GLACIATED COASTS

GLACIATED COASTS

Edited by *DUNCAN M. FITZGERALD*

Department of Geology
Boston University
Boston, Massachusetts

PETER S. ROSEN

Department of Geology
Northeastern University
Boston, Massachusetts

ACADEMIC PRESS, INC.
Harcourt Brace Jovanovich, Publishers
San Diego New York Berkeley Boston
London Sydney Tokyo Toronto

Photograph on cover by Duncan M. FitzGerald of the drumlin coast northeast of Halifax, Nova Scotia.

ACADEMIC PRESS, INC.
1250 Sixth Avenue, San Diego, California 92101

United Kingdom Edition published by
ACADEMIC PRESS INC. (LONDON) LTD.
24–28 Oval Road, London NW1 7DX

Library of Congress Cataloging in Publication Data

Glaciated coasts.

Includes index.
1. Glacial landforms. 2. Coasts. I. FitzGerald, Duncan M. II. Rosen, Peter S.
GB582.G57 1987 551.3'13 87–11378
ISBN 0–12–257870–8 (alk. paper)

PRINTED IN THE UNITED STATES OF AMERICA

87 88 89 90 9 8 7 6 5 4 3 2 1

Contents

List of Contributors

Numbers in parentheses indicate the pages on which the authors' contributions begin.

CHRISTOPHER T. BALDWIN (327), *Department of Geology, Boston University, Boston, Massachusetts 02215*

DANIEL F. BELKNAP (177), *Department of Geological Sciences and Oceanography Program, University of Maine, Orono, Maine 04469*

JON C. BOOTHROYD (115), *Department of Geology, University of Rhode Island, Kingston, Rhode Island 02881*

A. J. BOWEN (87), *Department of Oceanography, Dalhousie University, Halifax, Nova Scotia, Canada B3H 3J5*

R. BOYD (87), *Centre for Marine Geology, Dalhousie University, Halifax, Nova Scotia, Canada B3H 3J5*

BENNO M. BRENNINKMEYER (251), *Coastal Research Institute, Department of Geology and Geophysics, Boston College, Chestnut Hill, Massachusetts 02167*

KENNETH FINKELSTEIN (1), *Virginia Institute of Marine Science, Gloucester Point, Virginia 23062*

JOHN J. FISHER (279), *Department of Geology, University of Rhode Island, Kingston, Rhode Island 02281*

DUNCAN M. FITZGERALD (327), *Department of Geology, Boston University, Boston, Massachusetts 02215*

D. L. FORBES (51), *Geological Survey of Canada, Atlantic Geoscience Centre, Bedford Institute of Oceanography, Dartmouth, Nova Scotia, Canada B2Y 4A2*

R. K. HALL (87), *Centre for Marine Geology, Dalhousie University, Halifax, Nova Scotia, Canada B3H 3J5*

ALBERT C. HINE (115), *Department of Marine Science, University of South Florida, St. Petersburg, Florida 33701*

NOOR AZIM IBRAHIM (327), *Department of Geology, Boston University, Boston, Massachusetts 02215*

JOSEPH T. KELLEY (151, 177), *Maine Geological Survey, Augusta, Maine 04333, and Department of Geological Sciences and Oceanography Program, University of Maine, Orono, Maine 04469*

R. A. KOSTASCHUK (33), *Department of Geography, McMaster University, Hamilton, Ontario, Canada L85 4K1*

KENNETH LEACH (233), *Department of Geology, Northeastern University, Boston, Massachusetts 02115*

STEPHEN P. LEATHERMAN (307), *Laboratory for Coastal Research and Department of Geography, University of Maryland, College Park, Maryland 20742*

S. B. MCCANN (33), *Department of Geography, McMaster University, Hamilton, Ontario, Canada L85 4K1*

THOMAS F. MOSLOW (1), *Louisiana Geological Survey, Coastal Geology Program, Baton Rouge, Louisiana 70893*

DAG NUMMEDAL (115), *Department of Geology and Geophysics, Louisiana State University, Baton Rouge, Louisiana 70803*

ADAM F. NWANKWO (251), *Coastal Research Institute, Department of Geology and Geophysics, Boston College, Chestnut Hill, Massachusetts 02167*

PETER S. ROSEN (233), *Department of Geology, Northeastern University, Boston, Massachusetts 02115*

DAVID R. SANDS (327), *Department of Geology, Boston University, Boston, Massachusetts 02215*

R. CRAIG SHIPP (177, 209), *Oceanography Program, University of Maine, Orono, Maine 04469*

STEPHANIE A. STAPLES (209), *Department of Geological Sciences, University of Maine, Orono, Maine 04469*

R. B. TAYLOR (51), *Geological Survey of Canada, Atlantic Geoscience Centre, Bedford Institute of Oceanography, Dartmouth, Nova Scotia, Canada B2Y 4A2*

LARRY G. WARD (1, 209), *Center for Environmental and Estuarine Studies, Horn Point Environmental Laboratories, University of Maryland, Cambridge, Maryland 21613*

Preface

Glaciated Coasts results from the recognition that while the morphology and processes of shorelines south of the glacial limit in North America have been rigorously investigated and regionally compared, glaciated northern shorelines have not received a similar amount of regional analysis. Since the southern shore morphologies are largely secondary, or marine-modified shorelines, the variations are smaller compared with the glaciated shores to the north. A symposium was organized at the northeast section meeting of the Geological Society of America in 1983 in an attempt to assemble a view of the range of shoreline morphologies of glaciated coasts and the processes responsible for producing these coastlines. This volume is the result of this effort and is indicative of evolving interest in nonsandy shorelines. The chapters presented in this volume have been reviewed by at least two professionals in the field.

Diversity is the common theme to the glaciated coasts. North of the glacial limit, shorelines are most often primary in form and in various stages of modification by marine agents. Bedrock shorelines and coarse sediment, derived from eroding glacial deposits, are typical shore materials. Shoreline reaches, or distances of similar processes, become shorter as terrestrial processes control shore outlines. This more complex shoreline form results in a large percentage of the coast consisting of various types of embayments, such as estuaries and fjords. The prevalence of embayments increases the range of wave energy along the coast, which creates further small-scale variations in forms. Extreme differences in isostatic and tectonic setting produce large variations in relative sea level, both areally and in the Holocene history of shorelines. Both emergent and submergent coasts commonly occur.

The 13 chapters in this volume cover a range of geologic and geographic coastal settings and are roughly arranged in a north to south order. At least four chapters are involved with characterizing shorelines associated with embayments (Chapter 2 by McCann and Kostaschuk; Chapter 6 by Kelley; Chapter 7 by Belknap, Kelley, and Shipp; and Chapter 8 by Shipp, Staples, and Ward). Two chapters describe baymouth barriers that enclose embayments (Chapter 4 by Boyd, Bowen, and Hall and Chapter 13 by

FitzGerald, Baldwin, Ibrahim, and Sands). With the exception of areas consisting of reworked outwash deposits (i.e., Chapters 11 and 12 by Fisher and Leatherman, respectively), most chapters deal with beaches having coarse or mixed sediment populations (Chapter 3 by Forbes and Taylor; Chapter 9 by Rosen and Leach; Chapter 10 by Brenninkmeyer and Nwankwo; and Chapter 4 by Boyd, Bowen, and Hall). Brenninkmeyer and Nwankwo noted that approximately half the world's beaches are composed of gravel clasts that have been partially rounded and sorted by marine processes. Compared with sandy beaches, coarse sediment beaches have received little attention in the past.

Ward, Moslow, and Finkelstein's examination of the southeastern coast of Alaska (Chapter 1) shows the influence of active tectonism on a mountainous shoreline. The region contains both emergent and submergent shorelines, with a glacial imprint that is being reworked by modern processes. McCann and Kostaschuk define the sedimentation processes in a fjord in British Columbia. In this area, the glacial activity not only excavated the basin but also controlled the depositional sequence preserved in the fjord.

Forbes and Taylor summarize the effects of glaciation on the Canadian Atlantic Coast with examples of gravel beach environments in differing coastal settings. Boyd, Bowen, and Hall's investigation of the Eastern Shore of Nova Scotia identifies submerging coastal drumlins as the primary source of coastal sediments. The sand and gravel accumulates as valley–mouth barrier systems, which undergo cyclic progradation and subsequent landward translation.

A unique glaciated coastal environment is discussed by Nummedal, Hine, and Boothroyd in Chapter 5. The evolution of the south–central coast of Iceland is controlled by a combination of glacial, fluvioglacial, volcanic, and marine processes. Coastal sedimentation patterns are dominated by catastrophic "glacier bursts."

Three different aspects of coastal sedimentation along the Maine coast are presented. Kelley developed a detailed inventory of coastal environments along the Maine shoreline. This shore is characterized by a series of estuaries that have been excavated in bedrock by glacial flow. Glacial sediments eroded on the estuary margins have resulted in mudflat deposition as a dominant accumulative environment. Belknap, Kelley, and Shipp present a Quaternary stratigraphy of estuarine deposits, which has been affected by the extreme fluctuations in sea level in that area. Shipp, Staples, and Ward focus on a single embayment, identifying geomorphic elements and variation, which is primarily related to varying exposure to wave energy.

Rosen and Leach identify accumulative processes resulting from the erosion of drumlins in Boston Harbor, Massachusetts. Distinct progradational and recessional gravel spit forms correlate with the volumes of longshore inputs. Overwash in this environment is a nonstorm event, which results in a distinct depositional sequence. Brenninkmeyer and Nwankwo traced the sources of shingle on a barrier on Massachusetts Bay. While the material was derived from eroding ground moraines, the bedrock sources were local. The shingle making up the barrier accumulates from landward transfers along the nearshore bottom.

Two chapters on the shoreline development of Cape Cod, near the glacial terminus, show that glaciated coasts in some environments can be similar to shores on the coastal plain to the south. Fisher describes the evolution and form of the outer shore of Cape Cod, Massachusetts, which has evolved from glacial outwash deposits. The regional shoreform is analogous to coastal compartments to the south. Leatherman details the reworking of these deposits to form Provincetown Spit, which is the result of late Holocene northerly drift that followed the submergence of George's Bank.

FitzGerald, Baldwin, Ibrahim, and Sands show that the shore outline in Buzzards Bay, near the base of Cape Cod, is dominated by a series of peninsulas and embayments, in many ways similar to the coast of Maine. On the western half of the shore, there is an abundance of glacial sediments, which have been reworked to form barrier beaches, spits, and inlet systems.

PETER S. ROSEN

CHAPTER 1

Geomorphology of a Tectonically Active, Glaciated Coast, South–Central Alaska

LARRY G. WARD

Center for Environmental and Estuarine Studies
Horn Point Environmental Laboratories
University of Maryland
Cambridge, Maryland

THOMAS F. MOSLOW

Louisiana Geological Survey
Coastal Geology Program
Baton Rouge, Louisiana

KENNETH FINKELSTEIN

Virginia Institute of Marine Science
Gloucester Point, Virginia

Significant variations in the coastal geomorphology of the predominantly bedrock shoreline from Lower Cook Inlet to Prince William Sound, Alaska, occur due to crustal movements (submergence versus emergence), glaciations, and wave energy. In areas that underwent submergence during the Good Friday, 1964, earthquake, the coastline is dominated by steep bedrock slopes with extremely narrow intertidal zones. The primary depositional features associated with these bedrock coasts are relatively steep (~6–10°), coarse-grained talus fan beaches or pocket beaches. Conversely, areas uplifted during the 1964 earthquake have wider intertidal zones, with wave-cut platforms being common. Also, depositional features occur more frequently along emergent shorelines with linear, sand to boulder beaches and talus fan beaches dominating. Bedrock shorelines exposed to the Gulf of Alaska, which has one of the most severe wave climates in the world, are actively undergoing erosion.

River valley and glacial outwash fan deltas occur along both submergent and emergent shorelines where sediment supply is abundant (usually glacially derived). Lobate fan deltas with relatively large tidal flat deposits dominate in low-wave-energy environments. Conversely, where wave energy is high, the fan deltas are arcuate–cuspate in shape with beach ridge plains and recurved spits occurring frequently.

The morphologic characteristics of beach environments vary according to sediment texture and exposure to wave energy. Sediments comprising the beaches range from fine sand to boulders, although modes occur in the medium sand and pebble range. Coarser grained beaches tend to be narrower and steeper than finer grained beaches. Beach slope also decreases with increasing wave energy.

On the basis of the geomorphic and sedimentologic processes observed in the study area, four principle types of coastal units are delineated: (1) mountainous coastline, (2) fjord coastline, (3) exposed embayment coastline, and (4) linear island coastline.

I. Introduction

The Kenai Peninsula, Alaska, region is a tectonically active, mountainous environment characterized by shear cliffs exceeding 300 m in height, deep embayments, and a narrow continental shelf (Fig. 1). Adding to the ruggedness and complexity of this area is the presence of relatively large ice fields, the remnants of the last major glaciation that completely covered the southern coast of Alaska (Karlstrom, 1964). Previous glaciations helped shape the landscape by forming features such as fjords and covered much of the area with a blanket of till. Presently, meltwater runoff from the glacial fields provides large volumes of sediment to the coastal zone, where it is shaped by marine processes (waves and tides). Large stretches of southern Alaska are directly exposed to the storm-wave environment of the Gulf of Alaska (Davies, 1973), which creates one of the highest energy coasts in the world (Nummedal and Stephen, 1978). Consequently, the geomorphology of the southern coast of Alaska is controlled by the interactions of tectonic processes, glacial processes (past and present), and hydrographic regime (primarily wave action).

Excellent descriptions of the geomorphologic characteristics and sedimentary processes of a large portion of the southern coast of Alaska have

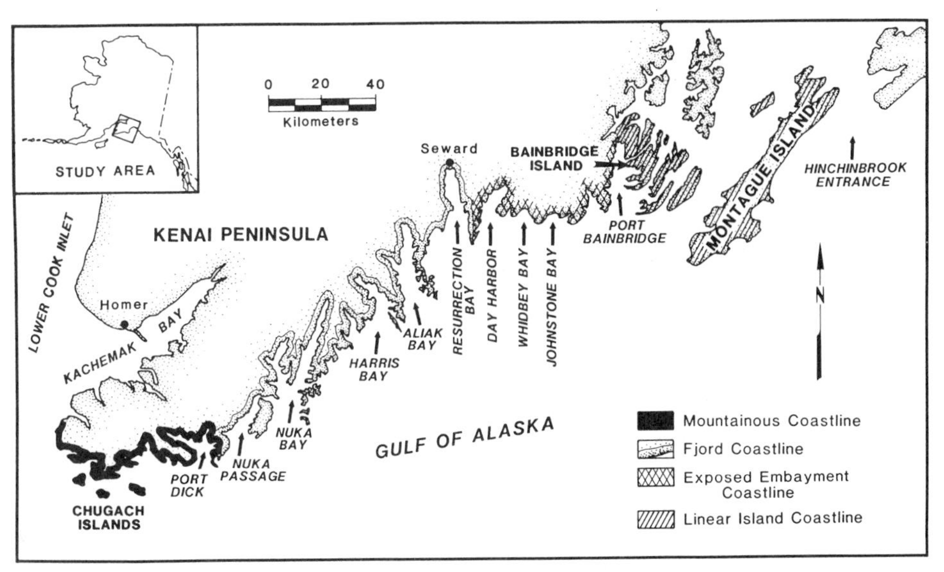

FIG. 1. Location map of the study area. Principal bays and islands referenced in the text are labeled. The study area is divided into four coastal physiographic regions as labeled.

been previously presented by a number of investigators. Hayes and Michel (1982) described in detail the coastal region of Lower Cook Inlet and developed a depositional model for a forearc embayment. Much of this coastal region, at least in recent times, has been undergoing submergence due to crustal subsidence. Boothroyd and Ashley (1975), Hayes *et al.* (1976), and Ruby (1977) described coastal segments east of Prince William Sound and developed a coastal classification scheme for this heavily glaciated, sediment-rich region. The sedimentary characteristics of the intertidal zone for the entire southern coast of Alaska from Yakutat to Cook Inlet have been reported on by Sears and Zimmerman (1977).

Hayes (1979) presented a model that describes a hierarchy of controls on coastal environments in general. Of major importance is the location of a coastal segment with respect to global tectonics (leading or trailing edge). Leading-edge coastlines are typically very rugged with great relief and relatively few major sedimentary deposits, while trailing-edge coastlines are more subdued topographically and have vast sediment accumulations. Once the tectonic setting is determined, the morphology of the coastline is a function of climate, sediment input, and hydrographic regime. Hayes (1979) and Nummedal and Fisher (1979) ultimately developed conceptual models for temperate, trailing-edge coastlines based on relative inputs of wave and tidal energy.

Although these and other investigations provide important insights to the overall geomorphology and dominant sedimentary processes of bedrock coasts, no one of the regions studied exhibited all the various tectonic, glacial, and hydrographic processes that actively control the coastal geomorphology of southern Alaska. In general, detailed descriptions of the coastal geomorphology of bedrock coasts are lacking. In this chapter, we report the results of a study concerned with the coastal geomorphology of the Kenai Peninsula area (Fig. 1). The purpose of this study is to describe the geomorphic characteristics of this leading-edge coast and assess the impact of tectonics, glaciation, and marine processes on both a regional (shore type) and local (intertidal) level.

The field study was conducted during the summer of 1979 using a modification of the zonal method described by Hayes *et al.* (1973). The survey included geomorphic mapping of the entire 2160 km of shoreline by aerial reconnaissance. Coastal morphology, depositional environments, and sediments were described in detail. Ground reconnaissance was conducted at 50 representative sites, 29 of which were surveyed using the method described by Emery (1961). Sediment samples were collected at survey stations for textural analyses.

II. Physical Setting

The study area extends from the entrance of Lower Cook Inlet to Hinchinbrook Entrance of Prince William Sound (Fig. 1) and is located on the leading edge of the North American plate (Inman and Nordstrom, 1971; Davies, 1973). Consequently, the study area is tectonically active and subjected to intense earthquake activity. The most recent significant crustal movement occurred during the 1964 Good Friday earthquake. Regional deformation by this powerful earthquake, which measured ~8.6 on the Richter scale, affected an area of 256,000 km^2 (Eckel, 1970). Vertical displacements occurred from the southern tip of the Kodiak Islands through Prince William Sound (Fig. 2), with roughly four-fifths of the outer Kenai Peninsula undergoing crustal subsidence. The axis of maximum subsidence within this zone trended northeast along the crest of the Kodiak–Kenai–Chugach Mountains, intersecting the coastline in the study area along the western margin of Nuka Passage. A maximum crustal downwarp of 2.3 m was recorded on the southwest coast of the Kenai Peninsula (Plafker, 1969). Conversely, the region east of Resurrection Bay to Prince William Sound was uplifted. Montague Island, elevated as much as 11.5 m (Plafker, 1969), lies along the axis of maximum uplift. Thus,

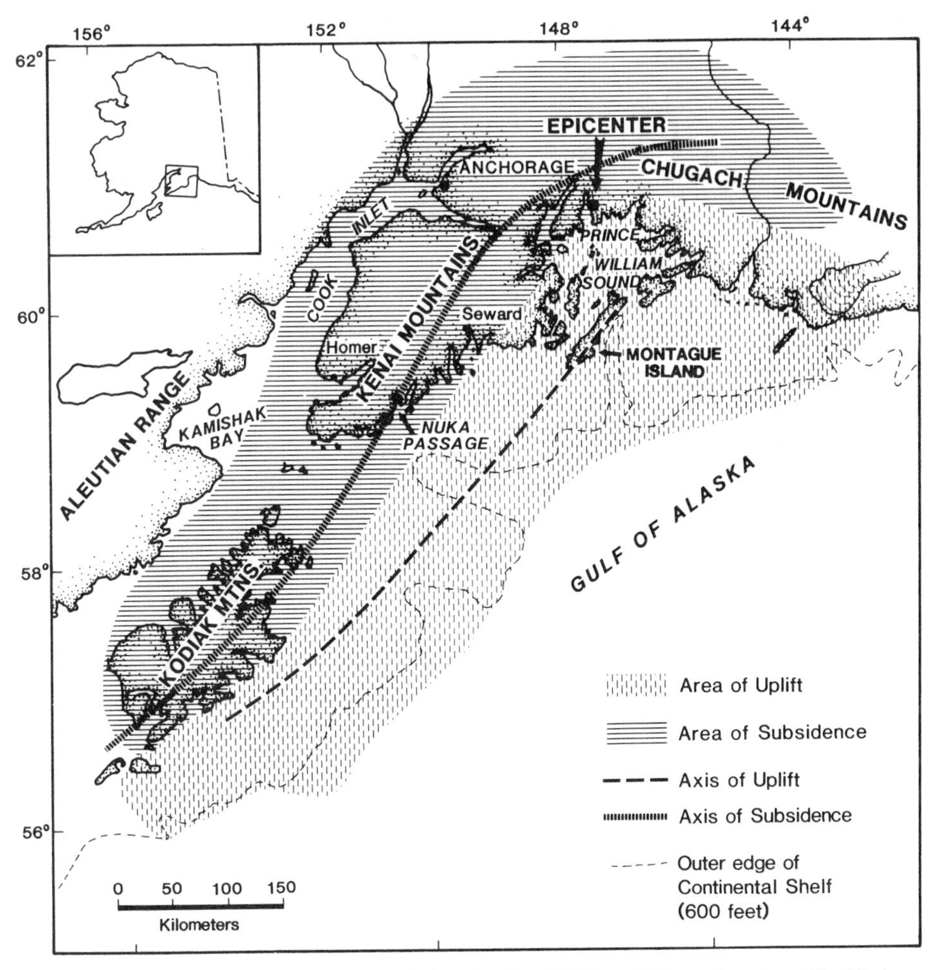

FIG. 2. Vertical crustal movements during the Good Friday, 1964, earthquake. [Modified from Plafker (1966).]

within the study area, this one instantaneous tectonic event produced vertical crustal movements resulting in 13.8 m of net elevation change (−2.3 m to +11.5 m).

Typical of leading-edge or young mountain range coastlines, this region displays great relief, having a hinterland composed of mountain systems (Kenai and Chugach Mountains) with ridges exceeding 1200 m in elevation within 10 km of the shoreline. The Kenai–Chugach Mountains are composed chiefly of dark-gray metasandstones, slates, and argillites of Mesozoic and Tertiary age (Wahrhaftig, 1965; Clark, 1972). Almost all of the sedimentary rocks in the Kenai Mountains are mildly metamorphosed. Granitoid intrusions are exposed in a few places along the southern peninsular coast. The islands forming the southern boundary of Prince William Sound (Bainbridge to Montague) contain large bodies of greenstone in association with the argillite and graywacke.

The geologic formations in the Kenai Peninsula and Prince William Sound area include the Valdez Group, a sequence of eugeosynclinal deposits that comprise the vast majority of the outer Kenai Peninsula, and the Orca Group, a sequence of early Tertiary age rocks that can be divided into a lower volcanic unit and an upper sedimentary unit (Case *et al.*, 1966). The elongated islands that form the southern boundary of Prince William Sound are included in the Orca Group. These rocks are faulted into contact with the Valdez Group of the outer Kenai Peninsula along a north–south line that divides Bainbridge Island into two halves (Case *et al.*, 1966).

The Kenai Peninsula region has a maritime climate with one of the highest frequencies of winter storms in the Northern Hemisphere (Petterssen, 1969). Numerous low-pressure centers enter the Gulf of Alaska in winter, creating significant storm activity (Nummedal and Stephen, 1978). In addition, storm waves generated further south in the Pacific impact the area. Ship wind and wave observations reported in the Summary of Synoptic Meteorological Observations (SSMO) (U.S. Naval Weather Service Command, 1970) for the area offshore of Kenai Peninsula indicate that the prevailing and dominant winds are out of both the east and the west (Fig. 3). The SSMO wind velocities for the Seward data square over the period from 1963 to 1970 were between 60 and 85 km/hr ~7% of the time; deep-water wave heights were between 3.9 and 4.9 m ~5% of the time. Although waves affecting the exposed coastal areas of the Kenai Peninsula are significantly altered due to shoaling and refraction, a high-energy wave climate is indicated. The mean wave energy flux (regardless of direction) computed from the SSMO data is $\sim 30 \times 10^{10}$ ergs/m s (Fig. 3), which is the highest for the Gulf of Alaska (Nummedal and Stephen, 1978). This agrees with geomorphic evidence observed in the field. Gravel berms several meters in height are common on beaches exposed to the Gulf of Alaska. In sharp contrast, wave energy within the protected embayments and fjords can be very low due to limited fetches.

Tidal range is less variable than wave energy, increasing steadily from a mean of 2.3 to 3.2 m in a westward direction along the study area (U.S. Department of Commerce, 1979). There is minor amplification of tidal range within some fjords, but this is limited due to their relatively great water depths and steep submarine slopes.

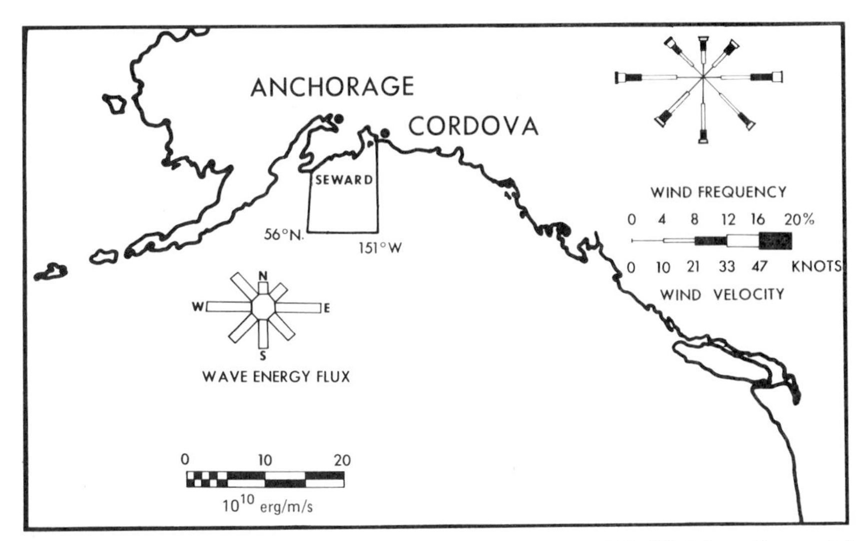

FIG. 3. Distribution of wind and wave energy in study area. [Modified from Nummedal and Stephen (1978).] Observations are for the area enclosed by solid lines and labeled Seward.

III. Coastal Geomorphology

The study area can be divided into four large physiographic units that display distinct coastline morphologies determined by tectonic movements (emergence versus submergence), proximity to active glacial fields, and presence of fjords. These segments are (1) the rugged, mountainous coastline with both exposed bedrock cliffs and sheltered embayments found in the far western Kenai Peninsula from the Chugach Islands to Port Dick; (2) the large fjords from Port Dick to Day Harbor; (3) the exposed embayments from Day Harbor to Port Bainbridge; and (4) the linear islands from Bainbridge Island to Montague Island (Fig. 1). The first two regions (mountainous coastline and fjords) lie within the area that has undergone crustal subsidence, at least in historic times (i.e., 1964 earthquake). Conversely, the exposed embayments and linear islands are located in the area uplifted during the same tectonic event. Although there are a number of similar coastal geomorphic features that occur in all four of these regions (i.e., bedrock scarps), each region is unique in its overall morphologic framework. In order to synthesize the type and distribution of coastal geomorphic features found in the study area, morphologic models based on representative sections of coastline have been developed and are described in the following.

A. Mountainous Coastline

The mountainous coastline found along the far western portion of the Kenai Peninsula extending from the Chugach Islands to Port Dick (Fig. 1) is characterized by rugged mountains exceeding 850 m in elevation within 2 km of the shore (Fig. 4). During the Good Friday, 1964, earthquake, this region instantaneously subsided ~2 m (Plafker, 1969). Most of the region is exposed to the storm-wave environment of the Gulf of Alaska, although a significant amount of shoreline is sheltered from wave erosion in protected embayments (Fig. 4).

The combination of rugged terrain coupled with the high wave energy of the areas open to the Gulf of Alaska has led to much of the region (~73%) being composed of steeply sloping bedrock scarps with elevations exceeding 450 m. These scarps, referred to here as exposed bedrock, are actively being eroded, as indicated by slopes often exceeding 45° (Fig. 5a), lack of vegetation, and the frequent presence of broad talus fans (Fig. 5b). Sea caves and stacks are common features. Depositional intertidal features are largely absent, with the exception of pocket or talus fan beaches. Sediments comprising the pocket beaches in the exposed bedrock coast are coarse-grained, ranging from coarse sand to cobbles. These beaches are relatively steep (~6–10°), short in length (<0.5 km), and narrow (<40 m) (Fig. 5c). Similarly, talus fan beaches are steep, extremely coarse-grained consisting of cobble- to boulder-size material, and are rarely more than a few hundred meters in length or a few tens of meters in width. However, larger depositional features are found at the mouths of small river valleys incised into this mountainous coastline (Fig. 4). These river valleys are normally less than 5 km long and are lined with a veneer of glacial till. Short, high-gradient streams drain the valleys, transporting coarse detritus to the littoral zone where sediment is deposited in large volumes and molded by marine processes. In some locations, till deposits are being eroded by wave action and form low scarps (<10 m) fronted by sand and gravel beaches.

An interesting contrast to the exposed bedrock areas, which are relatively barren of depositional features, occurs in the more sheltered embayments, which are protected from the storm-wave climate of the Gulf of Alaska. These sheltered bedrock shorelines tend to have much lower slopes than their exposed counterparts, which are subjected to significantly higher wave energy. Also, the sheltered bedrock is covered by vegetation close to the water's edge, and talus slopes are rare; both morphologic features are suggestive of slower rates of coastal erosion. Where river valleys intersect the shoreline and discharge their bedload into the shallow

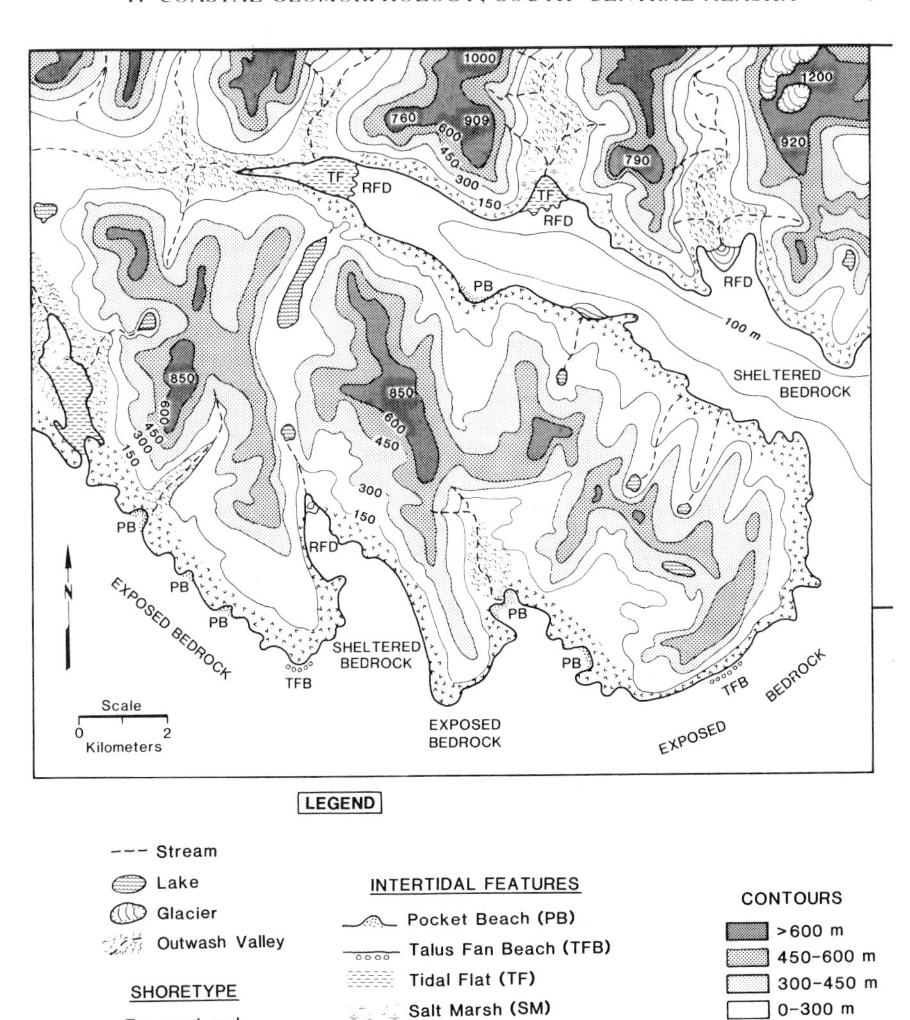

Fig. 4. General geomorphic setting for the mountainous coastline region from the entrance to Lower Cook Inlet to Port Dick (see Fig. 1). The model is based on a segment of coastline found west of Port Dick. "Shore type" describes the overall geomorphic character of a coastal region, while "intertidal features" refers to the smaller scale, local morphology.

FIG. 5. Typical geomorphic features associated with mountainous coastlines. (a) Exposed bedrock shoreline located just east of Port Dick. Bedrock scarps exposed to storm waves from the Gulf of Alaska are typically steep, unvegetated, and have little intertidal area. (b) Talus fan beach located along the seaward margin of the Chugach Islands.

FIG. 5. (c) Pocket beach found along exposed bedrock shoreline. (d) Tidal flat and salt marsh found at the head of the western embayment of Port Dick.

waters at the head of these sheltered embayments, lobate fan deltas are found in association with extensive sandy tidal flats and salt marshes (Figs. 4 and 5d).

B. Fjord Coastline

The coastline from Port Dick to Resurrection Bay consists of predominantly large north–south trending fjords with numerous smaller embayments located along their perimeters (Fig. 6). This entire coastal region was partially submerged during the Good Friday, 1964, earthquake, with the axis of maximum crustal subsidence (2.3 m; Plafker, 1969) trending northeast–southwest near Nuka Passage (Fig. 2). Morphology and sedimentation in this area are still influenced by active glaciation, with large glacial fields and valley glaciers affecting Nuka, Harris, Aliak, and Resurrection Bays.

At the entrances to some of the fjords are island groups that range in elevation from 150 to over 300 m, and have steep bedrock scarps with little to no intertidal area (Fig. 7a). The 100-m bathymetric contour is often within 1 km of the shoreline. These island groups have the appearance of a submerged mountain system.

Although this entire coastal region is dominated by large fjords, dramatic variability exists in terms of associated depositional features. Approximately 65% of the coastline in the region from Port Dick to Nuka Passage is composed of bedrock scarps, while ~35% is associated with depositional features, primarily pocket beaches and small river fan deltas. In the larger fjords to the east with active glaciation, bedrock scarps comprise approximately the same proportion of the shoreline as in the unglaciated areas. However, the sedimentary deposits are typically much larger. Glacial outwash fans, reworked morainal deposits, and long, wide arcuate beaches fed by glacial streams are common (Fig. 6).

River fan deltas are found along the margins and landward extremes of both glaciated and unglaciated fjords where till-lined fluvial valleys intersect the bedrock scarp coast. As with the mountainous coastline region, fan deltas vary in morphology depending on rate of sediment supply and exposure to wave reworking. Within the fjords there is a progression of fan delta morphologies. In areas sheltered from high wave energy, lobate fan deltas that exhibit digitate margins and have large, sandy tidal-flat and salt-marsh deposits are most common (Fig. 7b). Where wave energy is significant, arcuate–cuspate fan deltas with sand to gravel berms and beach ridges at their seaward margin (Fig. 7c) are dominant. This progression of fan types has been described by Hayes and Michel (1982) during their study of Lower Cook Inlet. Within the study area, arcuate–cuspate fans

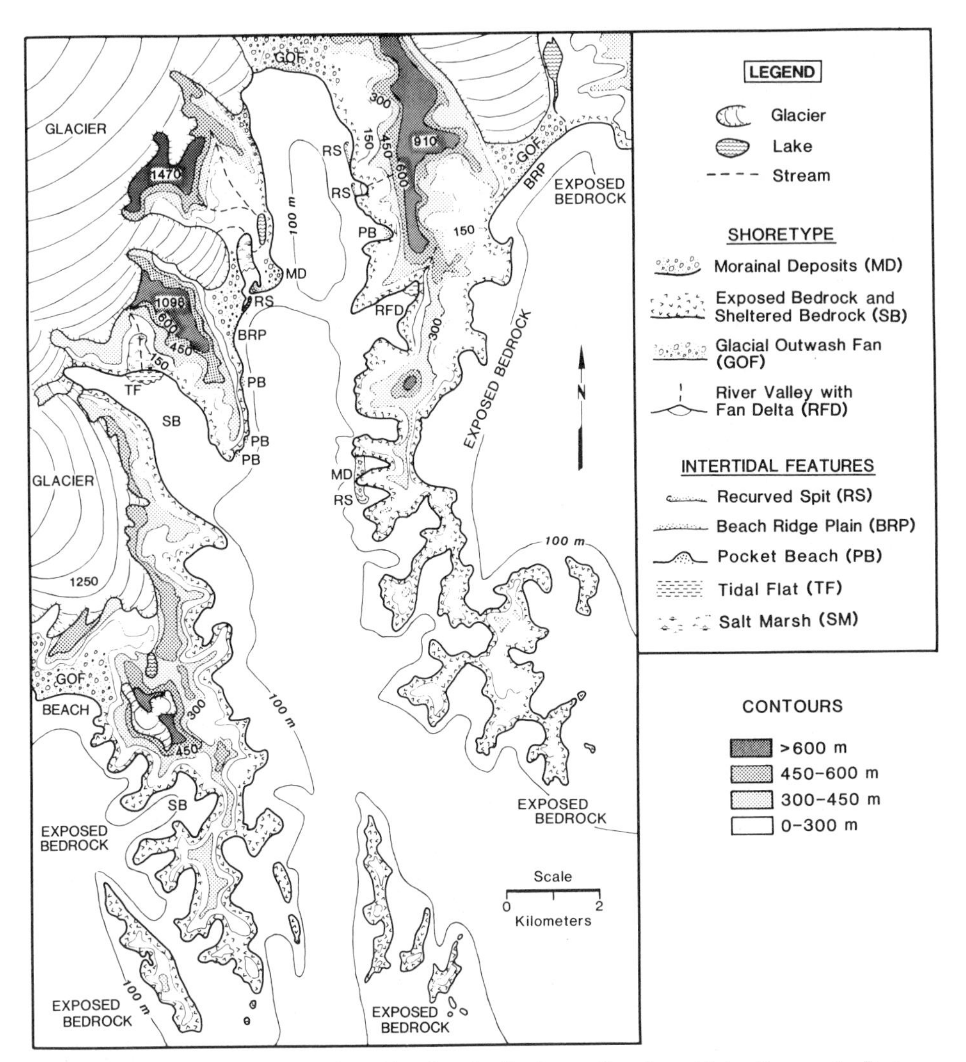

FIG. 6. General geomorphic setting for the fjord coastline from Nuka Passage to Day Harbor (see Fig. 1). The model is based on the Aliak Bay area. "Shore type" describes the overall geomorphic character of a coastal region, while "intertidal features" refers to the smaller scale, local morphology.

are most common along the flanks or near the entrance of the fjords where wave energy is higher, while lobate fan deltas dominate in the landward reaches of the fjords or in sheltered embayments.

In the glaciated fjords, glacial outwash fans (sandurs) are common. Again, depending on rate of sediment input and exposure to wave action,

FIG. 7. Typical geomorphic features associated with fjord coastlines. (a) Typical island group found at the entrance to Nuka Bay. Normally these islands have almost no intertidal features, due to their extremely steep slopes. Depositional features, such as the one shown in this photograph, occasionally occur due to the presence of moraines. (b) Lobate fan delta located in Resurrection Bay near Seward.

FIG. 7. (c) Arcuate–cuspate fan delta located in Resurrection Bay. (d) Glacial outwash fan terminating in a lobate fan delta located in Nuka Bay.

these outwash fans have a range of morphologic characteristics. In low-wave-energy environments, the outwash fans terminate in tidal-flat–salt-marsh systems (Fig. 7d), similar to lobate fan deltas previously described. In locations exposed to higher wave energy, the terminus of the outwash fans tends to be composed of beach ridges with a morphology similar to that of arcuate–cuspate fan deltas. These beach ridge plains are typically composed of sand and gravel, are relatively wide, and have several shore-parallel berms or ridges.

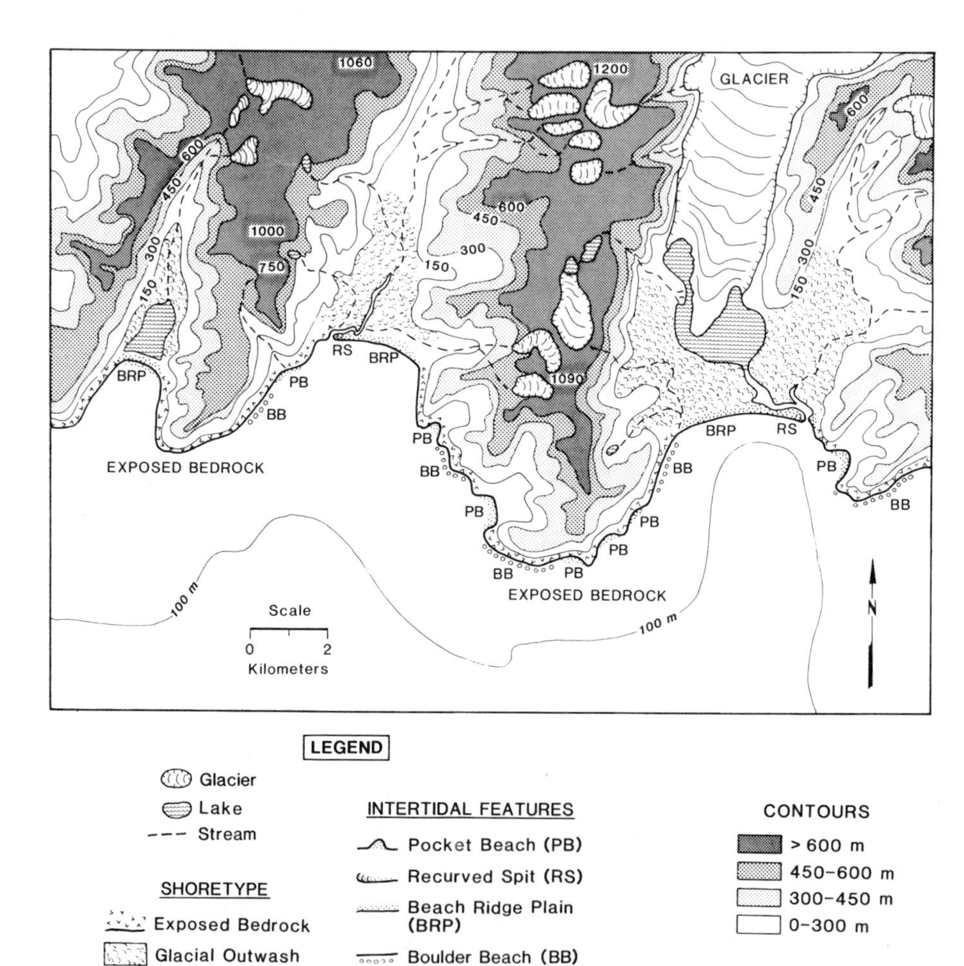

FIG. 8. General geomorphic setting for the exposed embayment coastline located from Day Harbor to Port Bainbridge (see Fig. 1). The model is based on the Whidbey and Johnstone Bay areas. "Shore type" describes the overall geomorphic character of a coastal region, while "intertidal features" refers to the smaller scale, local morphology.

C. Exposed Embayment Coastline

The coastal region from Day Harbor to Port Bainbridge is composed of relatively small embayments that face into the Gulf of Alaska (Fig. 8). The coast has great relief, with mountain peaks exceeding 1000-m elevation within 4 or 5 km of the shore. The embayments, which typically extend less than 10 km inland, have steep bedrock scarps near their entrances and large glacial outwash fans at their landward margins. These outwash fans are exposed to relatively high wave energy that forms large beach ridges composed of sand and gravel (Fig. 9). The beach ridges along with pocket beaches form ~33% of the exposed embayment coastline.

The bedrock scarps, which constitute ~65% of the shoreline within the exposed embayment coastline, are typical of the exposed bedrock environments found throughout the Kenai Peninsula area in that they are very steep, often unvegetated, and have talus fan beaches. Unlike their counterparts in the mountainous coastline region (western Kenai), which

Fig. 9. Typical beach ridge located at the head of an exposed embayment (Day Harbor). This type of beach is normally composed of multiple, coarse-grained sand and gravel ridges with numerous washover features.

underwent recent submergence, these scarps frequently have boulder beaches at their base. Bedrock scarps more sheltered from wave erosion have lower slopes than the exposed scarps and are covered by forest vegetation. These sheltered bedrock scarps are fronted by narrow, relatively continuous, sand to boulder beaches. The texture of these beaches depends on proximity to glacial outwash fans.

Although the axis of maximum uplift during the Good Friday, 1964, earthquake occurred further to the east at Montague Island (Fig. 2), a number of morphologic features indicate this area has been uplifted and is out of equilibrium with the present hydrographic regime. Notable are narrow wave-cut platforms at the base of some bedrock scarps, and gravel ridges on uplifted bayhead beaches, formerly intertidal features that are now vegetated.

D. Linear Island Coastline

The far eastern segment of the study area consists of a series of elongated islands of variable length (20–90 km) forming the southern border of Prince William Sound (Fig. 1). Similar to the rest of Kenai Peninsula, these islands exhibit great relief with peaks exceeding 900 m (Fig. 10). Each island has a mountainous ridge along the longitudinal axis. However, unlike the mountainous coastal areas to the west, some of these islands are flanked by a flat, narrow band of rolling lowlands (Fig. 11a), possibly reflecting previous tectonic uplifts that exposed the flatter, nearshore shelf. During the Good Friday, 1964, earthquake, the nearshore shelf southwest of Montague Island may have been uplifted >15 m as indicated by changes in bottom soundings (Malloy, 1964).

The linear island coastline region dramatically contrasts with the rest of the Kenai Peninsula. Two of the more important differences are the relative abundance of depositional features and the appearance of very large, wave-cut platforms. Both of these reflect the episodic uplift. The wave-cut platforms, which constitute ~25% of this coastline, are typically wide (>0.5 km) and are covered with scattered cobbles and boulders (Figs. 11b and 11c). In their coastal geomorphic study of Kodiak Island, Alaska, Ruby *et al.* (1979) showed that the composition and structure of the bedrock exerted strong control on the morphology of wave-cut platforms. They noted that in locations where bedding planes are very steep or near vertical, as with Montague and Latouche Islands, wave-cut platforms are very irregular and contain numerous tidal pools (Fig. 11d). Conversely, where bedding planes are more horizontal, wave-cut platforms are flatter and more uniform. Both of these types occur in the study area.

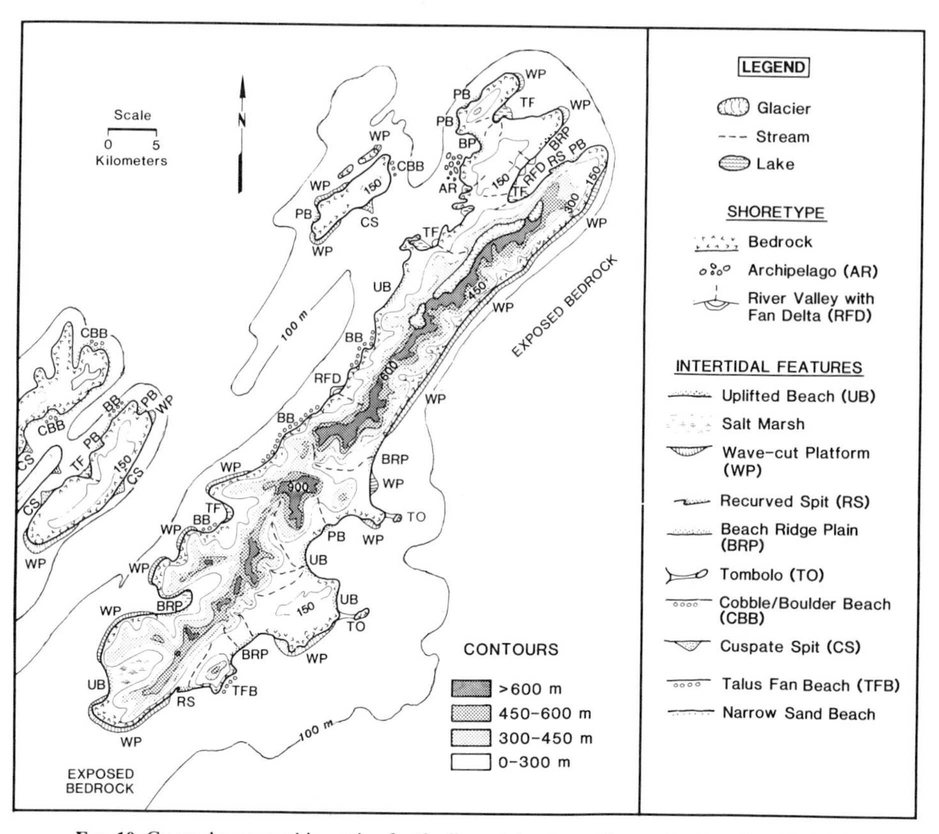

Fig. 10. General geomorphic setting for the linear island coastline region from Port Bainbridge to Montague Island (see Fig. 1). This model is based on the Montague Island area. "Shore type" describes the overall geomorphic character of a coastal region, while "intertidal features" refers to the smaller scale, local morphology.

Perhaps the most striking example of the influence of uplift on the geomorphology of this coastline is raised or uplifted beaches. These beaches typically have a large depositional berm, now well above the reach of wave activity, separated from the active beachface by a wide, flat platform. An abandoned lagoon may be located landward of the berm. For example, San Juan Bay, which is located on the southwestern end of Montague Island (Fig. 1), was mapped as an open lagoon on a 1953 United States Geological Survey topographic sheet (Fig. 12a). During this study, conducted in 1979, the lagoon had completely filled with sediment or had been abandoned (Fig. 12b). A pebble to cobble berm, 3–4 m in height, was located just seaward of the infilled lagoon. Separating the cobble berm,

FIG. 11. Typical features of the linear island coastlines. (a) Terraces flanking western shore of Latouche Island.

Fig. 11. (b, c, d) Examples of wave-cut platforms found along the southern and eastern shore of Montague Island. This area was uplifted ~11 m during the Good Friday, 1964, earthquake. Note the irregular relief and numerous tidal pools.

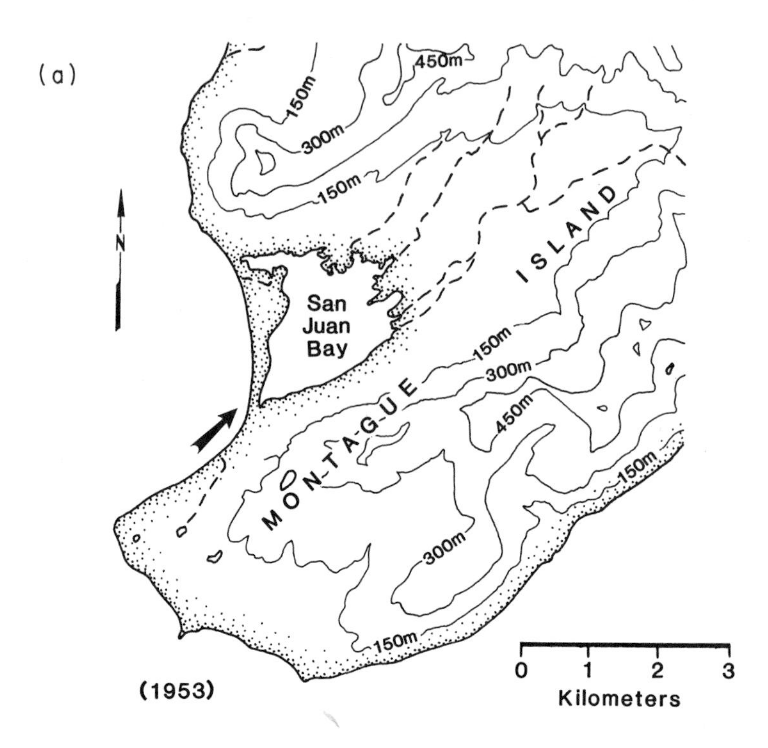

FIG. 12. (a) Map of San Juan Bay, Montague Island, based on a 1953 U.S. Geological Survey topographic sheet. This map was produced prior to the Good Friday, 1964, earthquake, which uplifted the region > 11 m. Note San Juan Bay was an open lagoon connected to the ocean by a narrow inlet (arrow). (b) Photograph of San Juan Bay taken in July 1979. Note the bay has either completely filled or was abandoned due to being uplifted above the maximum high tide level.

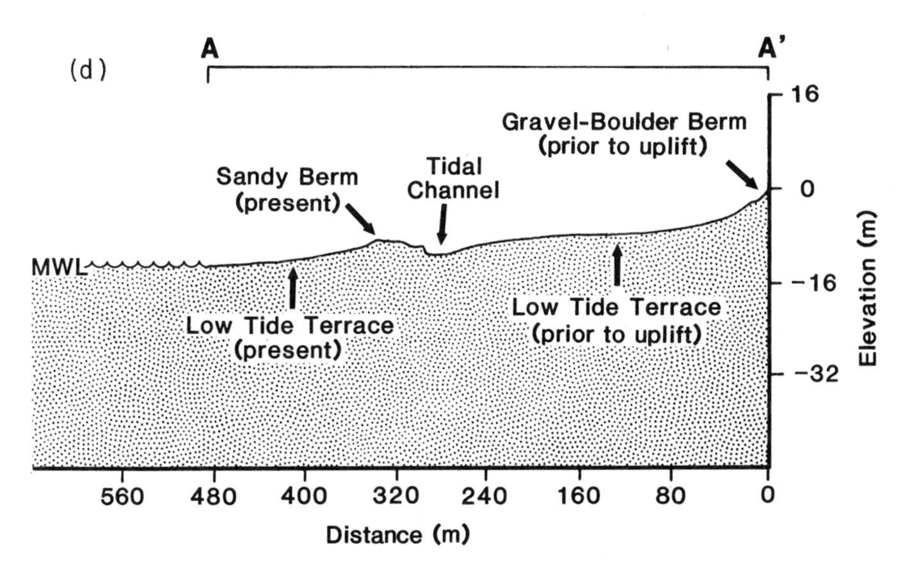

FIG. 12. (c) Aerial photograph of the beach fronting San Juan Bay in July 1979. The line running across the photograph shows position of beach profile shown in (d). (d) Topographic profile showing morphologic features of present and uplifted beaches at San Juan Bay.

which presently is above tidal influence, from an active recurved spit system is an ~300-m-wide terrace composed of medium to coarse sand (Figs. 12c and 12d). The morphologic and sedimentologic characteristics of this terrace indicate that it was the beachface (low-tide terrace?) prior to the most recent (1964) episode of uplift. The present low-tide terrace is relatively wide (~100 m) and is composed of fine to medium sand. The fine-grained nature of this sediment suggests that this material was deposited on the shoreface seaward of San Juan Bay prior to uplift.

IV. Coastal Sedimentary Environments

Although Kenai Peninsula is dominated by bedrock, nearly 35% of the coastline is composed of depositional features, including pocket beaches, talus fan beaches, cuspate spits, recurved spits, boulder beaches, uplifted beaches, tidal flats, and salt marshes. These depositional features vary in their morphologic and sedimentologic characteristics depending on source and rate of sediment supply and exposure to wave energy. The sediments comprising these depositional features cover an extremely large range.

A. Sediment Texture

The mean grain size [M_z from Folk (1974)] of the sediment samples analyzed during this study varies from nearly -5ϕ to 2.5ϕ (Fig. 13). Both field observations and textural analyses indicate the sediments composing many of the beaches are bimodal, with modes in the pebble and in the fine to medium sand ranges. In their study of Lower Cook Inlet, Alaska, which has a source of sediments similar to that of Kenai Peninsula, Hayes and Michel (1982) documented the occurrence of two primary modes (-4.0ϕ to -5.0ϕ and 1.5ϕ to 3.0ϕ) and a secondary mode (0.5ϕ to 1.5ϕ) in the beach sediments. Although Hayes and Michel used a much greater sampling density than was used during this study, similar sediment textural characteristics are indicated.

Sorting (σ_I from Folk, 1974) values of the samples analyzed also vary widely, ranging from very well sorted ($<0.35\phi$) to very poorly sorted ($>2\phi$) (Fig. 13a). In general, the coarsest (-3ϕ to -5ϕ) and finest (1ϕ to 3ϕ) mean grain sizes are the best sorted, while the poorest sorting occurs in the granule-size material. Folk and Ward (1957) attributed this type of distribution to the mixing of end members or natural modes that occur in the pebble- and sand-size ranges.

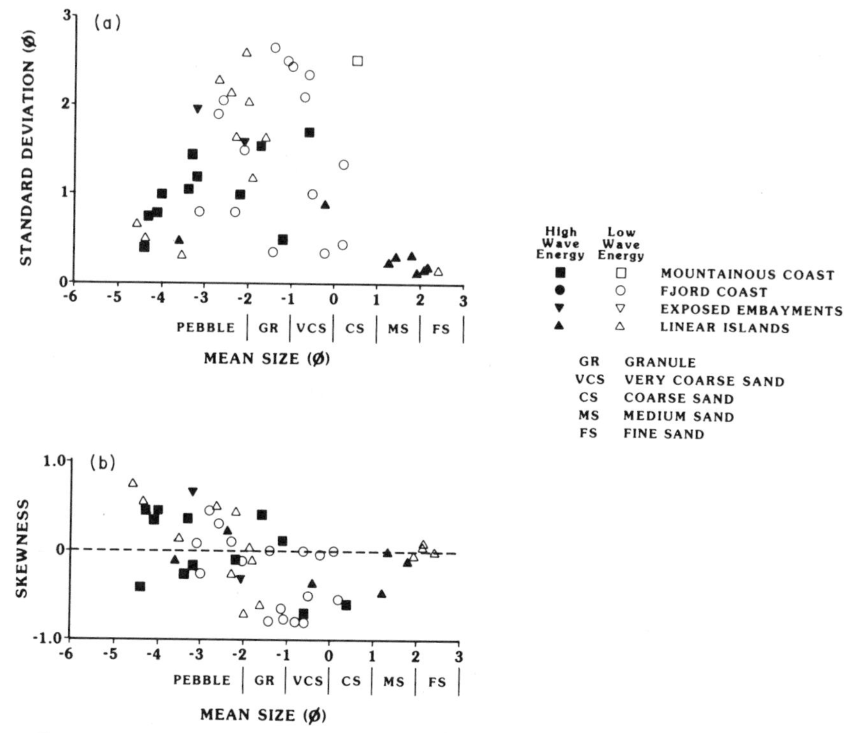

FIG. 13. Relationship between (a) graphic mean and inclusive graphic standard deviation and (b) graphic mean and inclusive graphic skewness for representative sediment samples taken along the Kenai Peninsula coast. Symbols refer to the physiographic units shown in Fig. 1. Most high-wave-energy beaches have direct exposure to the storm-wave environment of the Gulf of Alaska.

Similarly, the scatter plot of mean grain size and graphic skewness (Sk_I from Folk, 1974) varies over a wide range and indicates the mixing of two end-member populations (Fig. 13b). In general, the pebble size fraction is positively skewed while the sand size fraction is negatively skewed. Hayes and Michel (1982) found a similar relationship for the beaches in Lower Cook Inlet, Alaska.

Although a wide range of sediment texture occurs throughout the study area, some general relationships to the morphologic units described in the previous section exist. Coastal depositional features (primarily pocket beaches) sampled along the exposed mountainous coastline tend to be composed of granule to pebble-size material (-1ϕ to -5ϕ) with standard deviations less than 1.5ϕ (Fig. 13a). A major source of sediment to this region is the erosion of bedrock cliffs, which provides abundant coarse material. The relatively good sorting of these beaches results from the exposure to high wave energy. In general, most beaches exposed to the Gulf of Alaska tend to be better sorted. The finest sediments (1ϕ to 3ϕ) occur on the uplifted beaches exposed to high wave energy found along the linear island coastline. At least part of the reason for the fine-grained nature of these sediments is related to the most recent episode of uplift, which introduced shoreface sediment into the intertidal zone. Subsequent winnowing by wave action has led to these beach deposits being well to very well sorted. Conversely, sediments found in the lower wave energy regions of the linear island coast are poorly sorted and relatively coarse grained (-1ϕ to -3ϕ). The fjord and exposed embayment coastline show a wide range of grain sizes (-3ϕ to 1ϕ) and tend to be poorly sorted. The glacial source of much of the sediment composing the depositional features in both of these shoreline types accounts for this variability. The size and sorting of the coastal deposits in areas affected by glaciation largely depend on proximity to source and wave energy.

B. Beach Slope

The morphologic characteristics of the beaches along the Kenai Peninsula are strongly influenced by sediment texture and exposure to wave energy. Examination of representative profiles of the beaches indicates that coarse-grained beaches tend to be narrower and steeper than the finer grained beaches, although numerous exceptions exist. In addition, beaches exposed to high wave energy tend to have lower slopes than beaches with similar sediments in lower wave energy, sheltered environments. Figure 14 shows the relationship between slope, mean sediment size, and relative wave energy (high versus low) for selected beaches along the Kenai Peninsula coast. Some of the scatter in Fig. 14 is due to difficulties in identifying a representative slope or grain size where variability occurs. For the purposes of this study, the slope at midbeach was used (normally the steepest slope). When a bimodel sediment population was present, the smaller mean grain size (matrix) was used. The mean grain size of sediments less than $\sim -4.5\phi$ was determined by sieve analysis (Folk, 1974). For coarser sediment populations the modal grain size was estimated from vertical photographs taken in the field.

The relationship between grain size, beach slope, and wave energy has been described previously for sandy beaches along the east and west coasts of the United States (Bascom, 1951; Wiegal, 1964) and for sand–gravel beaches in South Island, New Zealand (McLean and Kirk 1969), and Kodiak Island, Alaska (Finkelstein, 1982). McLean and Kirk (1969) dem-

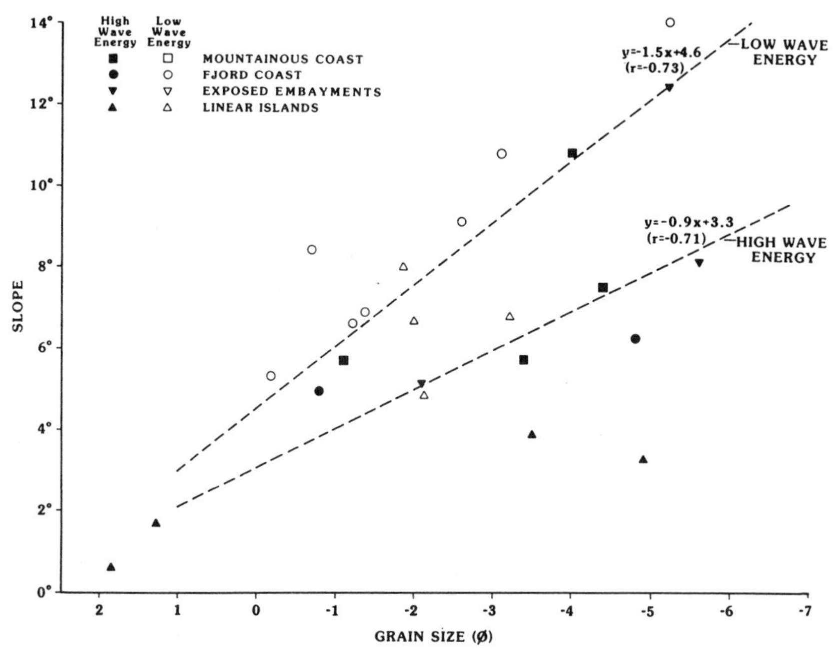

FIG. 14. Relationship between slope and grain size for high-and low-wave-energy beaches along the Kenai Peninsula. Symbols refer to physiographic units shown in Fig. 1. Most high-wave-energy beaches have direct exposure to the storm wave environment of the Gulf of Alaska. Beach slope and sediment grain size were measured at approximately midbeach, normally the steepest gradient. Mean grain size of sediments less than ~ − 4.5φ determined from sieve analysis [M_z from Folk (1974)]. Model grain size of sediments greater than ~ − 4.5φ estimated from vertical photographs taken in the field.

onstrated that both sediment size and sorting have an influence on beach slope, with better sorting leading to steeper slopes. Although there is considerable variability, the beaches examined in this study have a relationship between sediment size and slope similar to that of the beaches along the New Zealand coast. Finkelstein (1982) found that the gradients of sand–gravel beaches located on Kodiak Island, Alaska, had somewhat lower slopes than those reported by McLean and Kirk (1969) and that beach slope was strongly influenced by general erosional or depositional trends. Beaches undergoing accretion were steeper than erosional beaches, despite similarity in sediment size. Some of the beaches along the Kenai Peninsula may have been undergoing recent erosion in response to storm activity or general low sediment supply, which causes lower beach slopes (Bascom, 1951).

V. Summary

The coastal geomorphology of the study area is a product of the interactions of several major geologic processes. The overall ruggedness of this young mountain range coast results from its position at a convergent plate boundary (Inman and Norstrom, 1971; Hayes and Michel, 1982). Typical of collision coasts, this area is characterized by great vertical relief, a predominance of bedrock shorelines, and a narrow continental shelf. Within this geologic setting, the coastal morphology of a particular stretch of shoreline is primarily controlled by vertical crustal movements (emergence or submergence), glaciation (influencing sediment supply), and exposure to wave energy (Table I). Approximately two-thirds of the study area underwent recent (and probably long-term) submergence. Conversely, the rest of the coast has been uplifted. In areas of submergence, where

TABLE I. MAJOR SHORE TYPES AND ASSOCIATED INTERTIDAL FEATURES FOR THE COASTLINE FROM COOK INLET TO PRINCE WILLIAM SOUND RELATIVE TO WAVE ENERGY AND SEDIMENT SUPPLY

Shore type[a]	Associated intertidal features		Wave energy	Sediment supply
	Submergent coast[b]	Emergent coast[c]		
Sheltered bedrock	Bedrock (C) Pocket beach (C) Cuspate spit (R)	Wave-cut platform (C) Boulder beach (C) Cuspate spit (O)	Low	Low
Exposed bedrock	Steep bedrock (C) Talus fan beach (C) Pocket beach (O)	Wave-cut platform (C) Talus fan beach (O) Pocket beach (O) Boulder beach (O)	High	Low
River valley/glacial outwash fan delta (lobate)[d]	Tidal flat/salt marsh (C)		Low	High
River valley/glacial outwash fan delta (arcuate–cuspate)[d]	Beach ridge plain (C) Recurved spit (O)		High	High

[a]River valley fan deltas and glacial outwash fan deltas are combined as their geomorphic characteristics are similar.

[b]Submergence and emergence are based on vertical crustal movements during the Good Friday, 1964, earthquake. Only the most common features are presented.

[c]C, occurs commonly; O, occurs occasionally; and R, occurs rarely. These divisions are based on aerial photographs.

[d]Emergent and submergent coasts have similar shore types and intertidal features in areas of high sediment supply. The main differences are the presence of uplifted beaches or abandoned lagoons in areas of maximum uplift.

sediment supply is limited, much of the coast has the appearance of a flooded mountain system. In these areas, especially in the absence of any glaciation, the coast is largely composed of steeply sloping, bedrock cliffs with little intertidal area. Conversely, in uplifted regions, the intertidal areas are substantially larger, often with wide wave-cut platforms. In areas of maximum uplift (recent), the entire intertidal and nearshore subtidal zones have been raised to supertidal levels.

The second dominant control on the coastal geomorphology in this region is glaciation. Previous glaciations during the Quaternary (Karlstrom, 1964) carved great fjords and deposited a blanket of till in many areas. Presently, active glaciers provide abundant sediment to limited areas. Consequently, the major source of sediments to much of the Kenai Peninsula and Prince William Sound area is glacially derived. Many of the larger depositional features in the study area are directly related to this sediment source.

The final factor influencing the coastal geomorphology is marine processes. Although tidal action is undoubtedly important, the influence of waves has a greater effect on the coastal geomorphology. Mean tidal range is relatively consistent throughout most of the study area, varying less than a meter (2.3–3.2 m). However, wave energy varies from negligible in sheltered areas to extremely high in regions exposed to the Gulf of Alaska. The exposure to wave energy has dramatic effects on depositional features and appears to have subtle effects on bedrock coasts. Bedrock coasts facing the Gulf of Alaska tend to be steeper, having near-vertical slopes in many areas, in comparison to sheltered bedrock coasts in local embayments.

Although the dominant characteristic of the expanse of coastline between the entrance to Cook Inlet and Prince William Sound is its rocky nature, changes in geologic processes cause variations in coastal geomorphology. This generalization is applicable to both the depositional features and the bedrock shorelines. Coastal geomorphic studies in the past have largely concentrated on depositional environments along trailing-edge coastlines, and have produced eloquent conceptual models that relate the effects of physical processes and resultant geomorphic forms (Hayes, 1979; Nummedel and Fisher, 1979). Although this study was too limited in scope to produce clear process–response models for the rocky Kenai Peninsula coast, it does indicate that bedrock coastlines respond to geologic processes in a predictable manner. However, the role of other factors, such as bedrock lithology, degree of metamorphism, and structural patterns, in controlling geomorphic and sedimentologic characteristics of bedrock shorelines is, as yet, undetermined. In order to develop clear process–response models for rocky shorelines, much more geomorphic and sedimentologic work needs to be done along young mountain range or leading-edge coasts.

ACKNOWLEDGMENTS

Financial support for this project was given by the Bureau of Land Management, Offshore Continental Shelf Environmental Assessment Program (OCSEAP), contract RU-59, Miles 0. Hayes, principal investigator. We are indebted to Miles O. Hayes, Dag Nummedal, and Christopher H. Ruby for many valuable discussions on the coastal geomorphology of Alaska, and glaciated coastlines in general. The field work was conducted under the direction of Miles O. Hayes. Figures were drafted by Karen Westphall and Toni McLaughlin. The manuscript was typed by Anna Ruth McGinn and Jane Gilliard. This article is contribution no. 1740 from the University of Maryland Center for Environmental and Estuarine Studies.

REFERENCES

Bascom, W. M. (1951). The relationship between sand size and beach-face slope. *Trans. Am. Geophys. Union* **32,** 866–874.

Boothroyd, J. C., and Ashley, G. M. (1975). Processes, bar morphology and sedimentary structures on braided outwash fans, northeastern Gulf of Alaska. *In* "Glacialfluvial and Glaciolacustrine Sedimentation" (A. V. Jopling and B. C. McDonald, eds.), Spec. Publ. 23, pp. 193–222. Soc. Econ. Palentol. Mineral., Tulsa, Oklahoma.

Case, J. E., Barnes, D. F., Plafker, G., and Robbins, S. L. (1966). Gravity survey and regional geology of the Prince William Sound epicentral region. *U.S. Geol. Surv. Prof. Pap.* **543-C.**

Clark, S. H. B. (1972). Reconnaissance bedrock geologic map of the Chugach Mountains near Anchorage, Alaska. *U.S. Geol. Surv. Misc. Geol. Invest. Map* **MF-350.**

Davies, J. L. (1973). "Geographical Variation in Coastal Development." Hafner, New York.

Eckel, E. B. (1970). The Alaskan earthquake March 27, 1964: lessons and conclusions. *U.S. Geol. Surv. Prof. Pap.* **546.**

Emery, K. O. (1961). A simple method of measuring beach profiles. *Limnol. Oceanogr.* **6,** 90–93.

Finkelstein, K. (1982). Morphological variations and sediment transport in crenulate-bay beaches, Kodiak Island, Alaska. *Mar. Geol.* **47,** 261–281.

Folk, R. L. (1974). "Petrology of Sedimentary Rocks." Hemphill, Austin, Texas.

Folk, R. L., and Ward, W. C. (1957). The Brazos River Bar: a study in the significance of grain size parameters. *J. Sediment. Petrol.* **27,** 3–26.

Hayes, M. O. (1979). Barrier island morphology as a function of tidal and wave regime. *In* "Barrier Islands from the Gulf of St. Lawrence to the Gulf of Mexico" (S. Leatherman, ed.), pp. 1–27. Academic Press, New York.

Hayes, M. O., and Michel, J. (1982). Shoreline sedimentation within a forearc embayment, Lower Cook Inlet, Alaska. *J. Sediment. Petrol.* **52,** 251–263.

Hayes, M. O., Owens, E. H., Hubbard, D. K., and Abele, R. W. (1973). Investigation of form and processes in the coastal zone. *In* "Coastal Geomorphology," Proceedings, Third Annual Geomorphology Symposia Series (D. R. Coates, ed.), pp. 11–41. Dep. Geol., State Univ. of New York, Binghamton.

Hayes, M. O., Ruby, C. H., Stephen, M. F., and Wilson, S. J. (1976). Geomorphology of the southern coast of Alaska. *Proc. Int. Coastal Eng. Conf., 15th* pp. 1992–2008. Am. Soc. Civ. Eng., New York.

Inman, D. L., and Nordstrom, C. E. (1971). On the tectonic and morphologic classification of coasts. *J. Geol.* **79,** 1–21.

Karlstrom, T. N. V. (1964). Quaternary geology of the Kenai lowlands and the glacial history of the Cook Inlet region, Alaska. *U.S. Geol. Surv. Prof. Pap.* **443.**

Malloy, R.J. (1964). Crustal uplift of Montague Island, Alaska. *Science* **146,** 1048–1049.

McLean, R. F., and Kirk, R. M. (1969). Relationships between grain size, size-sorting, and foreshore slope on mixed sand-shingle beaches. *N.Z. J. Geol. Geophys.* **12,** 138–155.

Nummedal, D., and Fisher, I. A. (1979). Process–response models for depositional shorelines: the German and the Georgia Bights. *Proc. Int. Coastal Eng. Conf., 16th,* pp. 1215–1231. Am. Soc. Civ. Eng., New York.

Nummedal, D., and Stephen, M. F. (1978). Wave climate and littoral sediment transport, northeast Gulf of Alaska. *J. Sediment. Petrol.* **48,** 359–371.

Petterssen, S. (1969). "Introduction to Meteorology." McGraw-Hill, New York.

Plafker, G. (1966). Surface faults on Montague Island associated with the 1964 Alaska earthquake. *U.S. Geol. Surv. Prof. Pap.* **643-G.**

Plafker, G. (1969). Tectonics of the March 27, 1964 Alaska earthquake. *U.S. Geol. Surv. Prof. Pap.* **543-1.**

Ruby, C. H. (1977). "Coastal Morphology, Sedimentation and Oil Spill Vulnerability—Northeast Gulf of Alaska, Tech Rep. 15-CRD. Dep. Geol., Univ. of South Carolina, Columbia.

Ruby, C. H., Hayes, M. O., Finkelstein, K., and Reinhart, P. J. (1979). "Oil Spill Vulnerability, Coastal Morphology, and Sedimentation of the Kodiak Archipelago," Final report to the Outer Cont. Shelf Environ. Assess. Program. Natl. Oceanic Atmos. Adm., Washington, D.C.

Sears, H. S., and Zimmerman, S. T. (1977). "Alaska Intertidal Survey Atlas." U.S. Dep. Commer., Natl. Oceanic Atmos. Adm., Natl. Mar. Fish. Serv., Northwest Alaska Fish. Cent., Auke Bay.

U.S. Department of Commerce (1979). "Tide Tables, West Coast of North and South America." U.S. Dep. Commer., Natl. Oceanic Atmos. Adm., Natl. Ocean Surv., Rockville, Maryland.

U.S. Naval Weather Service Command (1970). "Summary of Synoptic Meteorological Observations, North American Coastal Marine Areas." Natl. Clim. Cent., Asheville, North Carolina.

Wahrhaftig, D. (1965). Physiographic divisions of Alaska. *U.S. Geol. Surv. Prof. Pap.* **482.**

Wiegel, R. L. (1964). "Oceanographic Engineering." Prentice-Hall, Englewood Cliffs, New Jersey.

CHAPTER 2

Fjord Sedimentation in Northern British Columbia

S. B. McCANN
AND
R. A. KOSTASCHUK

Department of Geography
McMaster University
Hamilton, Ontario, Canada

The character of fjord sediments in northern British Columbia is determined by glacial and glacial–marine events during the last glaciation, in particular the retreat phase, and by a variety of Holocene processes, of which the most important are related to fluvial inputs of sediment and delta progradation. Analysis of seismic profiles from Burke Channel–North Bentinck Arm, a 120-km-long fjord penetrating the Coast Mountains, identified two seismic sequences. The lower, late Pleistocene, sequence contains tills, glacial–marine deposits, and subaqueous moraines. For the most part it is overlain by a thinner, Holocene sequence, deposited from suspension from the plume of Bella Coola River, which enters the head of the fjord, and by turbidity currents that originate on the Bella Coola delta front.

I. Introduction

Attention has begun to focus on the sediments within the fjords that penetrate deeply into the Coast Mountains of British Columbia. This chapter presents an introductory review that stems, in part, from a study of sedimentation in the Burke Channel fjord system between 52° and 53°N in northern British Columbia (Fig. 1).

This research was originally concerned with recent sedimentation in the Bella Coola delta and North Bentinck Arm (Kostaschuk and McCann, 1983; Kostaschuk, 1985). The scope of the work was extended in 1982 when a seismic profiling survey was undertaken of late Quaternary sediments in following channels of the system: Burke Channel, North Bentinck Arm, South Bentinck Arm, Labouchere Channel, Dean Channel, and Fisher Channel (Fig. 2). The impetus for the broader study came from two sources. The first was a proposal to examine the notion that deltaic processes, including subaqueous processes, provide the key to understanding Holocene sedimentation in many British Columbian fjords. The second was the indication, from Bornhold's (1983) study of Douglas Arm, 180 km north of Bella Coola (Fig. 1), that the study fjords would probably contain a thick and varied sequence of late Pleistocene sediments of glacial and glacial–marine origin.

There are two sections to this chapter. The first considers the factors that have influenced sedimentation in Coast Mountain fjords—the physical setting, the glacial and sea-level history, and Holocene processes. The second presents an analysis, based on seismic profiles, of the late Quaternary sediments and sedimentary history of Burke Channel–North Bentinck Arm.

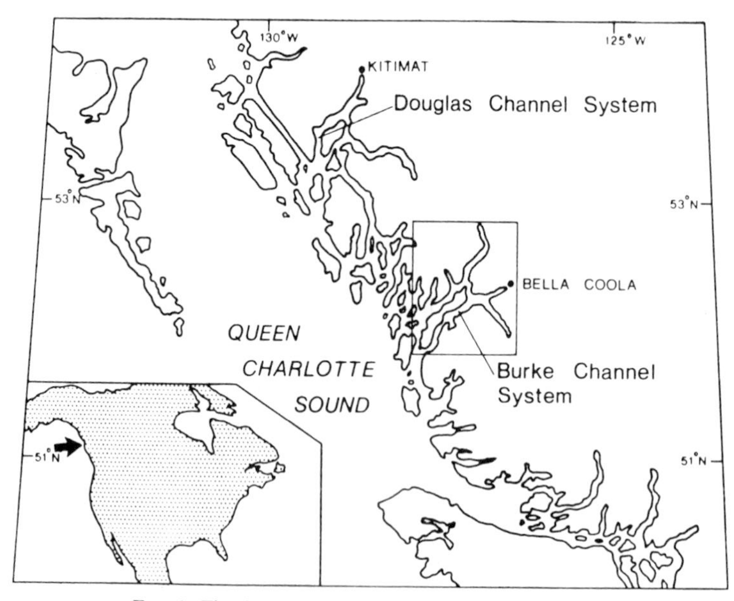

FIG. 1. Fjord coast of northern British Columbia.

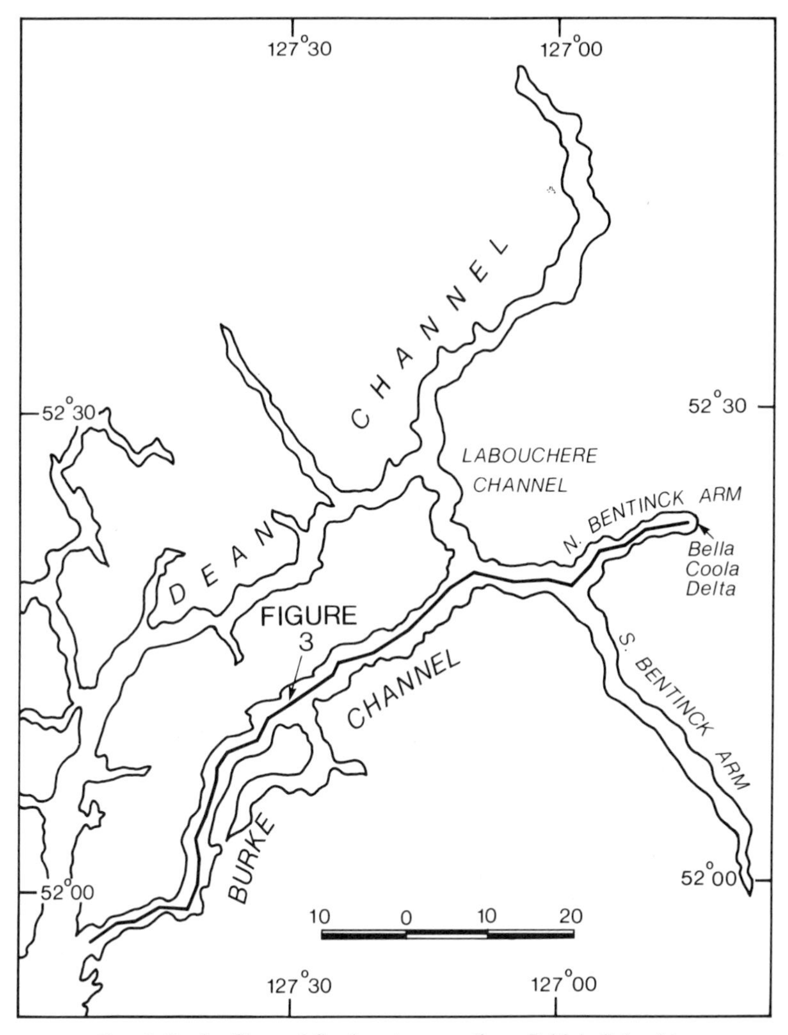

FIG. 2. Burke Channel fjord system, northern British Columbia.

II. Controls on Fjord Development and Sedimentation

A. Physical Setting

The British Columbia Coast is a leading-edge continental margin backed by rugged mountain ranges more than 3000 m in elevation, the higher areas of which maintain small ice caps and valley glaciers. The predominantly rocky coastline is extremely complex in outline, but the presence of numerous fjords, not only along the mainland coast but also on the west coast of Vancouver Island and the Queen Charlotte Islands, imparts a basic unity of landforms and depositional environments to the British Columbia coastal zone. This is continued northward into Alaska, so that the Pacific fjordland of North America extends through 10° of latitude. The fjords illustrated in Fig. 1, together with the associated narrow mountain valleys and marine channels, have developed in association with marked NW–SE and NE–SW trending structural alignments, and there is an overall rectilinear pattern to the coastal outline. The headward reaches are deep within the Coast Mountains. They exhibit a typical fjord morphology characterized by steep, bedrock walls, smooth sediment floors, deep basins, and shallow bedrock or bedrock and morainic sills.

B. Glacial and Sea-Level History

The last major episode of glaciation in British Columbia, the Fraser glaciation of late Wisconsin age, reached its maximum extent about 14,500 years ago, when a thick cover of Cordilleran ice reached the outer coasts of Vancouver Island and the Queen Charlotte Islands. Very rapid glacial retreat followed. The outer mainland coast of northern British Columbia was free of ice by 12,700 yr B.P. (J. J. Clague, cited in Bornhold, 1983), and some of the fjord heads were clear of ice by 11,000 yr B.P. Bornhold (1983) indicates that the ice front in outer Douglas Channel was retreating at an average rate of 8–10 km/100 yr during the period 12,700 to 12,000 yr B.P. Sea level was high during and immediately after deglaciation. The late Pleistocene marine limit, represented by raised deltas, is 150 m above present sea level at Bella Coola (Andrews and Retherford, 1978) and about 200 m at Kitimat (Clague *et al.*, 1982). Thus, the fjords were the sites of rapidly retreating tidewater glaciers for a 1500-yr period toward the end of the Pleistocene. In other words, there was a brief period toward the end of the Pleistocene when the fjord environment was glacial–marine (sensu, Molnia, 1983; Powell, 1984). None of the British Columbia fjords, unlike some of their Alaskan counterparts, contain tidewater glaciers today. However, the presence of glaciers in the mountain catchments had, and continues to have, a marked effect on the hydrological and sedimentological regimes of most of the major rivers that enter the fjords. Early Holocene recession from the late Pleistocene marine limit was rapid, and along the northern mainland coast sea level had stabilized a few meters below present level by 8,000 B.P. Sea level had risen slightly during the past few thousand years.

C. Holocene Sedimentation

Contemporary sedimentation in Coast Mountain fjords is controlled by the characteristics of both the contributing basins, supplying new sediment, and the receiving basins into which it is deposited. The most important contributing basin characteristics are the hydrological and sedimentological regimes of the principal rivers; the most important receiving-basin characteristics are morphology and circulation patterns. The main locus of sedimentation is the delta that commonly occurs in a fjord head setting.

1. River Regimes

Discharge characteristics of some Coast Mountain streams are summarized in Table I. All of the streams follow a similar pattern through the year. Streamflows are low in winter, increasing in the spring due to snowmelt, remaining high in the summer, the period of peak glacier melt, and declining in the autumn. In autumn and winter, short-lived, intense floods can occur as a result of rain storms on ripe snowpack (Kostaschuk and McCann, 1983). Rivers in the Coast Mountains are bedload dominated, in response to high sediment yields in the basins, but considerable quantities of suspended load, primarily glacial flour, are also transported. Bed textures are in the sand and gravel range and most bedload is transported during floods. The glacially derived suspended load is at a maximum during summer glacier melt discharges. The bedload is concentrated into deltas, but the suspended sediment plumes have the potential to extend long distances down the fjords.

2. Fjord Morphology and Circulation

The main morphological influence (deep, narrow receiving basins) is in confining the rapidly prograding deltas. Subaqueous slopes are steep.

TABLE I. DISCHARGE CHARACTERISTICS OF SOME COAST MOUNTAIN STREAMS

River	Mean annual discharge (m^3/s)	Mean of maximum month (m^3/s)	Season of maximum discharge	Drainage area (km^2)
Bella Coola	119	267	Summer	4170
Dean	140	340	Spring–summer	7850
Homathko	283	697	Summer	5720
Kitimat	135	294	Summer	1990
Klinaklini	325		Summer	6462
Squamish	242	501	Summer	2340

In the Bella Coola delta the delta front extends to a depth of 300 m in a linear distance of 4 km (mean slope 4°): the slope to the 100-m depth is in excess of 9°, and slopes immediately below distributary channels are much steeper. The combination of rapid progradation and steep slopes results in subaqueous slope instabilities and the development of submarine slumps and slides. A major translational submarine slide occurred in 1875 on the Kitimat delta (Luternauer and Swan, 1978; Prior *et al.*, 1983). It began as a number of shallow rotational slides and became translational downslope, extending for a distance of 4 km. Submarine slope failures lead to the generation of subaqueous debris flows and turbidity currents. Turbidity currents provide a mechanism for transporting material, originally deposited in the delta front, out to the deep fjord basins and redepositing it as a distinctive facies of basin sediments. Turbidity currents have been invoked to explain sedimentation patterns in Norwegian and New Zealand fjords (Holtedahl, 1965; Glasby, 1978), and Bornhold (1983) considers them to be important in Douglas Channel–Kitimat Arm.

The complimentary sediment flux on the fjord surface is, of course, the suspended sediment plume introduced by the major rivers. During the summer period of high river discharge and high suspended sediment concentration, the upper reaches of the fjords exhibit a marked density stratification, with a layer of turbid, fresh water overlying saline basin water. Just how far this extends downfjord under different circumstances has not been documented. However, Syvitski and Murray (1981) record, in their study of floccing and settling processes of the Squamish River plume in Howe Sound in southern British Columbia, that in the summer months river water skims across the surface of the fjord in their 17-km study reach with very little mixing. They suggest that 60% of the suspended load introduced at this time bypasses the proximal 17 km of the Sound. The turbid, freshwater surface layer developed throughout North Bentinck Arm was 2.5–3.0 m thick in 1981 and 6.0 m thick in 1982. With declining river discharge in the fall, wave turbulence and tidal mixing break down the stratification.

Within the area of mainland coast shown on Fig. 1, tides are mixed semidiurnal in type, and mean tidal range increases from 3.5 m in the south to more than 5.0 m in the north. Mean and large tidal ranges at Bella Coola and Kitimat are 3.8 and 5.9 m, and 4.4 and 6.7 m, respectively. Tidal currents are important in the vicinity of shallow sills, where, with accelerated velocities due to the confining of the tidal prism, they may scour the bed and inhibit sedimentation. Additionally, there is enhanced turbulent mixing of the stratified fjord waters at these locations. Tidally induced currents and mixing are particularly well documented in Knight

Inlet (Pickard and Rodgers, 1959; Freeland and Farmer, 1979). At deltas it appears that the main effect of the tide is to change the location of the distributary outlets (Kostaschuk and McCann, 1983).

Pickard (1961) evaluated wave conditions on Coast Mountain fjords, concluding that waves were all generated within individual fjords and tended to be steep and short-period (2–4 s) with maximum heights in the range 0.9–1.5 m. For over 80% of the time in summer, wave heights did not exceed 0.9 m. This is our experience in North Bentinck Arm (Kostaschuk and McCann, 1983). Syvitski and Farrow (1983) indicated that storm wave heights may approach 2.4 m in Knight and Bute Inlets. Wave action influences the character of some of intertidal sediments at deltas, and wave turbulence helps break down the summer stratification of surface waters, but, overall, waves are an unimportant factor in fjord sedimentation.

III. Burke Channel–North Bentinck Arm

A. Methods of Study

The primary data sources for the analysis of the fjord basin sediments are the seismic profiles surveyed in an October 1982 cruise on *C.S.S. Parizeau*. Two types of profiles were obtained: high-penetration, low-frequency airgun records and low-penetration, high-frequency 3.5-kHz records. The seismic stratigraphy approach, described in the series of papers by Vail *et al.* (1977) in A.A.P.G. Memoir 26 (Payton, 1977) and also by Sheriff (1980), was used to provide a systematic analysis of the data. This involves the identification of seismic sequences, seismic units, and seismic facies and subfacies, using a hierarchy of seismic characteristics (Table II). The separation into two sequences was based largely on the occurrence of a major reflector, which is interpreted as an unconformity separating Pleistocene glacigenic sediments and Holocene sediments. The separation into units is based on the three-dimensional shape (external form) of groups of reflectors (Mitchum *et al.*, 1977): the separation into facies utilizes primary reflector characteristics (reflector-free, parallel–subparallel, divergent, complex reflectors), and the separation into subfacies is based on secondary reflector characteristics (amplitude, smoothness, continuity, frequency, abundance, etc.). The last two sets of criteria are particularly useful in making assessments of the depositional environments of sediments (Sheriff, 1980, p. 86). Additional information on

TABLE II. SEISMIC STRATIGRAPHY AND SEISMIC SIGNATURE OF QUATERNARY SEDIMENTS IN BURKE CHANNEL–NORTH BENTINCK ARM

Sequence	Unit (external form)	Facies (reflector configuration)	Subfacies (modifying terms and other criteria)
Upper sequence	7. Bank	Opaque,[a] chaotic[b]	Hummocky, hyperbolic[b]
	6. Sheet	a. Sheet parallel–subparallel[a,b]	Continuous, subparallel to wavy: lateral variations in cycle breadth and amplitude
		b. Sheet–drape parallel–subparallel	Discontinuous, wavy[a,b]
	5. Lens	Opaque,[a] subparallel[b]	Hummocky, discontinous
Lower sequence	4. Mound	Opaque,[a] chaotic to subparallel[b]	i. Surface form ii. Subsurface form
	3. Sheet	Opaque,[a] subparallel[b]	Disrupted, hummocky[b]
	2. Chaotic fill	Chaotic[a,b]	i. Opaque;[a] disrupted, parallel[b] ii. Disrupted, wavy[a,b] iii. Disrupted, contorted[a,b]
	1. Parallel fill	Parallel[a,b]	Wavy, V-shaped perturbations

[a]From 3.5-kHz record.
[b]From airgun record.

delta front sediments and processes comes from echo sounding, sidescan sonar, and bottom sampling surveys (Kostaschuk and McCann, 1983; Kostaschuk, 1984).

B. MORPHOLOGY

Burke Channel extends 120 km northeastward from its junction with Fitzhugh Sound to the Bella Coola delta at its head (Fig. 2). It usually maintains a uniform width of about 3 km and is never more than 4 km wide. The surrounding mountains increase in height from 1500 m in the southwest to 2000 m in the northeast, and much of the channel is more than 400 m deep. In places there is a vertical drop of 2000 m in a horizontal distance of 4 m from mountain top to fjord floor. The fjord is contained in the southwest by an 8-km-wide bedrock sill that comes within 36 m of sea level. The outer bedrock slopes of the sill extend to a depth of 130 m before they are covered by sediment: the steeper, inner bedrock slope extends down to 240 m before it is covered. There are three small sediment-filled bedrock basins on the inner flank of the sill that become progressively deeper toward the northeast. Depths to the fjord floor in these basins are 240, 370, and 490 m; and to bedrock, 460 and 650 m. The main Burke Channel Basin has a relatively uniform depth of 600 m for a distance of 30 km before it begins to rise slowly toward the Bella Coola delta in North Bentinck Arm. The depth to bedrock in the main basin approaches 980 m. There are three tributary fjords to Burke Channel–North Bentinck Arm: Kwatna Inlet and South Bentinck Arm on the south, and Labouchere Channel on the north. As far as can be ascertained, each of these channels represents a separate depositional system.

C. Seismic Stratigraphy

Seven seismic units occur within the system, and they can be grouped into two distinct sequences separated by a pronounced and continuous reflector, which is interpreted as an unconformity (Table II and Fig. 3). The lower sequence of four units, below the unconformity, is interpreted as glacial or glacial–marine in origin and is considered to be late Pleistocene in age. The upper sequence of three units consists for the most part of deltaic and delta-related sediments and is Holocene in age.

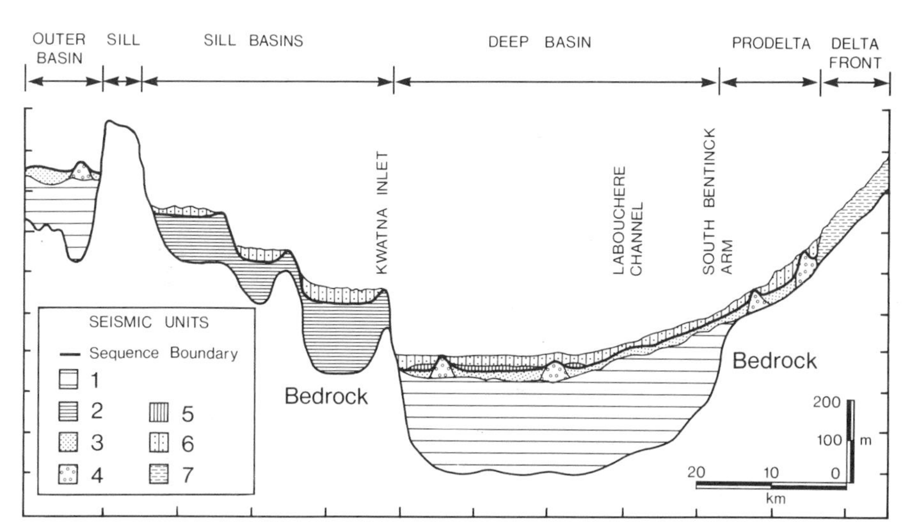

Fig. 3. Schematic representation of bedrock topography and seismic units of Quaternary sediments in Burke Channel–Northern Bentinck Arm.

1. Lower Sequence

a. Unit 1: Parallel Fill. This is a thick (200–250 m) basin fill unit, occupying the deepest parts of the main fjord basin. The airgun record shows continuous parallel, undulating reflectors, with V-shaped perturbations, which in places onlap onto bedrock. There are high-amplitude reflectors within the top few meters of the unit on the 3.5-kHz record, but at depth it becomes opaque (Fig. 4). By virtue of its position and

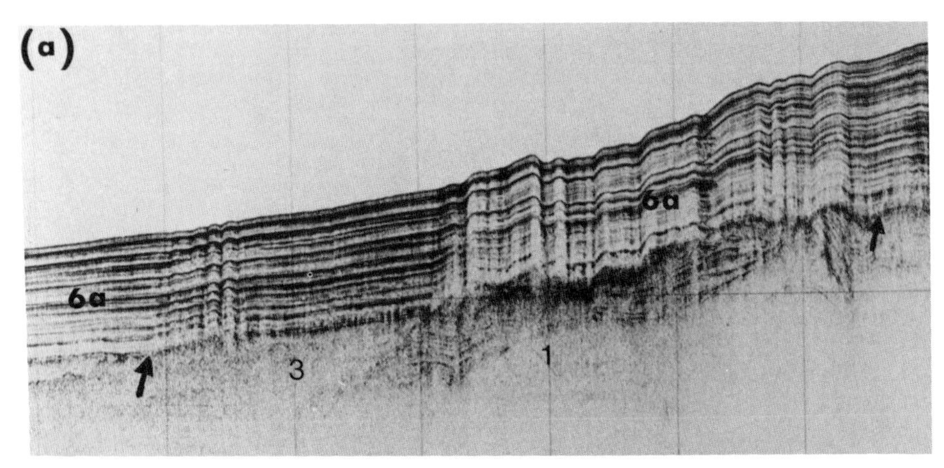

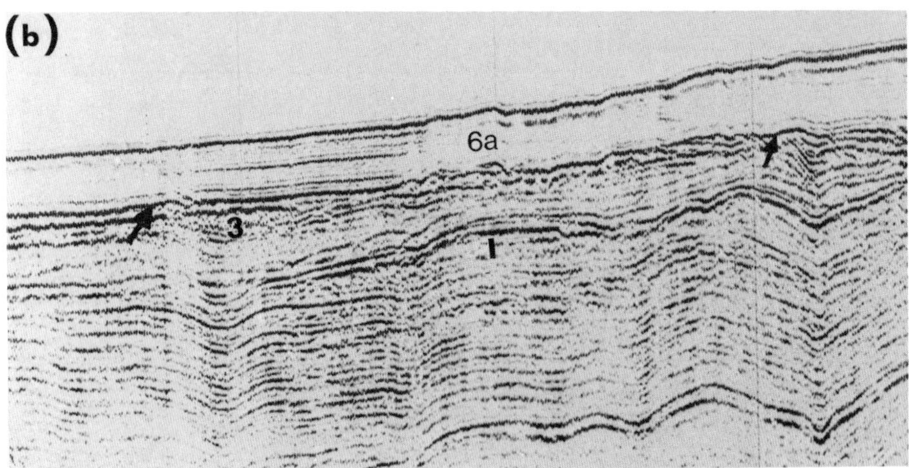

FIG. 4. (a) The 3.5-kHz and (b) airgun records from North Bentinck Arm showing upper-sequence unit 6a (sheet facies) overlying lower-sequence unit 1 (parallel fill) and unit 3 (sheet). The base of unit 6a is indicated by arrows. Both records show clearly the presence of two subfacies within unit 6a. In the 3.5-kHz record, wavy reflectors in the eastern half give way to straight reflectors in the west; in the airgun record, the unit is reflector-free in the east.

setting, and the interpretation offered below of the overlying unit, it is thought that unit 1 represents a thick sequence of tills deposited during the main phase of the Fraser glaciation. Some of it may date from earlier episodes of glaciation, but there is little indication of any stratigraphic breaks on the seismic records.

b. Unit 2: Chaotic Fill. This occurs as a basin fill in the three sill basins where it rests directly on bedrock and is 100–150 m thick (Fig. 5). It exhibits a chaotic array of reflectors, which are sufficiently similar throughout to suggest that only one facies occurs, though three subfacies are recognized (Table II). One subfacies, which is opaque on the 3.5-kHz record and has disrupted, parallel reflectors on the airgun record, occurs only in the outer sill basin. The other two subfacies, which exhibit disrupted, wavy, and disrupted, contorted reflectors on both records, occur in the middle and inner sill basins and appear to grade into each other. For the most part the unit is overlain by unit 6 (upper-sequence sheet), but it outcrops occasionally on the fjord floor with an opaque, rough surface. The chaotic reflectors and irregular surface of the unit indicate that it consists of a mixed assemblage of sediments deposited in a dynamic environment, where penecontemporaneous deformation was common. It is suggested that this was an ice contact–ice disintegration environment, developed during the rapid retreat of the tidewater glacier front across the irregular bedrock topography of the inner flank of the sill. Grounding, calving, and irregular intermittent retreat, with the deposition of subglacial till, push moraines, meltout debris, and subaqueous outwash, would all occur in such a setting.

c. Unit 3 Sheet and Unit 4 Mound. Though clearly part of the lower sequence, overlying unit 1, unit 3 does not have the configuration of a basin fill; rather, it is a thin (15 m), continuous sheet. It is opaque on the 3.5-kHz record and exhibits disrupted, hummocky, subparallel reflectors on the airgun record (Fig. 6). Laterally it appears to pass into the unit 4 mounds. There are two mounds in the deep basin, of which one outcrops on the fjord floor and the other is buried by upper-sequence sediments, and two in the approaches to the Bella Coola delta toward the head of the fjord, both of which outcrop on the fjord floor. The exposed mounds extend 40–60 m above the fjord floor sediments and are 2–3 km wide: the inner two have hummocky crests. They are opaque on the 3.5-kHz record and exhibit chaotic or subparallel reflectors on the airgun record. The sheet (unit 3) is interpreted as ice-rafted diamicton and bergstone mud deposited during the continuous rapid retreat of the glacier front across the smooth till floor of main basin and inner fjord. The mounds (unit 4) are interpreted as subaqueous moraines or morainal banks deposited during short-lived halts in the frontal retreat.

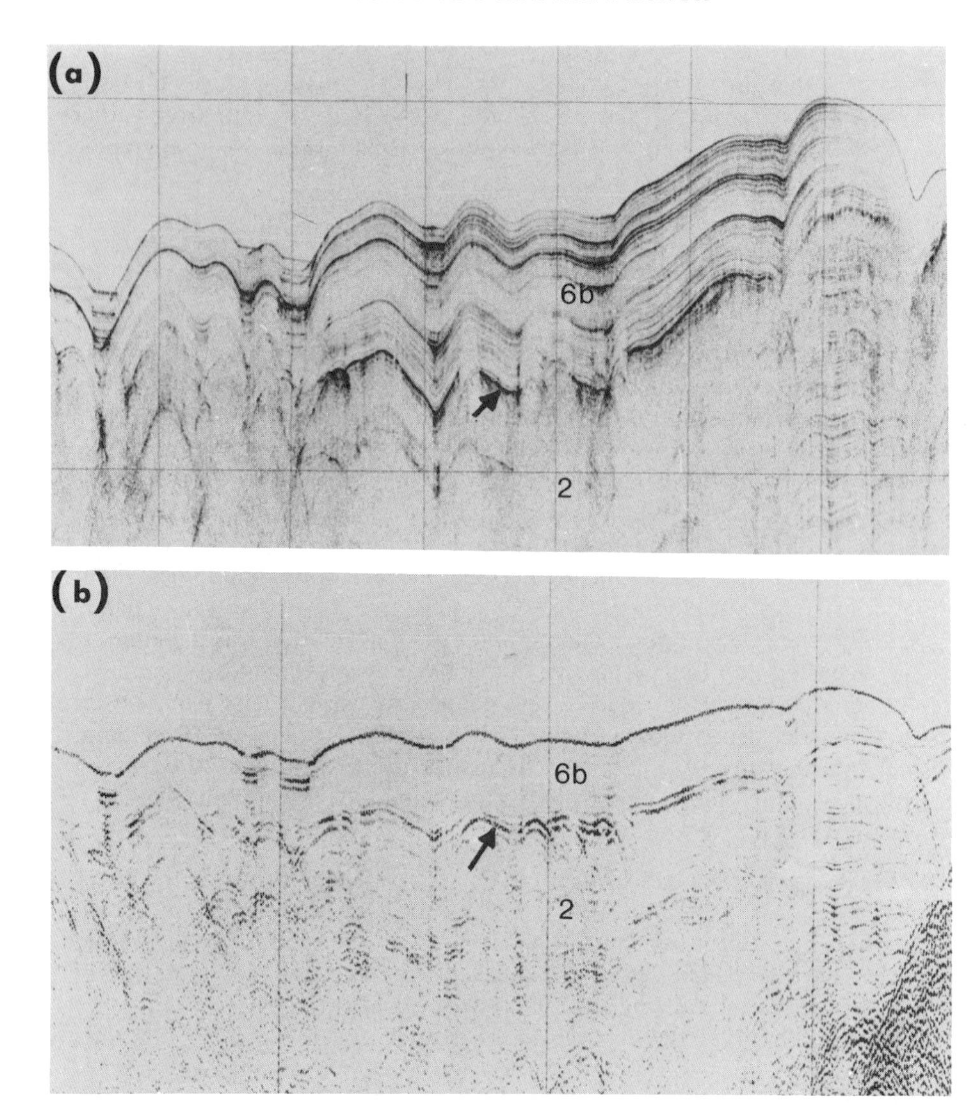

FIG. 5. (a) The 3.5-kHz and (b) airgun records from western Burke Channel showing upper-sequence unit 6b (sheet–drape facies) overlying lower-sequence unit 2 (chaotic fill). The base of unit 6b (indicated by arrows) is draped over the hummocky surface of the chaotic fill: reflectors within the unit and the fjord floor mirror this surface. Both records show variations within the chaotic fill that suggest that there are three subfacies (see Table II).

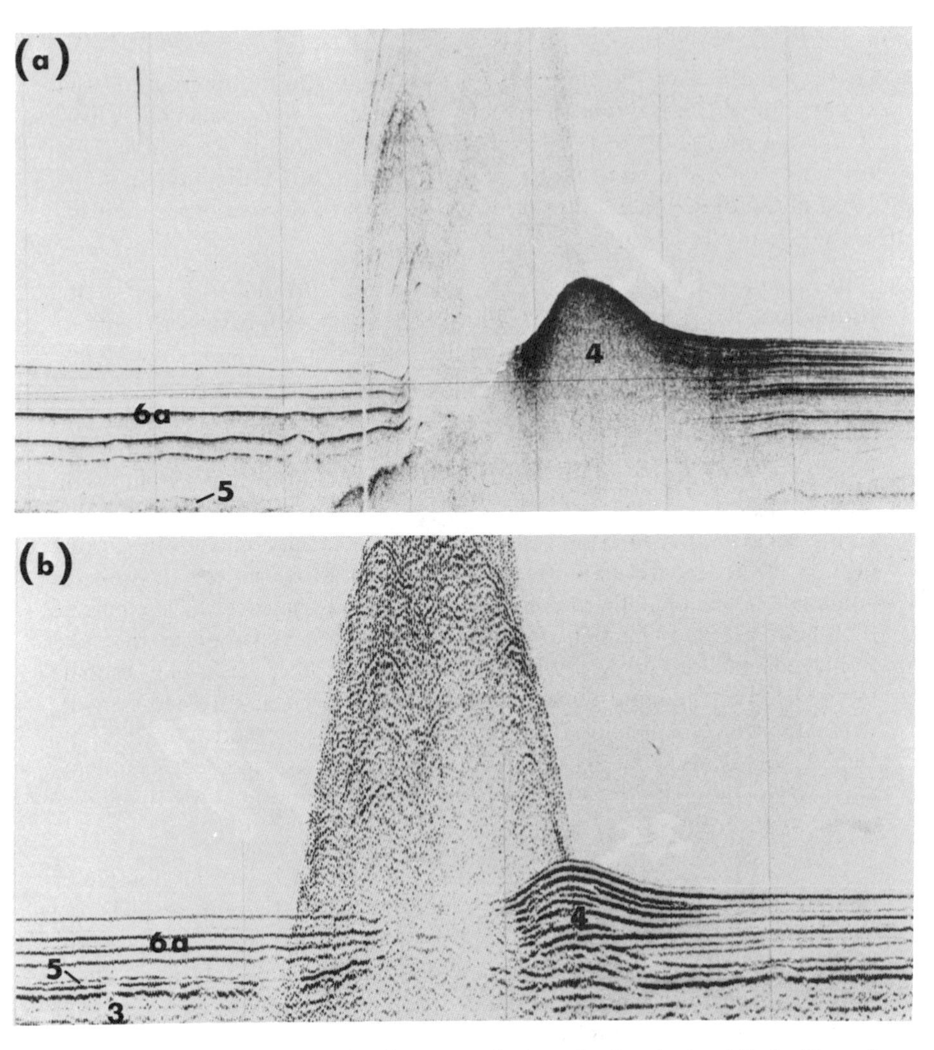

FIG. 6. (a) The 3.5-kHz and (b) airgun records from the deep basin of Burke Channel showing upper-sequence unit 6a (sheet facies) onlapping onto a lower-sequence unit 4 mound, which outcrops on the fjord floor. To the west of the mound, unit 6a overlies unit 5 (a thin lens). This in turn rests on lower-sequence unit 3 (sheet), which material appears to pass laterally into the base of the mound. The sharp peaked areas of reflectors just west of the mound are due to interference from the sidewalls of the fjord, due to the narrowness of the basin.

2. Upper Sequence

a. Unit 5: Lens. The lens is a thin (2 m), discontinuous unit that occurs above the prominent reflector in the deep fjord basin (Fig. 6). It is opaque on the 3.5-kHz record and shows discontinuous, subparallel reflectors on the airgun record. It clearly represents a distinctive, short-lived phase of sedimentation prior to the start of deposition in the early Holocene of the Bella Coola delta.

b. Unit 6: Sheet. This is the distinctive surficial unit, occurring throughout the fjord as a 20- to 40-m thick sheet with a smooth, slightly undulating surface, which makes up the fjord floor. It can be divided into two facies: a sheet, which occurs in the prodelta and deep basin (Fig. 4), and a sheet drape, which occurs in the three sill basins in the outer part of the fjord (Fig. 5). In the sheet, internal reflectors are generally parallel to subparallel but they do not mirror the surface of the underlying sediments. The base of the sheet may truncate reflectors in underlying units. In the sheet drape, internal reflectors and the surface mirror the configuration of the underlying units. Systematic lateral variations in reflector amplitude, cycle breadth, and configuration within the sheet facies indicate that it consists of a number of subfacies. In the zone adjacent to the delta front, reflectors are wavy or disrupted; toward the deep basin they become more orderly, continuous, and horizontal. The sheet is interpreted as muds deposited from suspension from the plume of sediment introduced by the Bella Coola River. The disrupted and wavy reflectors in sheet sediments adjacent to the delta are probably related to submarine slope instability on the delta front.

c. Unit 7: Bank. A steeply sloping, delta front bank unit completes this inventory. It is opaque on both the 3.5-kHz and airgun records and has a hummocky and mounded surface (Fig. 7). Mounds are of various shapes, and they tend to coalesce and overlap with a local relief of about 3 m. The seismic signature indicates that the sediments are sandy or coarser, and the topography provides evidence of slope movements and general instability. Further information on this delta front section of the fjord is available from echo sounder records and sidescan sonargraphs.

D. Sedimentary History

The genetic interpretations of the seismic units recognized within the fjord, inside the bounding bedrock sill, suggests that there have been four phases of deposition, each with a different mode of sedimentation. An attempt to place these events within the general chronology of late Wis-

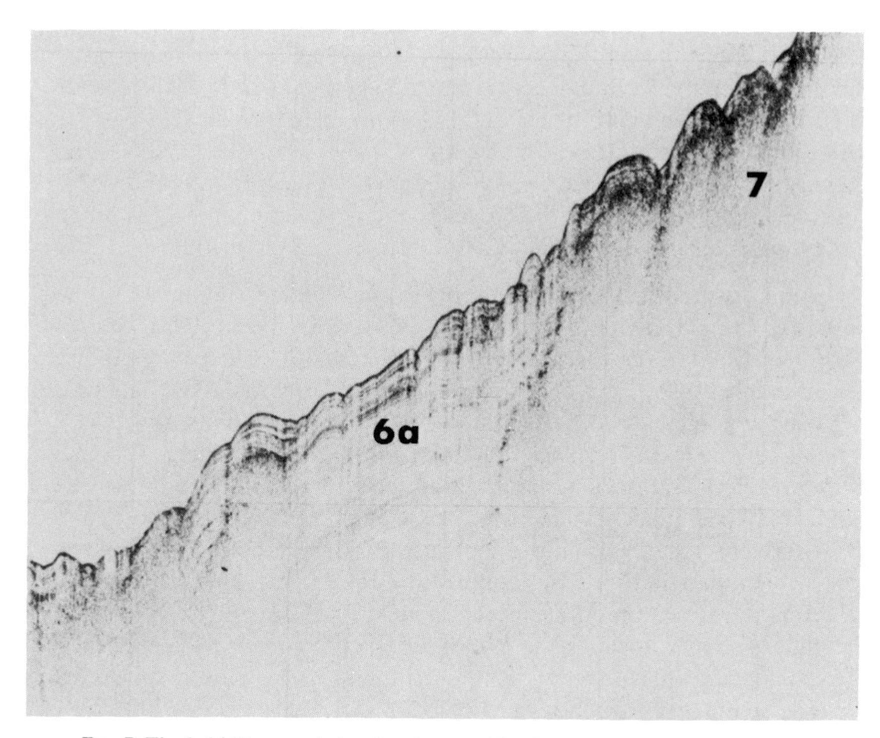

FIG. 7. The 3.5-kHz record showing the transition from upper-sequence unit 7 (bank) with hummocky, opaque reflectors, to upper-sequence unit 6a (sheet facies) in the Bella Coola delta. Alternatively, this could be termed the transition from the lower delta front to the prodelta zone.

consin events in coastal British Columbia suggests that the four depositional phases have been of unequal duration. From youngest to oldest (a–d) these phases have been:

(a) The Holocene phase, dominated by fluvial in puts of sediment from the Bella Coola River and the progradation of the Bella Coola delta at the head of the fjord; 9,000–10,000 yr of hemipelagic deposition interspersed with turbidity currents has produced the ubiquitous sheet unit, which is 45 m thick at the head of the fjord and thins to less than 10 m in distal areas.

(b) The final retreat phase of the Fraser glaciation, with rapid frontal recession of a tidewater glacier up the fjord; 1000–2000 yr of glacial–marine sedimentation, mostly by settling from suspension in very turbid water conditions, with the addition of ice-rafted debris, produced a continuous, though thin (15 m), sheetlike unit interrupted by small morainal mounds.

(c) A phase of slow retreat or stagnation of the tidewater glacier front against the inner flank of the bedrock sill: 1000–2000 yr of glacial marine sedimentation in the frontal, ice-grounding, ice-contact zone of the glacier infilled the sill basins with a thick (100–150 m), chaotic fill.

(d) The onset and full Fraser glaciation phase, when the fjord was the routeway for Cordilleran ice, which extended westward and southward across the continental shelf: 10,000 years of glacial sedimentation infilled the deep fjord basin with a thick (200–250 m) sequence of tills.

In comparison with the sequence of unconsolidated sediments in the Douglas Channel–Kitimat Arm, described by Bornhold (1983), the sequence described here is much thinner (300 m versus 600 m). There are other differences also, chief of which is the presence of a massive morainic sill, involving a thickness of 900 m of unconsolidated sediments, in Douglas Channel. However, the formation of this morainic sill in Douglas Channel and the development of the chaotic fill in the bedrock sill basin in Burke Channel may well represent the same phase of glacier front stability. They both clearly represent a major event in the retreat of the ice from the fjords, which was followed by rapid frontal retreat and a different mode of glacial–marine sedimentation. The phase of Holocene delta-related sedimentation that completes the record in Burke Channel–North Bentinck Arm also occurs in Douglas Channel–Kitimat Arm. In both fjords the Holocene sediments, which form the smooth fjord floor, constitute only about one-sixth of the total thickness of the late Quaternary fill in the deep bedrock basins.

Acknowledgments

We acknowledge the support of the Geological Survey of Canada (G.S.C.), and the good offices of the Pacific Geoscience Centre, where John Luternauer, Brian Bornhold, and Chris Yorath were very helpful. We thank Capt. B. L. Newton, the officers and crew of C.S.S. *Parizeau;* I. Frydecky, T. Forbes, and G. Jewsbury of G.S.C.; and M. T. Krawetz of McMaster University for their assistance during the cruise, and P. N. Matsushita for assistance with data analysis. The research was supported by N.S.E.R.C. Operating grant no. A5082 to S. B. McCann and an N.S.E.R.C. Post Graduate Scholarship to R. A. Kostaschuk.

References

Andrews, J. T., and Retherford, R. M. (1978). A reconnaissance survey of late Quaternary sea levels, Bella Bella/Bella Coola region, central British Columbia coast. *Can. J. Earth Sci.* **15,** 341–350.

Bornhold, B. D. (1983). Sedimentation in Douglas Channel and Kitimat Arm. *Can. Tech. Rep., Hydrogr. Ocean Sci.* **18,** 88–114.

Clague, J. J., and Bornhold, B. D. (1980). Morphology and littoral processes of the Pacific coast of Canada. *In* "The Coastline of Canada" (S. B. McCann, ed.), *Geol. Surv. Can. Pap.* **80–10,** 339–380.

Clague, J. J. Harper, J. R., Hebda, R. J., and Howes, D. E. (1982). Late Quaternary sea levels and crustal movements, coastal British Columbia. *Can. J. Earth Sci.* **19,** 597–618.

Freeland, H. J., and Farmer, D. M. (1980). The circulation and energetics of a deep strongly stratified inlet. *Can. J. Fish. Aquat. Sci.* **37,** 1398–1401.

Glasby, G. P. (1978). Sedimentation and sediment geochemistry of Caswell, Nancy and Milford Sounds. *In* "Fjord Studies: Caswell and Nancy Sounds, New Zealand" G. P. Glasby, ed., *N. Z. Oceanogr. Inst. Mem.* **79,** 19–37.

Holtedahl, H. (1965). Recent turbidites in Hardangerfjord, Norway. *In* "Submarine Geology and Geophysics—Coulston Papers" (W. P. Whitter and R. Bradshaw, eds.), pp. 107–141. Butterworth, London.

Kostaschuk, R. A. (1984). Sedimentation in a fjord-head delta, Bella Coola, British Columbia. Ph.D. Thesis, McMaster Univ., Hamilton, Canada.

Kostaschuk, R. A. (1985). Rivermouth processes in a fjord delta, British Columbia. *Mar. Geol.* **69,** 1–23.

Kostaschuk, R. A., and McCann, S. B. (1983). Observations on delta-forming processes in a fjord-head delta, British Columbia, Canada. *Sediment. Geol.* **36,** 269–288.

Luternauer, J. L., and Swan, D. (1978). Kitimat submarine slope deposits: a preliminary report. *Geol. Surv. Can. Pap.* **78-1A,** 327–332.

Mitchum, R. M., Vail, P. R., and Sangree, J. B. (1977). Seismic stratigraphy and global changes of sea level, Part 6: Stratigraphic interpretation of seismic reflection patterns. *In* "Seismic Stratigraphy—Applications to Hydrocarbon Exploration" (C. E. Payton, ed.), *Am. Assoc. Pet. Geol. Mem.* **26,** 117–133

Molnia, B. F. (1983). Subarctic glacial-marine sedimentation: a model. *In* "Glacial–Marine Sedimentation" (B. F. Molnia, ed.), pp. 95–143. Plenum, New York.

Payton, C. E., ed. (1977). "Seismic Stratigraphy—Applications to Hydrocarbon Exploration," *Am. Assoc. Pet. Geol. Mem.* **26.**

Peacock, M. A. (1935). Fjord-land of British Columbia. *Geol. Soc. Am. Bull.* **46,** 633–696.

Pickard, G. L. (1961). Oceanographic features of inlets in the British Columbia coast. *J. Fish. Res. Board Can.* **18,** 907–999.

Pickard, G. L., and Rodgers, K. (1959). Current measurements in Knight Inlet, British Columbia. *J. Fish. Res. Board Can.* **16,** 635–678.

Powell, R. D. (1984). Glacimarine processes and inductive lithofacies modelling of ice shelf and tidewater glacier sediments based on Quaternary examples. *Mar. Geol.* **57,** 1–52.

Prior, D. B., Coleman, J. M., and Bornhold, B. D. (1983). Results of a known sea floor instability event. *Geomar. Lett.* **2,** 117–122.

Sheriff, R. E. (1980). "Seismic Stratigraphy." Intern. Hum. Resour. Dev. Corp., Boston, Massachusetts.

Syvitski, J. P. M., and Farrow, G. E. (1983). Structures and processes in bay-head deltas, Knight and Bute Inlet, British Columbia. *Sediment. Geol.* **36,** 217–244.

Syvitski, J. P. M., and Murray, J. W. (1981). Particle interaction in fjord suspended sediment. *Mar. Geol.* **39,** 215–242.

Vail, P. R., Mitchum, R. M., Jr., Todd, R. G., Widmier, J. M., Thompson, S., III, Sangree, J. B., Bubb, J. N., and Hatlelid, W. G. (1977). Seismic stratigraphy and global changes of sea level (paper has II parts). *In* "Seismic Stratigraphy—Applications to Hydrocarbon Exploration" (C. E. Payton, ed.), *Am. Assoc. Pet. Geol. Mem.* **26,** 49–211.

CHAPTER 3

Coarse-Grained Beach Sedimentation under Paraglacial Conditions, Canadian Atlantic Coast

D. L. FORBES
AND
R. B. TAYLOR

Geological Survey of Canada
Atlantic Geoscience Centre
Bedford Institute of Oceanography
Dartmouth, Nova Scotia, Canada

Glaciation has left a strong imprint on development of the coast in eastern Canada. Postglacial changes in relative sea level have varied widely across the region, exerting a major control on shoreline migration and tidal range. Glacigenic deposits are the major sources of sediment supply to the coastal zone in most parts of the region. These sources include extensive thin covers of ground moraine, major drumlin fields, and localized thick deposits of ice-contact, glaciomarine, and glaciofluvial sediments.

The widespread occurrence of gravel beaches in glaciated areas can be attributed in part to high proportions of coarse clastic material in glacigenic sediments. Case studies presented in this chapter include sand–gravel and gravel beaches on the exposed sediment-deficient coast of southern Newfoundland, on the drumlin coast of southeast Cape Breton Island, and in the macrotidal Bay of Fundy. These examples illustrate the role of various geological and environmental factors in the shaping of beach morphology and facies characteristics.

Beach sedimentation is closely dependent on the location, volume, and sedimentology of source deposits and rates of supply from them. Beach characteristics are affected substantially by the relative proportions of sand and gravel in the system. Changes in relative sea level influence sediment supply and stratigraphic development. Other important factors include coastal planform, particularly the degree of compartmentalization, and aspects of the modern process regime, including wave climate, tidal range, and ice conditions.

I. Introduction

The Atlantic coast of Canada (Fig. 1) shows a great diversity of coastal geomorphology and sediments. The area has experienced repeated glaciation throughout late Cenozoic time, most recently in the late Wisconsinan. The imprint of this glacial heritage is evident in the abundance of glacial landforms and associated sediments and in the range of relative sea-level history experienced across the region. The variability of the coast provides an exceptionally good opportunity to examine the effects of various parameters such as sea-level change and source sediment properties on beach development in a range of geographical settings.

Use of the term *paraglacial* in this chapter follows its introduction by Ryder (1971) "to define nonglacial processes that are directly conditioned by glaciation" (Church and Ryder, 1972, p. 3059). The term has been applied to postglacial processes substantially influenced by the effects of former glaciation and to intervals of time during which paraglacial processes operate. Paraglacial fluvial systems experience initially high rates of proglacial sediment transport, followed by long postglacial relaxation trends (Fig. 2a) as the supply of glacial and proglacial source sediments is consumed. In coastal systems, paraglacial supply rates may be more variable and protracted. The coastal paraglacial cycle exhibits a marked spatial and temporal variability strongly influenced by the local geometry of sediment source deposits and the rate of change of relative sea level (RSL). In particular situations, such as marine transgression through a coastal drumlin field, the system may exhibit a repetitive sequence of input cycles (Fig. 2b) (Wang and Piper, 1982; Boyd *et al.*, this volume, Chapter 4).

This chapter is focused on coarse-grained (sand–gravel and gravel) beach sedimentation in a paraglacial setting. Although gravel beaches are found in many nonglaciated situations (see, e.g., Emery, 1955; Dobkins and Folk, 1970), the large volumes of gravel included in glacial sediments

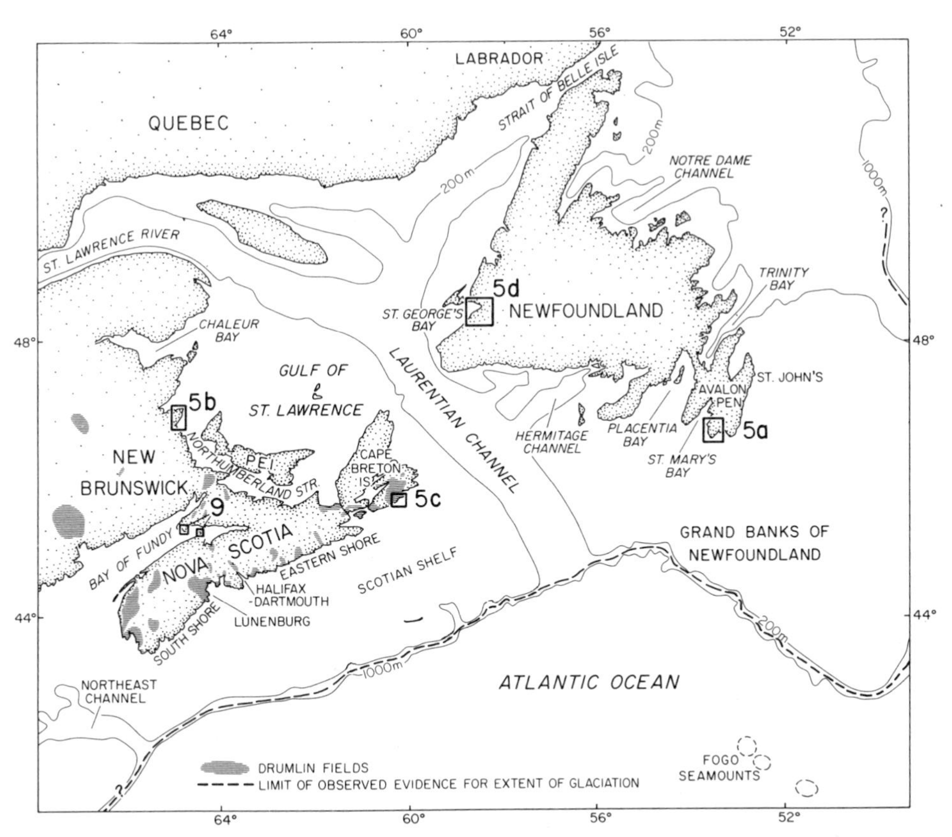

FIG. 1. Place names and locations of case study sites in Atlantic Canada. Drumlin fields are mapped from surveys by MacNeill (1974), Nielsen (1976), Munroe (1982), Stea and Fowler (1979, 1981), Stea (1983), and Rampton *et al.* (1984). The seaward limit of known evidence for glaciation is based on results of King and Fader (1986) and Piper *et al.* (in press).

contribute to the widespread occurrence of coarse-grained beaches in glaciated areas (Orford, 1979; Kirk, 1980; Carr, 1983; Carter and Orford, 1980, 1984; Carter *et al.,* 1987). The Canadian Atlantic coast is no exception. Throughout much of the region, glacial deposits constitute the major coastal sediment source. Coastal sands are common in some areas, notably in the southern Gulf of St. Lawrence where nonresistant sandstone is the predominant bedrock (Owens and Bowen, 1977; Forbes, 1987) and locally elsewhere (Bryant, 1983; Taylor *et al.,* 1985). Most overview papers dealing with coastal sedimentation in eastern Canada have emphasized sand-dominated coastal features (Owens, 1974a; McCann, 1979; Greenwood and Davidson-Arnott, 1979). Although certain aspects of coastal

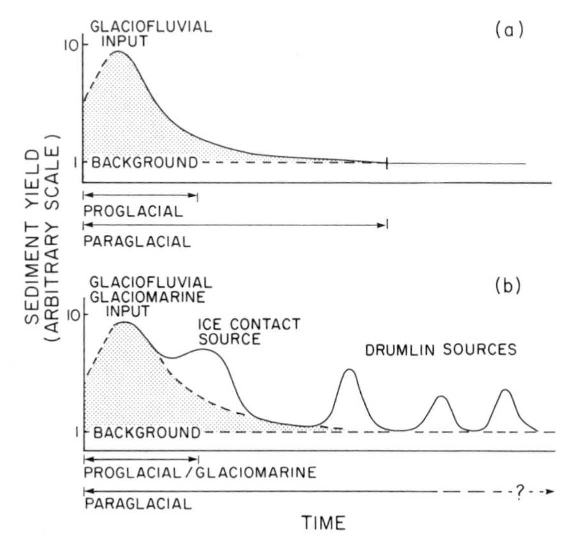

FIG. 2. Schematic representation of paraglacial sediment yield in (a) fluvial and (b) coastal environments. Both systems show long-term postglacial trend of decreasing sediment supply. Coastal system exhibits protracted paraglacial influence defined by the intersection of changing relative sea level and the distribution of glacial deposits. The figure illustrates a pattern of repetitive input cycles characteristic of postglacial marine transgression through a drumlin field. [Fig. 2a after Church and Ryder (1972), with permission.]

gravel sedimentation in the region have been reported (Johnson, 1925; Owens and Drapeau, 1973; Wightman, 1976; Rosen, 1979, 1980; Forbes, 1985; Taylor *et al.,* 1985; Shaw and Forbes, 1987), coarse-grained beach deposits are less well documented and remain relatively poorly understood. This chapter is an attempt to organize a summary of gravel beach characteristics as developed in a range of paraglacial settings throughout eastern Canada.

II. Glaciation in Atlantic Canada

Eastern Canada has been extensively glaciated on numerous occasions throughout the late Cenozoic (Grant and King, 1984). Cores from the Fogo Seamounts south of the Grand Banks (Fig. 1) suggest that the most extensive glaciation in the region may have been Illinoian, some time in the interval 200–125 ka B.P. (Piper, 1975; Alam *et al.,* 1983). Wisconsinan tills postdating an extensive raised shore platform, which has been referred to the last interglacial (Grant, 1976; Tucker and McCann, 1980), indicate that three major glacial advances occurred in eastern Canada since 75 ka B.P. Lithological evidence suggests that the oldest of these involved an extensive southeast-flowing ice sheet throughout the region, whereas the latter two involved more limited advances from local ice centers. Most of the modern coast was ice free by about 12 ka B.P. (Grant and King, 1984).

Many coastal and marine features of Atlantic Canada reveal the effects of glacial erosion, much of it pre-Wisconsinan. Examples include the fjords and other glacially deepened coastal embayments of Labrador, Newfoundland, and Nova Scotia; and major transverse troughs on the shelf, such as Northeast, Laurentian, Hermitage, and Notre Dame channels (Fig. 1). Although many of these features were defined initially by preglacial tectonics or fluvial erosion (Johnson, 1925; King, 1972; Piper *et al.*, 1986), substantial volumes of rock were removed by glacial processes (Alam and Piper, 1977).

Materials eroded by the ice were deposited elsewhere to form an extensive sedimentary record. Although in many areas these deposits form no more than a thin veneer over older rocks, thick accumulations at some coastal sites form important sources of sediment input to the littoral zone. Drumlin fields (Fig. 1) occur along much of the present coast of Nova Scotia. Other significant paraglacial sediment sources include complex stratigraphic sequences along the southwest coast of Nova Scotia (Grant, 1976), on the north shore of Minas Basin in the upper Bay of Fundy (Swift and Borns, 1967), and locally in southwest and southeast Newfoundland (MacClintock and Twenhofel, 1940; Brookes, 1974; Eyles and Slatt, 1977; Tucker and McCann, 1980). More restricted outwash valley-fill and deltaic deposits are present at many other locations (Flint, 1940; Tucker, 1974; Dubois, 1980; Taylor and Kelly, 1984).

III. Postglacial Relative Sea Level and Tidal Adjustments

Quinlan and Beaumont (1981, 1982) have presented a model that identifies four zones of postglacial relative sea level (RSL) adjustment in Atlantic Canada (Fig. 3). The model simulates the rheological response of the earth to differential late Wisconsinan ice loads, including the behavior of a glacial forebulge that migrated across the region following glacial recession. Zone A, closest to the center of late Wisconsinan ice, has experienced continuously falling RSL (from marine limits 75–200 m above present sea level), whereas zone D, the outermost zone, has undergone

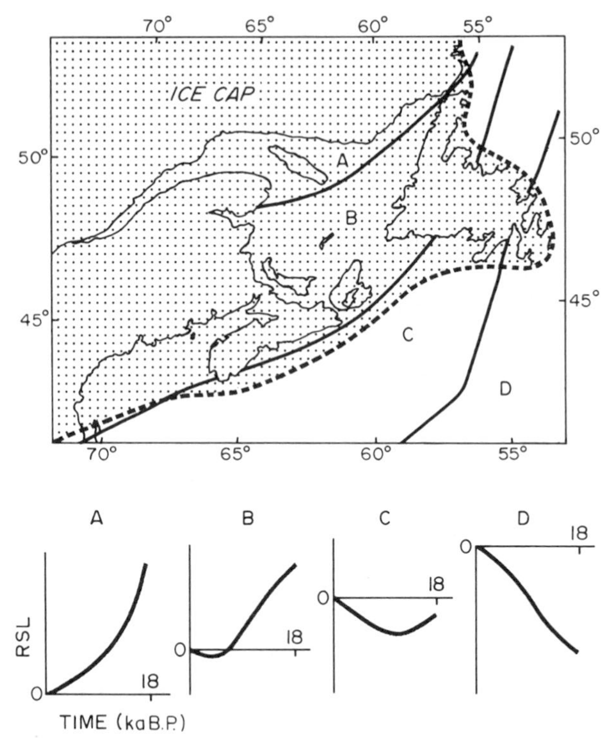

Fig. 3. Four zones of postglacial relative sea level (RSL) in Atlantic Canada, based on the model of Quinlan and Beaumont (1981), assuming a "maximum ice model" for 18 ka B.P. (stipple pattern). Zone A is a region of continuously falling postglacial RSL, zone D a region of continuously rising RSL, and zones B and C occupy the intermediate area where falling RSL in early postglacial time was followed later by rising RSL as the ice-marginal forebulge migrated across the region. Zone boundaries are schematic and depend on the deglaciation scenario selected for the computations. The maximum ice model shown here is no longer favored; see Quinlan and Beaumont (1982) for preferred model. In reality, the zone B–C boundary may lie farther north in Nova Scotia and Prince Edward Island and farther southeast in eastern Newfoundland (see Grant, 1977); location of the C–D boundary is not well known. [After Quinlan and Beaumont (1981), with permission.]

a continuous marine transgression throughout postglacial time. Between the two extremes (in zones B and C), falling RSL has been succeeded by rising RSL in varying proportions according to distance from the ice center.

Changes in RSL are reflected *inter alia* in the stratigraphy of coastal sediments (transgressive versus regressive sequences), in coastal morphology (including the distribution of former shore features and littoral deposits relative to present sea level), and in modern tide-gage records.

The preservation of specific regressive or transgressive sedimentary sequences is dependent on the rate of sea-level change, sediment supply, and wave climate (Sloss, 1962; Swift, 1975; Reinson, 1984; Boyd and Penland, 1984). In areas of falling RSL (zone A and early postglacial zone B), abandoned shoreline sequences have emerged above present sea level (Flint, 1940; Rosen, 1980; Dubois, 1980; Tucker, 1974). Where RSL is rising at present (zones B, C, D), transgressive shore-zone deposits may remain below sea level (Kranck, 1971; Grant, 1975; Barnes and Piper, 1978; Piper *et al.*, 1986). Much of the Canadian Atlantic coast exhibits a combination of regressive and transgressive sequences (zones B and C, Fig. 3; see, e.g., Wightman, 1976). In some areas, both old submerged transgressive sequences and older emerged regressive deposits are being reworked in the present coastal system. Other such combinations occur in zone D in response to factors other than changing relative sea level (Shaw and Forbes, 1987).

Changes in RSL can also force changes in tidal amplitude by altering basin geometry, particularly at the entrance to large embayments. This has been the case in the Bay of Fundy (Grant, 1975; Scott and Greenberg, 1983), where spring tide ranges as great as 16 m occur at present. Former fluctuations in tidal amplitude are recorded in the sediment sequences of emerged coastal features (Wightman, 1976) and in sublittoral deposits at the head of the bay (Amos, 1978; Amos and Zaitlin, 1985). Because the vertical range of wave action and the frequency of wave attack at given intertidal levels are controlled by tidal range, changes in tidal amplitude represent an important control on beach sedimentation processes.

IV. Modern Oceanographic Environment

The coast of Atlantic Canada is storm-wave dominated and characterized by sharp seasonal contrasts in wind, wave, and ice conditions. Storm waves are generated by cyclonic depressions passing northeastward across the region and also by occasional tropical storms moving north along the U.S. coast. Off the open Atlantic coast, wave-height isopleths roughly parallel the coast (Fig. 4), with annual deep-water significant wave heights in the 7–8-m range, and 10-yr significant wave heights from 10 to 13 m (Neu, 1982). Modal wave conditions observed on the Grand Banks are 2–4 m significant height and 6–8-s period; periods in the 8- to 14-s range are associated with larger waves (Neu, 1982). In the Gulf of St. Lawrence, annual wave heights of 5–8 m have been quoted for some sites (Ploeg, 1971; Owens, 1974b; Forbes, 1984).

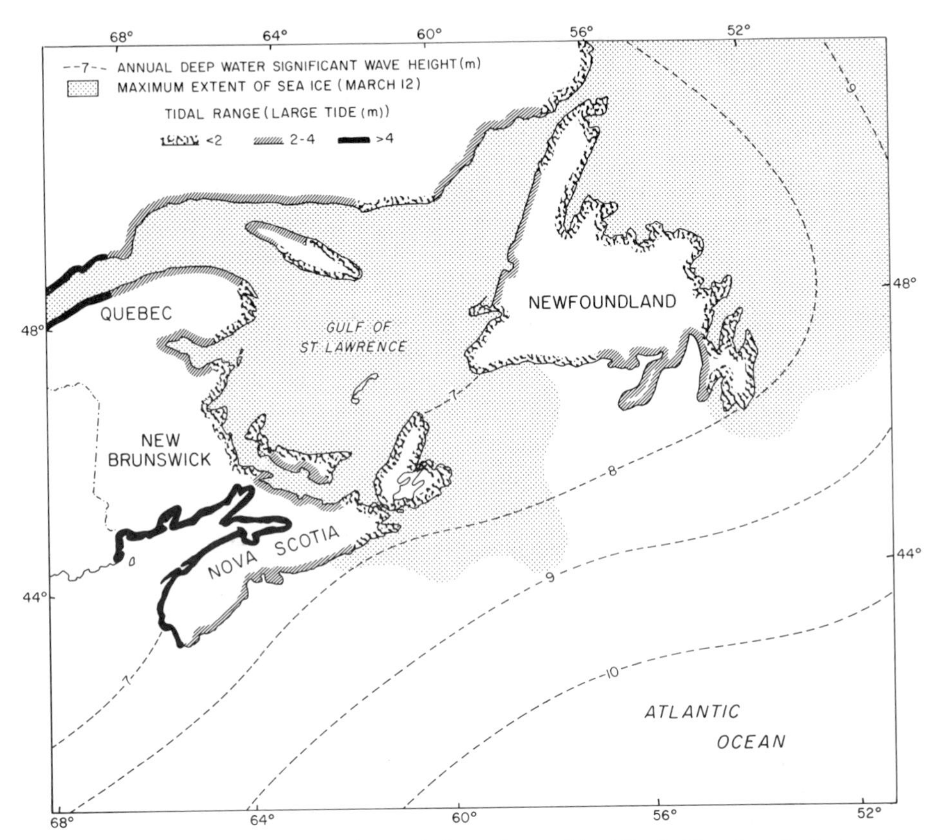

Fig. 4. The diversity of beach morphology observed in Atlantic Canada reflects in part the variety of tidal, wave, and sea-ice conditions in the region. Data compiled from Canadian Hydrographic Service (1984), Neu (1982), and Markham (1980).

Seasonal ice duration (Markham, 1980) varies from near zero on the coast of southern Nova Scotia to 6 months or more on the Labrador coast. In the southern Gulf of St. Lawrence, sea ice limits wave activity for up to 4 months each year. Although ice may be present in the Bay of Fundy for up to 4 months, the bay is never completely ice-covered (Owens and Bowen, 1977).

The tidal range increases from less than 2 m on much of the open Atlantic coast to 3–8 m northward along the Labrador coast; from almost 5 m near the entrance to the Bay of Fundy to over 16 m at the head of the bay; and from 0.8 m in parts of the Gulf of St. Lawrence to values as high as 3.3 m in Chaleur Bay (Canada Hydrographic Service, 1984) (Fig. 4).

V. Beach Morphology and Sediments—Selected Examples

The following three case studies include examples of beach development under a range of RSL, sediment supply, and oceanographic conditions. They have been selected from a much larger sample of observations at many sites throughout eastern Canada (see Forbes *et al.*, 1982; Forbes, 1984; Taylor *et al.*, 1985; Shaw and Forbes, 1987; among others). The examples presented here represent exposed, high-wave-energy, microtidal environments in RSL zone D (case 1: Figs. 3, 5a, 6, and 7b–d) and zone B or C (case 2: Figs. 5c, 7e, 7f, and 8) and a macrotidal setting in zone B (case 3: Figs. 9–10). They have been chosen also to highlight the effects of contrasting sediment source character, varying proportions of sand and gravel and RSL-controlled changes in tidal range on gravel beach development.

A. Case 1: Mutton and Holyrood Bays—Contrasting Sediment Sources on a Compartmented Coast

The Mutton Bay barrier (Fig. 5a) is an example of beach development in an area of restricted paraglacial sediment supply. Henderson (1972) mapped onshore surficial deposits in this part of the southern Avalon Peninsula as coarse till, 0–6 m thick (Fig. 7a). The pre-Quaternary rocks in the area are well-indurated Precambrian clastic sediments that represent a very limited sediment source. Mutton Bay bottom sediments are thin lag gravel over till or rock, except in the inner bay where some sand occurs (Table I) (Forbes, 1984). The coast is highly compartmented by prominent headlands that impede alongshore transport of sand and gravel. Although low till cliffs at both ends of the barrier supply a limited quantity of new material to the system, the beach is strongly supply-limited.

The barrier (Figs. 6, 7b, and 7c) is roughly 30–50 m wide at mean water level (MWL), with crest elevations ranging up to 4.5 m above higher high water at large tides (HHWLT). The barrier beachface is strongly reflective under almost all incident wave conditions. Nearshore sand (facies 1, Fig. 6, Table II) terminates below lower low water at large tides (LLWLT) against the base of a steep gravel step, 0.6–1.0 m high (facies 2). Above the step, the lower beachface (facies 3) exhibits a concave-up profile, rising to a very steep pebble berm face (facies 4). Above the berm, the upper beach face (facies 5), barrier crest (facies 6), and backbarrier slope (facies 7) consist of open-work pebble–cobble–boulder gravels (Tables I, II; Fig. 7c).

The profiles in Fig. 6 show changes in the Mutton Bay barrier at one

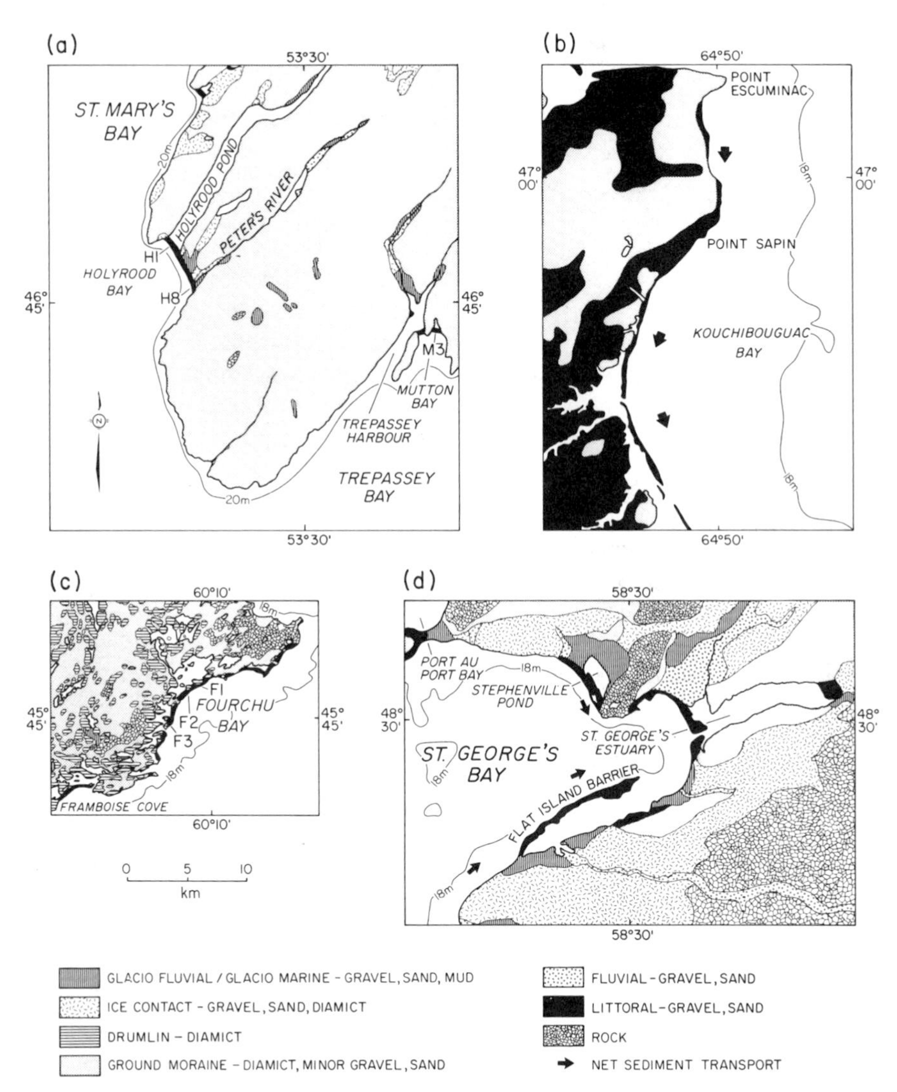

Fig. 5. The diversity and abundance of glacigenic sediments in Atlantic Canada are reflected in the size and geometry of present-day beach systems. Maps (all at the same scale) show inner-shelf bathymetry and onshore distribution of surficial deposits in four selected areas (see Fig. 1 for locations). (a) Highly compartmented coast of southern Avalon Penisula, Newfoundland, showing case-study examples in Mutton and Holyrood Bays. Surficial materials generalized after Henderson (1972), Rogerson and Tucker (1972), and Eyles and Slatt (1977). (b) Highly integrated coastal development on the northeast New Brunswick coast, where sandy barrier system in Kouchibouguac Bay is fed by longshore transport tapping littoral sheet sands, thin till, and underlying sandstone from Point Escuminac south. Surficial

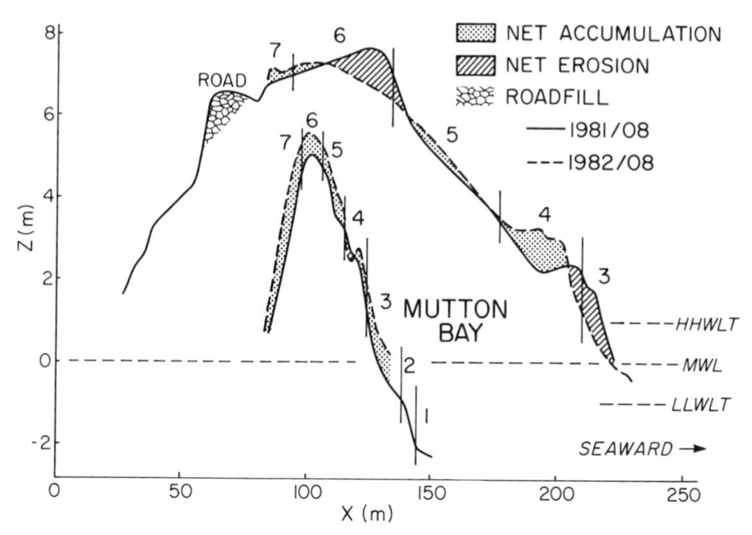

Fig. 6. Profiles across barriers at Holyrood Bay (H1) and Mutton Bay (M3) (see Fig. 5a for locations), showing net changes over a 12-month interval (August 1981 to August 1982) and cross-barrier distribution of surface facies (Tables II and III). Note contrasting width, height, and sediment storage volume, reflecting major differences in source materials and sediment budgets.

location (line M3) over a 12-month interval, during which two overtopping events occurred. The first caused erosion of the upper beachface and extensive washover deposition on the back-barrier slope. At line M3, the crest was built up about 0.4 m (Fig. 6); at an adjacent profile line (M4), the crest elevation was reduced some 0.5 m by scour of a 15-m-wide washover channel (Fig. 7c). During the second event, overtopping aggradation resulted in further build-up of the crest at line M3, but no deposition occurred on the back slope; location M4 experienced washover deposition on the landward slope and waning-stage sedimentation in the channel

---

materials generalized after Rampton *et al.* (1984). Note that, in this panel, organic deposits cover entire stippled area and dark-shaded area includes shallow marine facies. (c) Well-integrated transgressive beach development on a drumlin coast, southeast Cape Breton Island, showing case-study sites in Framboise Cove and Fourchu Bay. Irregular 18-m isobath reflects complex shoal bathymetry on inner shelf. Surficial materials generalized after Grant (1987). (d) Littoral and estuarine sediment sinks at head of St. George's Bay, southwest Newfoundland. Note large spit complex (Flat Island barrier) fed from extensive cliff exposures of glacigenic sediments along coast in southwest corner of map area. Surficial materials generalized after Brookes (1974).

FIG. 7. (a) Aerial oblique view of shore platform cut in steeply dipping Precambrian clastics, with thin gravel beach wedge backed by low till cliff (outhouse circled for scale), Mosquito Cove, Great Colinet Island, St. Mary's Bay (Fig. 1). [Geological Survey of Canada photo 190867, August 1, 1981 with permission.] (b) Aerial oblique view of Mutton Bay barrier (Fig. 5a), showing approximate location of profiles M3 and M4 (circle). Eye witnesses report barrier well seaward of foreshore rock (left of circle) about 50 years ago. [Geological Survey of Canada photo 190868, August 1, 1981, with permission.]

FIG. 7. (c) Ground view of Mutton Bay barrier, looking west along crest toward profile M3 (benchmark circled). Broken lines in foreground delineate washover channel cut across barrier crest at M4 in storm of January 10, 1982. Note open-work cobble–boulder gravel on barrier crest in foreground. Pack on crest beyond M3 and 1.5-m staff at far side of washover channel provide scale. [Geological Survey of Canada photo 190856, February 22, 1982, with permission.] (d) Beach at Holyrood Bay (Fig. 5a), looking northwest from top of berm at H8. Cliffs (large arrow) cut into major ice-contact deposit in middle distance provide the chief sediment source. Beach extends as barrier across mouth of fjord (Holyrood Pond) beyond cliffs, in vicinity of H1 (Figs. 5a and 6). Photo shows coarse end of beach: berm-crest sediment size grades from 18 mm at H1 to 43 mm at H8 (foreground). Note figure for scale. [Geological Survey of Canada photo 190855, February 21, 1982, with permission.]

Fig. 7. (e) Upper beachface and barrier crest at Fourchu Bay F2 (Figs. 5c and 8), showing featureless winter profile and ramp developed against the dune scarp. Drumlin cliffs in background. Helicopter and pack (circled) for scale. [Geological Survey of Canada photo 190862, February 17, 1981, with permission.] (f) Barrier crest at Fourchu Bay F3 (Figs. 5c and 8), showing closed-work (sand–pebble–cobble–boulder) gravel and extensive washover flats. Note contrast with site F2 and figures for scale. [Geological Survey of Canada photo 190858, August 7, 1980, with permission.]

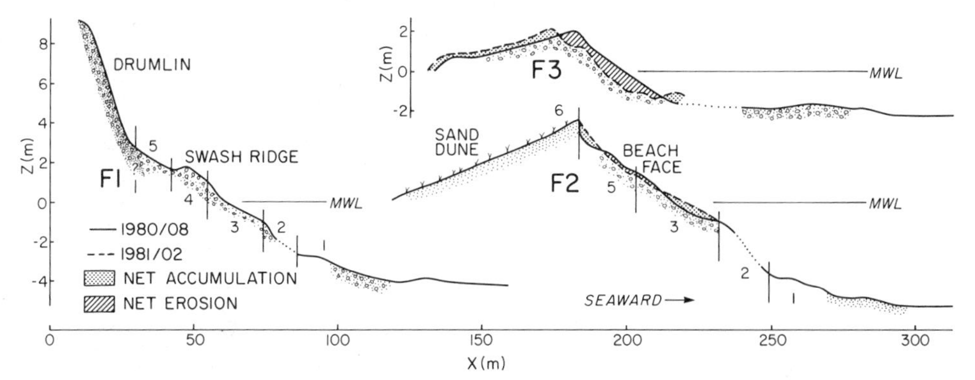

Fig. 8. Profiles across cliff-base beach (F1) and barrier (F2 and F3) in Fourchu Bay (see Fig. 5c for locations), showing net changes over a 6-month interval (August 1980 to February 1981) and cross-beach distribution of surface facies (Table IV). Note contrast in beach sediment volume between F2 and F3, indicating alongshore variability of sediment budget.

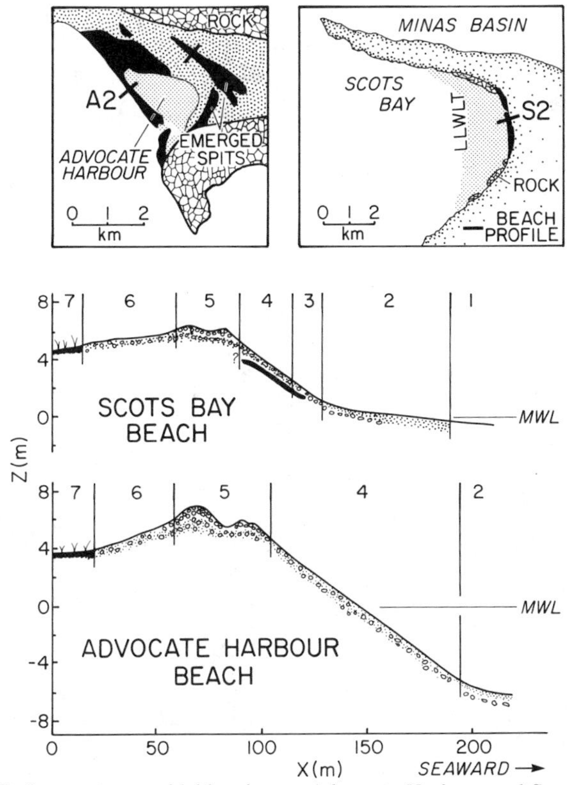

Fig. 9. Profiles across macrotidal barriers at Advocate Harbour and Scots Bay, Bay of Fundy, showing distribution of surface facies (Tables V and VI). Inset maps show profile locations (see Fig. 1 for map locations), intertidal flats (light stipple), and beach deposits (black). Profile line across raised-beach deposit at Advocate Harbour shows location of Wightman's (1976) section (Table VII). Distribution of glaciofluvial (heavy stipple) and littoral deposits (black) at Advocate Harbour generalized after Swift and Borns (1967).

Fig. 10. Beach sedimentation in a macrotidal setting, Scots Bay, Nova Scotia (Fig. 9). Distinct break in slope at mean sea level separates (a) extensive tidal flat, up to 1 km wide, from (b) cobble–boulder lag deposit.

Fig. 10. (c) Thin-bedded closed-work sand–gravel across the upper foreshore gives way at some locations to (d) open-work gravel on the barrier crest and backbarrier washover sequences and elsewhere to swash bars fronting low dune deposits. [Geological Survey of Canada photos 190861 (a), 190859 (b), 190857 (c), 190860 (d), August 28, 1984, with permission.]

TABLE I. SEDIMENT SIZE DATA—SELECTED SAMPLES FROM STUDY SITES

Locality/profile	Facies[a]	Size fractions (%)[b] m	s	p	c	b	Mean (phi)[c]	SD (phi)[c]
Mutton Bay								
—	1 Nearshore	0	100	0	0	0	2.90	0.18
—	1 Nearshore	0	100	0	0	0	2.77	0.67
M4	4 Berm	0	0	98	2	0	−4.65	0.50
M4	6 Crest	0	0	95	5	t	−5.14	0.56
M4	6 Crest	0	0	45	55	t	−6.09	0.62
Holyrood Bay								
H8	3 Lower beachface	0	0	100	0	0	−1.89	0.47
H8	4 Berm	0	0	94	6	0	−5.45	0.31
H1	4 Berm	0	0	100	0	0	−4.19	0.53
H1	5 Upper beachface	0	91	9	t	t	0.04	0.69
Fourchu Bay								
F1	s Drumlin till	3	24	73	t	t	−2.85	3.22
F1	1 Nearshore	0	10	35	55	0	−5.07	2.54
F1	1 Nearshore	1	99	0	0	0	3.10	0.44
F1	3 Lower beachface	0	83	17	0	0	0.81	1.54
F1	4 Swash ridge	0	0	100	0	0	−2.66	0.76
F2	1 Nearshore	1	99	0	0	0	3.28	0.33
F2	3 Lower beachface	0	94	6	0	0	1.18	1.22
F2	3 Swash ridge	0	1	99	0	0	−3.90	0.93
Framboise Cove								
—	s Drumlin till	54	24	22	t	0	0.85	3.50
B1	1 Nearshore	1	99	0	0	0	2.64	0.43
B1	3 Lower beachface	0	73	27	0	0	−0.35	0.82
B1	4 Swash ridge	0	17	83	0	0	−1.72	0.82
B1	5 Upper beachface	0	99	1	0	0	0.95	0.50
Scots Bay								
—	s Glaciofluvial	1	31	68	t	0	−1.88	2.05
—	s Till	28	34	38	t	t	1.47	4.59
S1	2 Lower beachface	1	90	9	0	0	1.15	1.41
S1	4 Upper beachface	1	58	41	0	0	−0.83	1.99
S1	5 Swash ridge	0	0	100	0	0	−2.96	0.94
S1	5 Crest	0	0	100	0	0	−4.22	0.67
S1	6 Dune	1	99	0	0	0	0.65	0.70
Advocate Harbour								
—	s Glaciofluvial	2	24	74	0	0	−1.70	1.87
—	s Till	73	21	6	t	t	6.51	4.02
A1	4 Lower beachface	0	3	97	0	0	−2.48	1.21
A1	4 Lower beachface	0	42	58	0	0	−1.65	1.84
A1	5 Crest	0	0	100	0	0	−4.26	0.70

[a]See Figs. 6, 8, and 9 for facies distribution; s refers to sediment source.

[b]Fraction abbreviations as follows: m, mud; s, sand; p, pebble; c, cobble; b, boulder; t indicates small proportion (not sampled).

[c]$d[\text{phi}] = -\log_2(d[\text{mm}])$ where d is grain size.

TABLE II. FACIES CHARACTERISTICS—MUTTON BAY BARRIER

Facies[a]	Zone	Slope[b]	Structure/texture
1. Nearshore shoreface	Subtidal	0.01	Thin veneer of parallel-laminated and ripple-cross laminated sand patches with gravel over diamict or rock
2. Step	Subtidal	0.25	Planar cross-stratified pebble gravel
3. Lower beachface	Intertidal	0.10/0.50	Imbricate pebble gravel (upper foreshore)/sand–pebble–cobble gravel (lower foreshore)
4. Berm	Supratidal	−0.08/0.13	Crudely stratified pebble–cobble gravel
5. Upper beachface	Supratidal	0.10/0.58	Crudely stratified pebble–cobble–boulder gravel with some large-scale cusps
6. Barrier crest	Supratidal	−0.18/0.09	Crude horizontally stratified weakly imbricated pebble–cobble–boulder gravel
7. Washover	Supratidal–lagoon	−0.43/−0.06	Crudely stratified landward-dipping pebble–cobble–boulder gravel

[a]See Figs. 6 and 7b,c.
[b]Expressed as tangent of slope angle, positive seaward; minimum and maximum values.

throat. The two events together resulted in up to 3-m net landward migration of the barrier crest. Preliminary airphoto analysis and information from local residents suggest that the barrier has migrated landward by as much as 50 m over the past 50 years (Forbes, 1984) without marked change in form. This indicates a sediment-starved condition in which washover transport exceeds the supply of new gravel to the beach.

Holyrood Bay (Figs. 5a, 7d), a 6-km-wide embayment in St. Mary's Bay (Fig. 1), is characterized by similar bedrock geology and is exposed to much the same wave environment as Mutton Bay. However, a large glacigenic sediment source, a small fluvial sediment supply, and a deep-fjord sediment sink—Holyrood Pond—set this site apart as a distinctive setting for beach and barrier development (Forbes, 1985). The complex sequence of glacial and proglacial sediments exposed in backshore cliffs up to 25 m high (Rogerson and Tucker, 1972; Eyles and Slatt, 1977) provides the major sediment source for the large volume of beach deposits in this system.

The beach has evolved to a smooth curvilinear planform that cuts into thick source deposits in the middle of the bay and forms a barrier across Holyrood Pond (Fig. 5a). It occupies a broad compartment confined by rock cliffs at both ends. The barrier segment fronting Holyrood Pond is approximately 200 m wide at MWL, with crest elevations over 6 m above HHWLT (Fig. 6). It is composed of sand–pebble and pebble gravel (Table I), up to 12 m thick, resting on the fjord sill (Forbes, 1984). Much of the body of the barrier is made up of thinly bedded landward-dipping washover sands and gravels (Table III). Sands are stored on the upper supratidal beachface and in washover facies, in minor dune deposits on the barrier crest, in deposits associated with an intermittent tidal channel at the mouth of Holyrood Pond, and in a thin shoreface veneer of sand and gravel out to depths of at least 30 m.

The beachface at location H1 is characterized by a steep reflective foreshore slope (facies 3, Fig. 6, Table III). A pronounced supratidal berm or swash ridge (facies 4) is a common feature of the middle beachface. Above the berm, the beach exhibits a broad upper slope (facies 5) that may function to promote higher swash excursions under extreme storm-surge and large-wave conditions, in a manner similar to that suggested by

TABLE III. FACIES CHARACTERISTICS—HOLYROOD POND BARRIER

Facies[a]	Zone	Slope[b]	Structure/texture
1. Nearshore shoreface	Subtidal	0.01	Thin veneer of patchy sand and gravel over glacial diamict or rock
3. Lower beachface	Intertidal	0.03/0.35	Fine pebble gravel
4. Berm	Supratidal	−0.03/0.07	Interstratified pebble and sand–pebble gravel
5. Upper beachface	Supratidal	0.06/0.08	Parallel-stratified sand and imbricate sand–pebble gravel (with occasional pebble gravel units and scattered cobbles and boulders)
6. Barrier crest	Supratidal	−0.02/0.04	Parallel-stratified sand and imbricate sand–pebble gravel (with occasional pebble gravel units)
7. Washover	Supratidal–lagoon	−0.40/−0.03	Parellel-stratified sand and imbricate closed-work sand–pebble gravel; distal planar cross-stratified sand–pebble gravel

[a]See Figs. 6 and 7d.

[b]Expressed as tangent of slope angle, positive seaward; minimum and maximum values.

Carter and Orford (1984). Following one such event, large-scale cusps with a mean wavelength of ~90 m were observed on the upper beachface (Forbes, 1985).

Further east in Holyrood Bay, midway between locations H1 and H8 (Fig. 5a), the beach forms a gravel wedge roughly 80 m wide, rising to the base of the cliffs at about 6 m above HHWLT. At the southeast end of the bay (H8, Figs 5a, 7d), the open-work pebble–cobble beach (Table I) forms a fringing barrier up to 100 m wide and 5.5 m above HHWLT, fronting hummocky ice-contact and alluvial terrace deposits.

The Holyrood Bay profiles plotted in Fig. 6 show changes in the barrier at profile line H1 (Fig. 5a) over the same 12-month interval described above for Mutton Bay. The barrier is sufficiently high that the crest was affected only by the first of the two major events during the year. During this storm, at one location near line H1, overwash flow scoured a broad washover channel and washed out the road. At line H1 (Fig. 6), it removed a substantial volume of material at the ridge crest and deposited a thin sand–gravel washover unit, with scattered cobbles and boulders, on the upper back-barrier platform (facies 7).

B. Case 2: Fourchu Bay and Framboise Cove—Drumlin Till Source on an Exposed Coast

Mixed sand–gravel barrier beaches extending between drumlin headland anchor points are characteristic of the low submergent coastline of southeastern Cape Breton Island (Figs. 1, and 5c) (Wang and Piper, 1982). At Framboise Cove the drumlin tills are as much as 15 m thick and contain a smaller proportion of gravel than the Fourchu drumlin deposits. This difference in source deposits is mirrored in the beaches (Framboise is the more sandy system), but given the difficulties in sampling mixed sand–gravel beach deposits and the large variance of grain size in both beaches, the contrast is somewhat obscure in sample data (Table I).

Although the coast in this area is exposed to the open North Atlantic, numerous rocky islets, reefs, and lag shoals cause considerable dissipation and refraction of incoming waves. The barriers in Framboise Cove and Fourchu Bay are less than 90 m wide at MWL (Fig. 8). They enclose brackish and freshwater lagoons with few tidal inlets. The barriers have a solitary dune crest (Figs. 7e and 8) that ranges in height from 1.3 to 3.5 m above HHWLT. In Fourchu Bay, landward of the low-gradient subtidal shoreface (facies 1, Fig. 8, Table IV), there is commonly a breakpoint step, 0.4–0.8 m high (facies 2), which can be associated with the presence of a stable berm or swash ridge (facies 4). The beach is largely reflective, as indicated by the presence of an erosional dune scarp, a steep concave

TABLE IV. FACIES CHARACTERISTICS—FOURCHU BAY BARRIER

Facies[a]	Zone	Slope[b]	Structure/texture
1. Nearshore shoreface	Subtidal	0.01/0.05	Parallel-/ripple-cross-laminated sand; patchy gravel over diamict adjacent to drumlins
2. Step	Subtidal	0.20/0.40	Planar cross-stratified pebble gravel (occasional ripple-cross-stratified coarse sand)
3. Lower beachface	Intertidal	0.02/0.11	Parallel-stratified sand–pebble gravel (lower foreshore); pebble gravel (upper foreshore)
4. Swash ridge	Inter- to supratidal	−0.07/0.21	Cross-stratified pebble gravel (with both seaward- and landward-dipping units); occasional cusp development
5. Upper beachface	Supratidal	0.08/0.15	Crudely stratified sand–pebble gravel and (locally) parallel-laminated sand (forming ramps against dune scarp); slump and mudflow diamict adjacent to drumlins
6. Barrier-crest dune	Supratidal	−0.50/0.09	Parallel-/cross-laminated sand with root channels, and scattered pebbles
7. Washover	Supratidal–lagoon	−0.14/−0.02	Crude parallel-stratified sand–pebble–cobble gravel with planar cross-stratified sand–gravel at distal margin

[a]See Figs. 7e,f and 8.
[b]Expressed as tangent of slope angle, positive seaward; minimum and maximum values.

beach face with cusps and swash ridges, and absence of nearshore bars. In Framboise Cove (Fig. 5c), where the sand supply is greater, there is a seasonal transition from a stable reflective beachface to a less reflective intermediate stage, characterized by lower beachface gradient and development of nearshore bars.

These barriers exhibit limited profile variability (Fig. 8). Crest and backbarrier sedimentation results from infrequent wave overtopping and wind transport of sands. Where the dune sand has has been removed in the course of commercial aggregate extraction (Taylor, 1982), sheet washover of gravel has occurred (with local minor channelization). As at other sites with high proportions of sand (Forbes *et al.*, 1982), wave runup and overtopping may be enhanced in winter by interstitial ice, which causes a reduction in permeability of mid- to upper-beachface sediments and restricts swash infiltration. Overtopping may be encouraged further by development of ramps where snow, wind-blown sediment, and swash deposits accumulate against the dune scarp.

The barriers in this area show pronounced alongshore trends in morphology and texture (Figs. 7e, 7f, and 8) that reflect the role of drumlin headlands as anchor points and discrete sources of sediment supply to the beach. The planform development of the barriers depends on a number of factors, including drumlin spacing and alignment in relation to the general trend of the coast (Wang and Piper, 1982; Carter *et al.*, 1987).

C. Case 3: Advocate Harbour and Scots Bay—Paraglacial Beaches in a Macrotidal Setting

Situated at the head of the Bay of Fundy, these sites (Fig. 9) are exposed to the southwest and west, the direction of prevailing winds and maximum fetch. Semidiurnal spring tidal ranges are 12.6 and 13.5 m, respectively (Canadian Hydrographic Service, 1984). Wave effectiveness in the littoral zone is influenced by the large vertical range over which wave energy is dissipated. The beaches consist largely of sediments reworked from late Wisconsinan glacial outwash deposits (Swift and Borns, 1967; Amos *et al.*, 1980). Both beaches are presently in a transgressive phase, retreating landward over salt marsh.

Shore-normal beach profiles at these sites reveal at least seven distinct facies, defined by changes in slope and texture (Figs. 9 and 10; Tables I, V, and VI). These include (1) lower-foreshore tidal flat (Scots Bay only), (2) beachface sand and gravel below a major break in slope, (3) middle-beachface cobble–boulder lag above break in slope, (4) mid- to upper-beachface sand–gravel, (5) barrier-crest gravel, (6) transgressive washover gravel, and (7) salt-marsh peat. Significant differences between the two sites include the extensive low-tide flat and the near obliteration of the backbarrier estuary at Scots Bay. At Advocate, the cobble–boulder lag and break in slope occur at or just below LLWLT, whereas at Scots Bay these features are found above MWL. Barrier crest elevations extend up to 3.0 m above HHWLT (15.6 m above LLWLT) at Advocate and less than 1.0 m above HHWLT (14.5 m above LLWLT) at Scots Bay. The extensive tidal flat at Scots Bay may reflect the near absence of an estuarine sediment sink: fine sediments remain on the lower shoreface, while

TABLE V. FACIES CHARACTERISTICS—SCOTS BAY BARRIER

Facies[a]	Zone	Slope[b]	Structure/texture
1. Lower beachface	Intertidal	0.01/0.02	Sandy mud with isolated cobbles and boulders
2. Lower beachface	Intertidal	0.01/0.05	Rillwash-scoured ripple cross-laminated sand over cobble–boulder gravel over mud and peat
3. Middle beachface	Intertidal	0.06/0.10	Weakly imbricated cobble–boulder gravel over mud and peat
4. Upper beachface	Intertidal	0.10/0.11	Crudely stratified sand–pebble gravel over peat
5. Barrier crest	Supratidal	0.08/0.22	Crude parallel-stratified pebble gravel (horizontal to landward-dipping) over sand–pebble gravel
5. Terrace	Supratidal	−0.50/0.00	Landward-dipping planar cross-stratified sand–pebble gravel over crude parallel-stratified sand–pebble gravel
6. Backbarrier dune	Supratidal	−0.13/−0.05	Parallel- and cross-laminated sand with root channels
6. Washover	Supratidal–lagoon	−0.13/−0.05	Crudely stratified landward-dipping sand–pebble gravel over peat
7. Saltmarsh	Supratidal		Peat

[a]See Figs. 9 and 10.
[b]Expressed as tangent of slope angle, positive seaward; minimum and maximum values.

the coarser material is transported landward onto the upper beach. The latter forms a thin veneer over old backbarrier organic muds that crop out at the base of facies 3 (Fig. 9).

The planform of the modern barrier at Advocate Harbour is similar to that of an early postglacial spit and lagoon complex 27 m above present sea level (Fig. 9). However, differences in shore-normal profile and sedimentology (Table VII), particularly the vertical limits of foreshore facies, indicate that the emerged beach was developed during a period of much smaller tidal range, roughly 3.4 m (Wightman, 1976). The difference in tidal range and the greatly enhanced vertical extent of foreshore deposits on the modern beach reflect basin-scale adjustments in the Bay of Fundy caused by postglacial changes in RSL (Grant, 1975; Scott and Greenberg, 1983).

TABLE VI. FACIES CHARACTERISTICS—ADVOCATE HARBOUR MODERN BEACH

Facies[a]	Zone	Slope[b]	Structure/texture
2. Nearshore lower beachface	Subtidal	0.01/0.03	Cobble–boulder gravel with sand–pebble matrix
4. Beachface	Intertidal	0.10/0.45	Poorly sorted sand–pebble gravel (lower foreshore); pebble gravel with low linear transverse ridges and cusps (upper foreshore)
5. Barrier crest	Supratidal	−0.45/0.14	Cross-stratified pebble–cobble gravel; sand and sand–pebble gravel, crude horizontal stratification at crest (some irregular structure associated with log debris)
6. Backbarrier	Supratidal	−0.50/−0.02	Crudely stratified landward-dipping sand–pebble gravel with planar-cross-stratified gravel over marsh facies.

[a]See Fig. 9.
[b]Expressed as tangent of slope angle, positive seaward; minimum and maximum values.

VI. Beach Development under Varying Sediment-Source Regimes

The location and scale of Holocene beach deposits in Atlantic Canada depend in large part on the distribution and character of glacigenic source sediments. Sediment storage volumes, facies characteristics, and beach stability are all substantially affected by sediment input conditions. The critical role of sediment supply in the development of coarse clastic beach deposits (Carter and Orford, 1980, 1984) is well illustrated by the contrast between Mutton Bay and Holyrood Bay (Figs. 5a and 6) as well as by temporal changes from regressive to transgressive behavior in several other barrier systems of the southern Avalon Peninsula (Shaw and Forbes, 1987), related to paraglacial reduction in sediment supply (Fig. 2).

Sediment production and dispersal from glacigenic sources is affected by coastal planform irregularity. In compartmented settings (Fig. 5a), large beach deposits are localized adjacent to major sources, as in Holyrood Bay. Where sediment sources are exposed along an open stretch of coast and longshore transport is unimpeded by projecting headlands, major depositional features can form downdrift (Fig. 5b, d). The large Flat Island

TABLE VII. FACIES CHARACTERISTICS—ADVOCATE HARBOUR RAISED BEACH[a]

Facies	Zone	Slope[b]	Structure/texture
1. Nearshore shoreface	Subtidal	−0.05/0.38	Thin parallel-stratified sand–pebble gravel; parallel-, shallow-trough, cross-laminated sand and pebbly sand
2. Lower beachface	Intertidal	−0.50/0.51	Parallel-bedded sand–pebble gravel with lenses of planar cross-laminated and channel-fill sand (lower foreshore); parallel-bedded sand–pebble gravel and interstratified open- and closed-work large-scale cross-stratified gravel
3. Berm	Supratidal	−0.05/0.09	Quasi-horizontal sand–pebble gravel, strongly imbricated
4. Upper beachface	Supratidal	0.05/0.12	Graded to interstratified sand to pebble gravel
5. Barrier crest and washover	Supratidal	−0.62/0.12	Parallel-bedded interstratified sand to pebble gravel; cross-stratified pebble gravel; cross-stratified sand–pebble gravel at distal margin

[a]Generalized from description of regressive unit by Wightman (1976); see Fig. 9 for location.

[b]Expressed as tangent of dip angle, positive seaward, minimum and maximum values.

barrier in St. George's Bay (Fig. 5d) (Shaw and Forbes, 1987) has developed downdrift from a long section of coastal cliffs cut into thick ice-contact, outwash, and glaciomarine deposits (MacClintock and Twenhofel, 1940; Brookes, 1974).

In some exposed coastal areas of low relief, where the sea has transgressed over resistant rocky terrain with a thin veneer of glacigenic sediment—as where a pre-Quaternary erosion surface intersects sea level along much of the Atlantic coast of Nova Scotia (Johnson, 1925; Piper *et al.*, 1986)—scattered erratic boulders may be found resting on wave-washed outcrop in the shorezone. In other areas of rock overlain by thin till, well-developed rock platforms (Fig. 7a) are common features, often associated with small-scale beaches or fringing barriers of locally derived sands and gravels. The nearshore zone may exhibit rock and lag-gravel bottom types, with thin patches of sand and fine gravel in shallow water off the major beaches (Kranck, 1967; Forbes, 1984, 1987). The latter are frequently coarse grained and supply limited, except adjacent to localized large sediment sources (Fig. 5a) or in downdrift sinks (Fig. 5b).

In contrast to areas of extensive thin till, coastal drumlin fields in Atlantic Canada provide sets of discrete sediment sources with characteristic drumlin volumes of the order of 10^5 to 10^7 m^3. Individual drumlins may be formed entirely in glacial drift or may be rock cored with a thin veneer of unconsolidated sediment. The amount of sediment input to the coastal zone depends on the sedimentology and size of the drumlins and on wave exposure. Erosion rates are highly variable alongshore. Areas of important drumlin–coast interaction in Atlantic Canada include the Lunenburg drumlin field (Johnson, 1925; Piper *et al.*, 1983, 1986), the Eastern Shore of Nova Scotia (Boyd *et al.*, this volume, Chapter 4), and southeast Cape Breton Island (Wang and Piper, 1982).

Marine transgression through a drumlin field (case 2, Fig. 5c) initiates barrier beach sedimentation cycles that are driven by erosion and subsequent depletion of individual drumlin deposits. Evolutionary models based on this concept (Wang and Piper, 1982; Boyd *et al.*, this volume, Chapter 4; Carter *et al.*, 1987) envisage barrier genesis and development during an initial period of high sediment supply, followed by barrier retreat and destruction as the drumlin source is depleted and sediment supply diminishes (Fig. 2b). Gravel barriers in such settings may incorporate recycled sediment from exhausted drumlin sources. They often respond to depleted sediment supply by accelerated landward migration and the continued integrity of a barrier "stretching" between drumlin headlands depends on the maintenance of an adequate sediment supply from the headland source (Carter *et al.*, 1987). On the Eastern Shore of Nova Scotia, the cycle is believed to have a period of the order of 1 ka, commensurate with typical drumlin length scales of the order of 1 km and erosion rates of 0.1–3 m/yr (Taylor *et al.*, 1985).

Facies characteristics of gravel barrier deposits (Tables II–VII) are particularly sensitive to proportions of sand in the system. In the absence of sand, or where small quantities are confined to the lower intertidal or subtidal zone (as in Mutton Bay, Tables I and II), barrier-crest dunes are absent and the crest, formed by gravel deposition from overtopping swash, may be remarkably uniform alongshore. In such cases, the beach gravel is open-work and beachface slopes are steep and strongly reflective. Even at sites with abundant sand supply, high wave-energy levels may disperse much of the sand to subtidal, lower intertidal, backshore (largely washover), or backbarrier sinks, leaving a gravel-dominated beachface (as in Holyrood Bay, Figs. 5a and 7d, Table I). Limited dune deposits may be present in the backshore, but washover sedimentation may still account for the bulk of sediment storage in the barrier (Forbes *et al.*, 1982). Lower beachface slopes may still be steep and reflective, but with increasing proportions of sand the morphodynamic range may increase and closed-work (mixed sand–gravel) deposits may predominate. Systems with very high proportions of sand may develop dissipative beachface configurations and store substantial sediment volumes in windblown dune deposits. Tidal inlet facies may account for much of the storage volume in sand or sand–gravel barriers (Reinson, 1984), but are rare in sand-starved gravel systems.

In areas of limited sediment supply, coarse-grained barrier systems tend to migrate onshore, and may do so even under stable sea-level conditions (Carter and Orford, 1984). This behavior results from the preferential landward transport of gravel clasts, due in part to high reflection coefficients, high infiltration rates, and swash dominance on steep rough beachface slopes. Landward migration of gravel-based barriers is also favored by high seepage rates and associated rarity of tidal inlets. In many cases, landward translation of the barrier increases planform compartmentalization of the coast (Mutton Bay, Fig. 5a). This impedes longshore transport and further diminishes sediment supply, particularly where resistant rock forms projecting headlands. Landward migration may also result in increased substrate control of nearshore profile adjustment, particularly where bedrock becomes exposed in the lower foreshore (Fig. 7a) or where cohesive glacigenic deposits are present on the shoreface (Davidson-Arnott and Askin, 1980; Dick and Zeman, 1983).

Where rates of sediment supply are sufficient, spit extension (Flat Island, Fig. 5d) or barrier expansion (Holyrood Bay, Fig. 5a) may occur. Barrier width may increase either by continued washover transport without significant foreshore recession or by beachface progradation, often resulting in deposition of multiple storm ridges. Such regressive sequences may develop under rising RSL with younger (more seaward) ridge crests higher than older ones (Forbes, 1985; Shaw and Forbes, 1987) or under falling RSL, when deposits of raised beach ridges may develop (Blake, 1975; Rosen, 1980; Taylor, 1980; Taylor and McCann, 1983).

VII. Beach Development under Varying Relative-Sea-Level Regimes

Changes in RSL exercise an important control on the evolution of beach systems. In Atlantic Canada, the postglacial history of RSL shows a marked regional variability related to the complex interaction between eustatic and isostatic adjustments following deglaciation. The time history of RSL shows four distinct zonal patterns (Fig. 3) (Quinlan and Beaumont, 1981). It is convenient to examine the effects of RSL changes on coastal morphology and facies characteristics in the context of these four zones.

The coast in zone A has experienced continuing marine regression throughout postglacial time. Early postglacial trends in zones B and C were similar to those in A. Relict shore features can be found above present sea level in zones A and B. Rates of change in RSL were initially very rapid but declined with time (Fig. 3). Under these conditions, as limits of wave uprush vary about a falling relative mean water level, older beach sediments are left above the reach of wave action and migration of the shore zone leads to accumulation of littoral deposits farther seaward. In areas of abundant supply, this process may result in deposition of thin sheets of littoral and shallow marine sediments or (particularly in gravel) "flights" of abandoned storm ridges (Blake, 1975; Rosen, 1980; Taylor, 1980). Where sediment sources are limited, or where steep slopes intervene, beaches may be left as isolated relict features stranded above sea level (Fig. 9) (Wightman, 1976; Stea, 1983). Zone-A and early zone-B deposits are predominantly regressive, but some examples show transgressive facies sequences that may reflect either anomalies of coastal dynamics and sediment supply or short-term reversals in the trend of RSL (Wightman, 1976). The extensive veneer of littoral and shallow marine deposits adjacent to the northeast New Brunswick coast (Fig. 5b) (Rampton *et al.*, 1984) is an example of early postglacial regressive sedimentation in zone B.

Coastal development in zones B to D has been strongly affected by the late Holocene marine transgression: see, for example, drowned valleys in Trepassey, Kouchibouguac, and St. George's Bays (Fig. 5a, 5b, and 5d) and flooded interdrumlin depressions on the southeast Cape Breton coast (Fig. 5c). At many sites, spits and barriers anchored on coastal promontories extend across flooded embayments. Although coastal deposits commonly mirror the general transgressive trend, prograded beach deposits with associated regressive facies sequences have developed in areas of sufficient sediment supply (Forbes, 1985; Shaw and Forbes, 1987; Boyd *et al.*, this volume, Chapter 4). The progress of the transgression in the Lunenburg drumlin field (Fig. 1) can be traced in a sequence of shore platforms developed at successively shallower depths (Piper *et al.*, 1983). There is evidence from the Eastern Shore of Nova Scotia to suggest that coarse sediment is moved preferentially landward on the shoreface, leaving only a thin veneer of transgressive coastal facies on the inner shelf, except where glaciofluvial or estuarine sediments occur in bedrock depressions (Boyd and Penland, 1984).

VIII. Beach Development under Varying Dynamic Regimes

Despite the emphasis in this chapter on paraglacial effects as they influence the beach sedimentation process, regional variability in oceanographic conditions (Fig. 4) is an important factor also. While the location and scale of beach deposits may be related primarily to source sediment distribution (Fig. 5), other characteristics are dominated by nearshore wave dynamics.

Barrier width, crest elevation, and volume are complex functions of sediment supply (Fig. 6) and incident wave energy. The examples described in this chapter represent exposed high-energy environments. Low-energy gravel barriers in the sheltered upper reaches of St. Mary's Bay (Fig. 1) have crest elevations and storage volumes significantly lower than those on the exposed outer coast (Forbes, 1985). A common feature of very protected sites is a poorly sorted mud–sand–gravel beach veneer, in which the gravel fraction may be lag or only rarely mobilized. Barrier-crest elevation is also a function of tidal range: macrotidal beaches in the Bay of Fundy display very extensive beachface slopes (Fig. 9) with a wide foreshore component. The vertical extent of foreshore facies can be used to estimate paleotidal range from preserved raised-beach deposits (Wightman, 1976). However, where fine sediments accumulate in the lower foreshore, as at Scots Bay (Figs. 9 and 10a), the vertical extent of exposed beachface gravels may be substantially reduced, despite very large tidal range.

In parts of the Atlantic coastal region, winter ice development (Fig. 4) is an important element of the coastal process set. Ice effects in the shore zone include both protection (by icefoot development or limitation of wave activity) and erosion (by direct ice thrusting or scour). Specific ice-related features include structural discontinuities due to sedimentation against ice, textural anomalies due to ice thrust or rafting, and distinctive geomorphological features such as boulder-strewn marshes and tidal flats, boulder mounds or clusters, and boulder barricades (Dionne, 1972a,b; Knight and Dalrymple, 1976; Rosen, 1979; Martini, 1980; Drake and McCann, 1982; Taylor and McCann, 1983; Forbes, 1984). Ice-formed features on beaches include stratified ice-sediment kaimoo deposits that thaw to produce mixed-structure facies, including both planar and contorted or disrupted lamination; ice-raft and ice-thrust sands and gravels with hummocky relief and poorly defined structure; and pitting due to thaw of ice floes incorporated in beach deposits (Nichols, 1961; Dionne and Laverdière, 1972; Reinson and Rosen, 1982; Taylor and McCann, 1983).

IX. Conclusions

The long-term impact of glaciation on coastal development, and in particular on beach sedimentation processes, is well displayed in Atlantic Canada. The region possesses a diversity of glacigenic source deposits, a wide spectrum of postglacial sea-level history, and a range of modern process environments, resulting in a great variety of coastal deposits with a preponderance of gravel and mixed sand–gravel beaches. The following conclusions related to coarse-clastic beach sedimentation in a paraglacial setting can be formulated from surveys of many coastal sites in eastern Canada, including the case-study examples presented above.

(1) The persistence of paraglacial effects in coastal sedimentary systems is controlled largely by sea level. In much of the Canadian Atlantic coast region, the time scale of paraglacial relaxation has been prolonged by a postglacial regressive–transgressive sequence associated with collapsing forebulge migration.

(2) Except in areas of nonresistant bedrock (notably the southern Gulf of St. Lawrence), glacial deposits provide the major source of sediment for beach development in the region. In the absence of glacigenic source materials, beach sedimentation is severely restricted.

(3) Paraglacial sediment source types can be broadly subdivided into three classes: thin till with extensive bedrock outcrop, localized thick ice-marginal and outwash deposits, and drumlin fields. Distinctive littoral sedimentation patterns can be associated with each source type. Beach location, morphology, and sediment storage volume are strongly dependent on the types and sizes of local sediment sources and rates of supply from them, particularly on highly compartmented sections of the coast.

(4) Because glacigenic deposits tend to be poorly sorted with large proportions of sand and gravel, coarse-grained beaches are particularly common in formerly glaciated regions such as Atlantic Canada, especially in areas of resistant bedrock.

(5) Coarse-grained beach deposits include a variety of sand–gravel and gravel beaches, including some multiple prograded beach–ridge sets, in areas of abundant sediment supply; sand-starved gravel beaches and barriers in exposed locations with limited sediment input; and less prominent features such as low-energy muddy-gravel veneers and ice-related bouldery tidal flats and boulder barricades.

(6) Many aspects of beach sedimentation are strongly affected by the relative proportions of sand and gravel in the system. Sand-starved gravel

beaches are characterized by a limited morphodynamic range with restricted beachface profile variability and steep reflective foreshore slopes. High seepage rates in coarse-grained barrier deposits are associated with limited tidal-inlet development and rapid swash infiltration. Tidal-inlet facies may be abundant in sandy barriers, but beachface and washover facies predominate in gravel systems. Windblown dunes may be rare or absent in sand-starved situations, where barrier-crest morphology is controlled primarily by swash and overwash dynamics and associated sediment transport.

(7) Where sediment supply is limited, gravel barriers migrate onshore. Where sediment input is high, beach progradation or spit extension may occur. The resulting facies architecture depends on the trend and rate of RSL change in relation to sediment supply and wave climate.

ACKNOWLEDGMENTS

This chapter draws on results of coastal projects in many parts of eastern Canada over several years. We therefore owe a debt to the many individuals who have worked with us in the field, in particular to David Frobel and to the Program Support group of the Atlantic Geoscience Centre. We also wish to thank David Piper, Brian McCann, Ron Boyd, Michael Church, Bill Carter, Julian Orford, John Shaw and the two anonymous reviewers for helpful discussions and constructive comments on earlier versions of the manuscript.

REFERENCES

Alam, M., and Piper, D. J. W. (1977). Pre-Wisconsin stratigraphy and paleoclimates off Atlantic Canada, and its bearing on glaciation in Québec. *Géogr. Phys. Quat.* **31,** 15–22.

Alam, M., Piper, D. J. W., and Cooke, B. S. (1983). Late Quaternary stratigraphy and paleo-oceanography of the Grand Banks continental margin, eastern Canada. *Boreas* **12,** 253–261.

Amos, C. L. (1978). The post-glacial evolution of the Minas Basin, N.S.: a sedimentological interpretation. *J. Sediment. Petrol.* **48,** 965–982.

Amos, C. L., and Zaitlin, B. A. (1985). Sub-littoral deposits from a macrotidal environment, Chignecto Bay, Bay of Fundy. *Geo-Mar. Lett.* **4,** 161–169.

Amos, C. L., Buckley, D. E., Daborn, G. R., Dalrymple, R. W., McCann, S. B., and Risk, M. J. (1980). "Geomorphology and Sedimentology of the Bay of Fundy," Fieldtrip Guidebook 23. Geol. Assoc. Can., Waterloo, Ontario.

Barnes, N. E., and Piper, D. J. W. (1978). Late Quaternary geological history of Mahone Bay, Nova Scotia. *Can. J. Earth Sci.* **15,** 586–593.

Blake, W., Jr. (1975). Radiocarbon age determinations and postglacial emergence at Cape Storm, southern Ellesmere Island, Arctic Canada. *Geogr. Ann.* **57,** 1–71.

Boyd, R., and Penland S. (1984). Shoreface translation and the Holocene stratigraphic record: examples from Nova Scotia, the Mississippi Delta and eastern Australia. *In* "Hydrodynamics and Sedimentation in Wave-Dominated Coastal Environments" (B. Greenwood and R. A. Davis, Jr., eds.), *Mar. Geol.* **60,** 391–412.

Brookes, I. A. (1974). Late-Wisconsin glaciation of southwestern Newfoundland (with special reference to the Stephenville map-area). *Geol. Surv. Can. Pap.* **73–40,** 1–31.

Bryant, E. (1983). Sediment characteristics of some Nova Scotian beaches. *Marit. Sediments Atl. Geol.* **18,** 1–27.

Canadian Hydrographic Service (1984). "Canadian Tide and Current Tables." Fish. Oceans, Ottawa.

Carr, A. P. (1983). Shingle beaches: aspects of their structure and stability. *In* "Shoreline Protection," pp. 97–104. Inst. Civ. Eng., Thomas Telford, London.

Carter, R. W. G., and Orford, J. D. (1980). Gravel barrier genesis and management: a contrast. *Proc. Coastal Zone 80* pp. 1304–1320. Am. Soc. Civ. Eng., New York.

Carter, R. W. G., and Orford, J. D. (1984). Coarse clastic barrier beaches: a discussion of the distinctive dynamic and morphosedimentary characteristics. *In* "Hydrodynamics and Sedimentation in Wave-Dominated Coastal Environments" (B. Greenwood and R. A. Davis, Jr., eds.), *Mar. Geol.* **60,** 377–389.

Carter, R. W. G., Orford, J. D., Forbes, D. L., and Taylor, R. B. (1987). Gravel barriers, headlands, and lagoons: an evolutionary model. *Proc. Coastal Sediments 87*, pp. 1776–1792. Am. Soc. Civ. Eng., New York.

Church, M., and Ryder, J. M. (1972). Paraglacial sedimentation: a consideration of fluvial processes conditioned by glaciation. *Geol. Soc. Am. Bull.* **83,** 3059–3072.

Davidson-Arnott, R. G. D., and Askin, R. W. (1980). Factors controlling erosion of the nearshore profile in over-consolidated till, Grimsby, Lake Ontario. *Proc. Can. Coastal Conf. 1980, Burlington, Ont.* pp. 185–199. Assoc. Comm. Res. Shoreline Erosion Sediment., Nat. Res. Counc. Can., Ottawa.

Dick, T. M., and Zeman, A. J. (1983). Coastal processes on soft shores. *Proc. Can. Coastal Conf. 1983, Vancouver, B.C.* pp. 19–35. Assoc. Comm. Res. Shoreline Erosion Sediment., Nat. Res. Counc. Can., Ottawa.

Dionne, J.-C. (1972a). Caractéristiques des schorres des régions froides, en particulier de l'estuaire du Saint-Laurent. *Z. Geomorphol., Suppl.* **13,** 131–162.

Dionne, J.-C. (1972b). Caractéristiques des blocs erratiques des rives de l'estuaire du Saint-Laurent. *Rev. Géogr. Montréal* **26,** 125–152.

Dionne, J.-C., and Laverdière, C. (1972). Ice formed beach features from Lake St. Jean, Quebec. *Can. J. Earth Sci.* **9,** 979–990.

Dobkins, J. A., and Folk, R. L. (1970). Shape development on Tahiti-Nui. *J. Sediment. Petrol.* **40,** 1167–1203.

Drake, J. J., and McCann, S. B. (1982). The movement of isolated boulders on tidal flats by ice floes. *Can. J. Earth Sci.* **19,** 748–754.

Dubois, J. M. M. (1980). Géomorphologie du littoral de la côte nord du Saint-Laurent: analyse sommaire. *In* "The Coastline of Canada" (S. B. McCann, ed.), *Geol. Surv. Can. Pap.* **80–10,** 215–238.

Emery, K. O. (1955). Grain size of marine beach gravels. *J. Geol.* **63,** 39–49.

Eyles, N., and Slatt, R. M. (1977). Ice-marginal sedimentary, glacitectonic and morphologic features of Pleistocene drift: an example from Newfoundland. *Quat. Res. (N.Y.)* **8,** 267–281.

Flint, R. F. (1940). Late Quaternary changes of level in western and southern Newfoundland. *Geol. Soc. Am. Bull.* **51,** 1757–1780.

Forbes, D. L. (1984). Coastal geomorphology and sediments of Newfoundland. Current research, Part B. *Geol. Surv. Can. Pap.* **84–1B,** 11–24.

Forbes, D. L. (1985). Placentia Road and St. Mary's Bay: field trip guide to coastal sites in the southern Avalon Peninsula, Newfoundland. *Proc. Can. Coastal Conf. 1985, St. John's, Newfoundland* pp. 587–605. Assoc. Comm. Res. Shoreline Erosion Sediment., Nat. Res. Counc. Can., Ottawa.

Forbes, D. L. (1987). Shoreface sediment distribution and sand supply at C^2S^2 sites in the southern Gulf of St. Lawrence. *Proc. Coastal Sediments 87*, pp. 694–709. Am. Soc. Civ. Eng., New York.

Forbes, D. L., Taylor, R. B., and Frobel, D. (1982). Barrier overwash and washover sedimentation, Aspy Bay, Cape Breton Island. *Marit. Sediments Atl. Geol.* **18,** 43.

Grant, D. R. (1975). Recent coastal submergence of the Maritime Provinces. *Proc. N. S. Inst. Sci.* **27,** Suppl. 3, 83–102.

Grant, D. R. (1976). Reconnaissance of early and middle Wisconsinan deposits along the Yarmouth-Digby coast of Nova Scotia. Report of activities, Part B. *Geol. Surv. Can. Pap.* **76-1B,** 363–369.

Grant, D. R. (1977). Glacial style and ice limits, the Quaternary stratigraphic record and changes of land and ocean level in the Atlantic Provinces, Canada. *Géogr. Phys. Quat.* **31,** 247–260.

Grant, D. R. (1987). Surficial geology and Quaternary history of Cape Breton Island, Nova Scotia. *Mem. Geol. Surv. Can.* (with A-Series Map at 1:125,000). In press.

Grant, D. R., and King, L. H. (1984). A stratigraphic framework for the Quaternary history of the Atlantic Provinces. *In* "Quaternary Stratigraphy of Canada—A Canadian Contribution to IGCP Project 24" (R. J. Fulton, ed.), *Geol. Surv. Can. Pap.* **84-10,** 173–191.

Greenwood, B., and Davidson-Arnott, R. G. D. (1979). Sedimentation and equilibrium in wave formed bars: a review and case study. *Can. J. Earth Sci.* **16,** 312–332.

Henderson, E. P. (1972). Surficial geology of Avalon Peninsula, Newfoundland. *Mem. Geol. Surv. Can.* **368.**

Johnson, D. W. (1925). "The New England-Acadian Shoreline." Wiley, New York.

King, L. H. (1972). Relation of plate tectonics to the geomorphic evolution of the Canadian Atlantic Provinces. *Geol. Soc. Am. Bull.* **83,** 3083–3090.

King, L. H., and Fader, G. B. J. (1986). Wisconsinan glaciation of the Atlantic continental shelf of southeast Canada. *Geol. Surv. Can. Bull.* **363.**

Kirk, R. M. (1980). Mixed sand and gravel beaches: morphology, processes and sediments. *Prog. Phys. Geogr.* **4,** 189–210.

Knight, R. J., and Dalrymple, R. W. (1976). Winter conditions in a macrotidal environment, Cobequid Bay, Nova Scotia. *Rev. Géogr. Montréal* **30,** 25–85.

Kranck, K. (1967). Bedrock and sediments of Kouchibouguac Bay, New Brunswick. *J. Fish. Res. Board Can.* **24,** 2241–2265.

Kranck, K. (1971). Surficial geology of Northumberland Strait. *Geol. Surv. Can. Pap.* **71-53.**

McCann, S. B. (1979). Barrier islands in the Southern Gulf of St. Lawrence, Canada. *In* "Barrier Islands" (S. P. Leatherman, ed.), pp. 29–63. Academic Press, New York.

MacClintock, P., and Twenhofel, W. H. (1940). Wisconsin glaciation of Newfoundland. *Geol. Soc. Am. Bull.* **51,** 1729–1756.

MacNeill, R. H. (1974). "Surficial Geology Maps of Nova Scotia, 1:50,000." Nova Scotia Res. Found., Dartmouth.

Markham, W. E. (1980). "Ice Atlas Eastern Canadian Seaboard." Atmos. Environ. Serv., Environ. Can., Toronto.

Martini, I. P. (1980). Sea ice generated features of coastal sediments of James Bay, Ontario. *Proc. Can. Coastal Conf. 1980, Burlington, Ont.* pp. 93–101. Assoc. Comm. Res. Shoreline Erosion Sediment., Natl. Res. Counc. Can., Ottawa.

Munroe, H. D. (1982). Regional variability, physical shoreline types and morpho-dynamic units of the Atlantic coast of mainland Nova Scotia (R. B. Taylor, D. J. W. Piper, and C. F. M. Lewis, eds.). *Geol. Surv. Can. Open File* No. 725.

Neu, H. J. A. (1982). 11-year deep water wave climate of Canadian Atlantic waters. *Can. Tech. Rep. Hydrogr. Ocean Sci.* **13.**

Nichols, R. L. (1961). Characteristics of beaches formed in polar climates. *Am. J. Sci.* **259,** 694–708.

Nielsen, E. (1976). The composition and origin of Wisconsinan till in mainland Nova Scotia. Ph.D. Thesis, Dalhousie Univ., Halifax, Nova Scotia.

Orford, J. D. (1979). Some aspects of beach ridge development on a fringing gravel beach, Dyfed, west Wales. *In* "Les côtes atlantiques d'Europe, évolution, aménagement, protection," *Publ. Cent. Natl. Exploit. Oceans, Ser. Actes Coll. (Fr.)* **9,** 35–44.

Owens, E. H. (1974a). Barrier beaches and sediment transport in the southern Gulf of St. Lawrence. *Proc. Coastal Eng. Conf., 14th, Copenhagen* pp. 1177–1193. Am. Soc. Civ. Eng., New York.

Owens, E. H. (1974b). A framework for the definition of coastal environments in the southern Gulf of St. Lawrence. *In* "Offshore Geology of Eastern Canada" (B. R. Pelletier, ed.), *Geol. Surv. Can. Pap.* **74-30,** 47–76.

Owens, E. H., and Bowen, A. J. (1977). Coastal environments of the Maritime Provinces. *Marit. Sediments* **13,** 1–31.

Owens, E. H., and Drapeau, G. (1973). Changes in beach profiles at Chedabucto Bay, Nova Scotia, following large scale removal of sediments. *Can. J. Earth Sci.* **10,** 1226–1232.

Piper, D. J. W. (1975). Upper Cenozoic glacial history south of the Grand Banks of Newfoundland. *Can. J. Earth Sci.* **12,** 503–508.

Piper, D. J. W., Letson, J. R. J., Delure, A. M., and Barrie, C. Q. (1983). Sediment accumulation in low sedimentation, wave-dominated, glaciated inlets. *Sediment. Geol.* **36,** 195–215.

Piper, D. J. W., Mudie, P. J., Letson, J. R. J., Barnes, N. E., and Iuliucci, R. J. (1986). The marine geology of the South Shore, Nova Scotia. *Geol. Surv. Can. Pap.* **85–19.**

Piper, D. J. W., Mudie, P. J., Fader, G. B., Josenhans, H. W., MacLean, B., and Vilks, G. (in press). Quaternary geology. *In* "Geology of the continental margin off Eastern Canada" (M. J. Keen and G. L. Williams, eds.) Chapt. 11. *Geol. Surv. Can. Geology of Canada* 2 [*also* Geol. Soc. of Am. *The Geology of North America* I–1].

Ploeg, J. (1971). Wave climate study, Great Lakes and Gulf of St. Lawrence. *Mech. Eng. Rep. (Natl. Res. Counc. Can.)* **MH-107A 1,** 1–160.

Quinlan, G., and Beaumont, C. (1981). A comparison of observed and theoretical postglacial relative sea level in Atlantic Canada. *Can. J. Earth Sci.* **18,** 1146–1163.

Quinlan, G., and Beaumont, C. (1982). The deglaciation of Atlantic Canada as reconstructed from the postglacial relative sea-level record. *Can. J. Earth Sci.* **19,** 2232–2246.

Rampton, V. N., Gauthier, R. C., Thibault, J., and Seaman, A. A. (1984). Quaternary geology of New Brunswick. *Mem. Geol. Surv. Can.* **416.**

Reinson, G. E. (1984). Barrier-island and associated strand-plain systems. *In* "Facies Models" (R. G. Walker, ed.), 2nd Ed., pp. 119–140. Geol. Assoc. Can., Waterloo, Ontario.

Reinson, G. E., and Rosen, P. S. (1982). Preservation of ice-formed features in a subarctic sandy beach sequence: geologic implications. *J. Sediment Petrol.* **52,** 463–471.

Rogerson, R. J., and Tucker, C. M. (1972). Observations on the glacial history of the Avalon Peninsula. *Marit. Sediments* **8,** 25–31.

Rosen, P. S. (1979). Boulder barricades in central Labrador. *J. Sediment. Petrol.* **49,** 1113–1124.

Rosen, P. S. (1980). Coastal environments of the Makkovik region, Labrador. *In* "The Coastline of Canada" (S. B. McCann, ed.), *Geol. Surv. Can. Pap.* **80-10,** 267–280.

Ryder, J. M. (1971). The stratigraphy and morphology of para-glacial alluvial fans in south-central British Columbia. *Can. J. Earth Sci.* **8,** 279–298.

Scott, D. B., and Greenberg, D. A. (1983). Relative sea-level rise and tidal development in the Fundy tidal system. *Can. J. Earth Sci.*, **20,** 1554–1564.

Shaw, J., and Forbes, D. L. (1987). Coastal barrier and beach-ridge sedimentation in Newfoundland. *Proc. Canadian Coastal Conference 1987*, Quebec, pp. 437–454. Assoc. Comm. Res. Shoreline Erosion Sediment., Natl. Res. Counc. Can., Ottawa.

Sloss, L. L. (1962). Stratigraphic models in exploration. *J. Sediment. Petrol.* **32,** 415–422.

Stea, R. R. (1983). Surfical Geology of the western part of Cumberland County, Nova Scotia. Current research, Part A. *Geol. Surv. Can. Pap.* **83-1A,** 197–202.

Stea, R. R., and Fowler, J. H. (1979). Minor and trace element variations in Wisconsinan tills, Eastern Shore region, Nova Scotia. *N. S. Dep. Mines Energy Pap.* No. **79-4.**

Stea, R. R., and Fowler, J. H. (1981). Pleistocene geology and till geochemistry of central Nova Scotia. *N.S. Dep. Mines Energy Map* No. 81-1 (1:100,000).

Swift, D. J. P. (1975). Barrier island genesis: evidence from the central Atlantic Shelf, eastern USA. *Sediment. Geol.* **14,** 1–43.

Swift, D. J. P., and Borns, H. W., Jr. (1967). A raised fluviomarine outwash terrace, north shore of the Minas Basin, Nova Scotia. *J. Geol.* **75,** 693–710.

Taylor, R. B. (1980). Coastal environments along the northern shore of Somerset Island, District of Franklin. *In* "The Coastline of Canada" (S. B. McCann, ed.), *Geol. Surv. Can. Pap.* **80-10,** 239–250.

Taylor, R. B. (1982). Seasonal shoreline change along Forchu Bay and Framboise Cove, Nova Scotia—a comparison between exploited and natural beaches. *Proc. Workshop Atl. Coastal Erosion Sediment., Halifax, N.S.* pp. 37–41. Assoc. Comm. Res. Shoreline Erosion Sediment., Nat. Res. Counc. Can., Ottawa.

Taylor, R. B., and Kelly, B. J. (1984). Beach observations along the east coast of Cape Breton Highlands National Park, Nova Scotia. *Geol. Surv. Can. Open File* No. 1119.

Taylor, R. B., and McCann, S. B. (1983). Coastal depositional landforms in northern Canada. *In* "Shorelines and Isostasy" (D. E. Smith and A. G. Dawson, eds.), Spec. Publ. Inst. Br. Geogr. 16, pp. 53–75. Academic Press, London.

Taylor, R. B., Wittmann, S. L., Milne, M. J., and Kober, S. M. (1985). Beach morphology and coastal changes at selected sites, mainland Nova Scotia. *Geol. Surv. Can. Pap.* **85-12.**

Tucker, C. M. (1974). A series of raised Pleistocene deltas, Halls Bay, Newfoundland. *Marit. Sediments* **12,** 61–73.

Tucker, C. M., and McCann, S. B. (1980). Quaternary events on the Burin Peninsula, Newfoundland, and the islands of St. Pierre and Miquelon, France. *Can. J. Earth Sci.* **17,** 1462–1479.

Wang, Y., and Piper, D. J. W. (1982). Dynamic geomorphology of the drumlin coast of southeast Cape Breton Island. *Marit. Sediments Atl. Geol.* **18,** 1–27.

Wightman, D. M. (1976). The sedimentology and paleotidal significance of a late Pleistocene raised beach, Advocate Harbour, Nova Scotia. M. S. Thesis, Dalhousie Univ., Halifax, Nova Scotia.

CHAPTER 4

An Evolutionary Model for Transgressive Sedimentation on the Eastern Shore of Nova Scotia

R. BOYD

Centre for Marine Geology
Dalhousie University
Halifax, Nova Scotia, Canada

A. J. BOWEN

Department of Oceanography
Dalhousie University
Halifax, Nova Scotia, Canada

R. K. HALL

Centre for Marine Geology
Dalhousie University
Halifax, Nova Scotia, Canada

The Eastern Shore of Nova Scotia consists of a series of linear embayments and intervening headlands. This region experiences mesotidal conditions and a relatively high-energy east-coast wave environment. Sea-level transgression is driven by a relative sea-level rise of 35–40 cm/100 yr. An evolutionary model has been developed to account for the wide range of variability present in coastal compartments and shoreline processes. Deposition during transgression is controlled by valley and ridge systems aligned SE to NW. Sediments are derived from eroding till cliffs and drumlins in exposed headland locations. Submergence of the valleys provides the only major site for deposition. Sand and gravel are transported alongshore from headland sources to infill the submerged valley mouths with barrier systems. Tidal inlets exchange the large tidal prisms that form after barrier progradation. As sediment supply diminishes from till and preexisting barriers, relative sea-level rise becomes the dominant factor and a period of barrier retreat and destruction begins. Eolian, overwash and tidal inlet processes remove sediment from the barrier and deposit it in flood tidal delta, washover, and estuarine environments. Later, landward translation of the shoreline causes these sediments to be exhumed on the retreating shoreface, reworked onshore, and accumulated in new barrier locations. Evolution of the Eastern Shore can be summarized as alternate periods of transgression and regression within an overall framework of coastal submergence.

I. Introduction and Objectives

The Eastern Shore of Nova Scotia faces the Atlantic Ocean along an east–northeast trending coastline from Halifax to Chedabucto Bay (Fig. 1). The area examined in detail in this chapter lies between Halifax Harbour and Jeddore Cape, 35 km to the east. This shoreline is characterized by a series of linear, shore-normal embayments (estuaries and lagoons) and intervening headlands. Barrier systems infill the seaward margin of most embayments. Many of the barrier systems contain tidal inlets and associated tidal delta deposits.

Strong contrasts exist between individual embayments and barrier systems along the Eastern Shore. Some estuaries, such as Porters Lake (Fig. 1), are predominantly fresh to brackish with strong salinity contrasts at the seaward end. Other embayments, such as Chezzetcook Inlet, are predominantly marine systems throughout. Correspondingly, some barrier systems such as those in the entrance to Chezzetcook Inlet are actively building, while others are experiencing serious erosion (Martinique Beach) or have recently been destroyed (Rocky Run Barrier). The objective in this chapter is to explain the genesis of the Eastern Shore and to provide an evolutionary model explaining its complex coastal processes and morphology.

II. Physical Processes

The two major physical parameters controlling sediment dispersal on the Eastern Shore are waves and tides. Tides are semidiurnal and mesotidal, with a maximum spring range of 2.1 m (Canadian Hydrographic

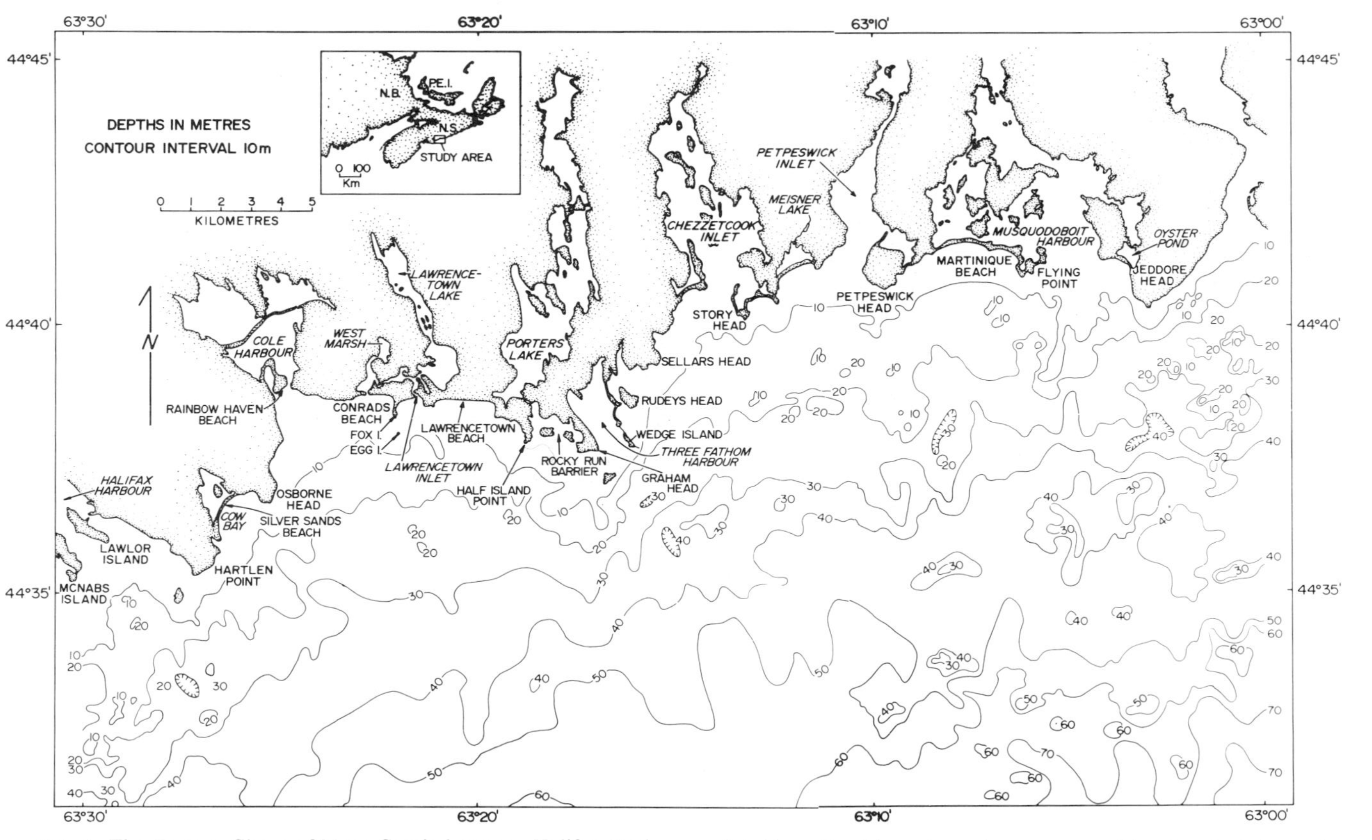

Fig. 1. The Eastern Shore of Nova Scotia between Halifax Harbour and Jeddore Head is an irregular coastline of headlands, intervening embayments, and coastal barrier systems. Holocene transgression of this coastline has resulted in a complex inner-shelf topography.

Service, 1984). The largest tidal current velocities are generated within inlet throats and estuarine tidal channels. Tidal currents provide the major mechanism for landward transport of sand- to mud-sized sediments into embayments.

Waves are responsible for reworking glacial sediments exposed along the coast and transporting sand- to cobble-sized material along adjacent barriers and into embayment mouths. The relatively high-energy wave climate received on the Eastern Shore is dominated by wind waves generated on the Scotian shelf superimposed on a component of lower amplitude, longer period Atlantic swell. The modal wave (for which the product of wave power with frequency of occurrence is at a maximum) is for waves of 1.5–2 m height and 9–10-s wave period. Average annual wave power is 2.14×10^4 W/m. Wave power displays a marked seasonality, with minimum values received from May to September and maximum values, representing frequent winter storms, occurring from October to April. These Eastern Shore features are characteristic of a Northern Hemisphere, east-coast wave environment (Davies, 1980).

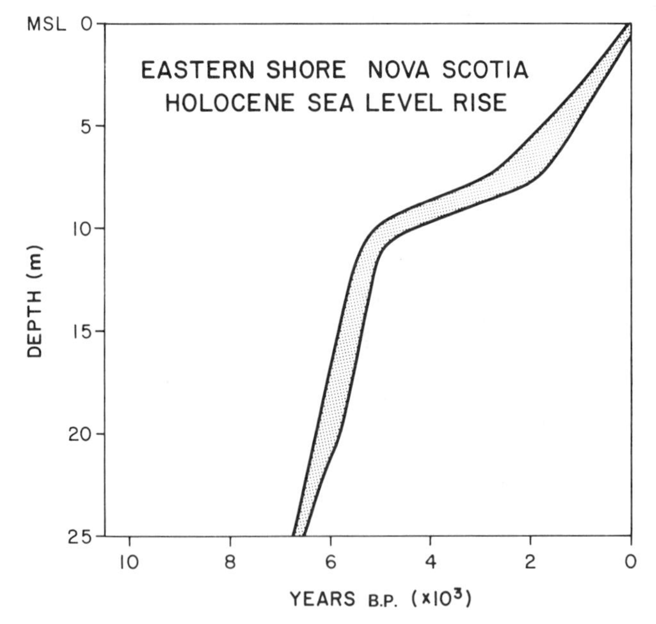

FIG. 2. Holocene sea level relative to the Eastern Shore. A transgression since 7000 yr B.P. has resulted in rates of RSL rise around 35 cm/100 yr. [After Scott and Medioli (1982).]

III. Relative Sea Level

Holocene relative sea level (RSL) in Atlantic Canada has been dominated by the ablation of continental ice masses. Coastal zones lying close to the former ice margin experience the passage of a glacial forebulge (Quinlan and Beaumont, 1981). Following ice recession, the glacial forebulge migrates inward through the marginal ice zone. This migration initially causes a fall in RSL as the forebulge approaches, followed by an RSL rise after passage of the forebulge crest.

The late Wisconsinan ice advance reached its maximum in Nova Scotia between 32000 and 12000 yr B.P. (thousand years before present). Late Wisconsinan crustal loading was insufficient during deglaciation to produce any higher RSL than that experienced at present. Following glacial forebulge passage, RSL has risen from depths of -27 m at 7000 yr B.P. to its present level (see Fig. 2) at an average rate of 35 cm/100 yr (Scott and Medioli, 1982). Tide gage records from Halifax Harbour (Grant, 1970) indicate that RSL is continuing to rise at current rates of about 40 cm/100 yr.

IV. Coastal Geology

Ordovician quartzites and slates of the Meguma Group are often exposed along the Eastern Shore as the headlands defining individual coastal systems and separating embayments. Relief of up to 40–60 m has been cut into the Meguma Group by glacial action during Pleistocene times. Linear scours and valley systems trend NW–SE, parallel to the main Pleistocene ice transport direction, and occasionally coincide, as at Porters Lake, with cross-cutting bedrock faults.

Unconsolidated Pleistocene glacial deposits on the Eastern Shore (Fig. 3) take the form of either thin basal lodgement or ablation till (<3 m thick), and thicker drumlin accumulations (often >10 m thick).

The occurrence of barrier systems in the coastal zone of Atlantic Nova Scotia is closely linked with an abundant sediment supply, usually resulting from marine erosion of glacial sediments and, in particular, drumlin fields. Piper *et al.* (1986) and Wang and Piper (1982) have described examples of coastal sediments accumulating when drumlin fields intersect the coastline around Lunenburg on Nova Scotia's South Shore and along the southeast coast of Cape Breton. In contrast, geomorphology is characterized by bare rock outcrop where glacial sediments are sparse, such as along the coastal granite cliffs southwest of Halifax. Between Halifax

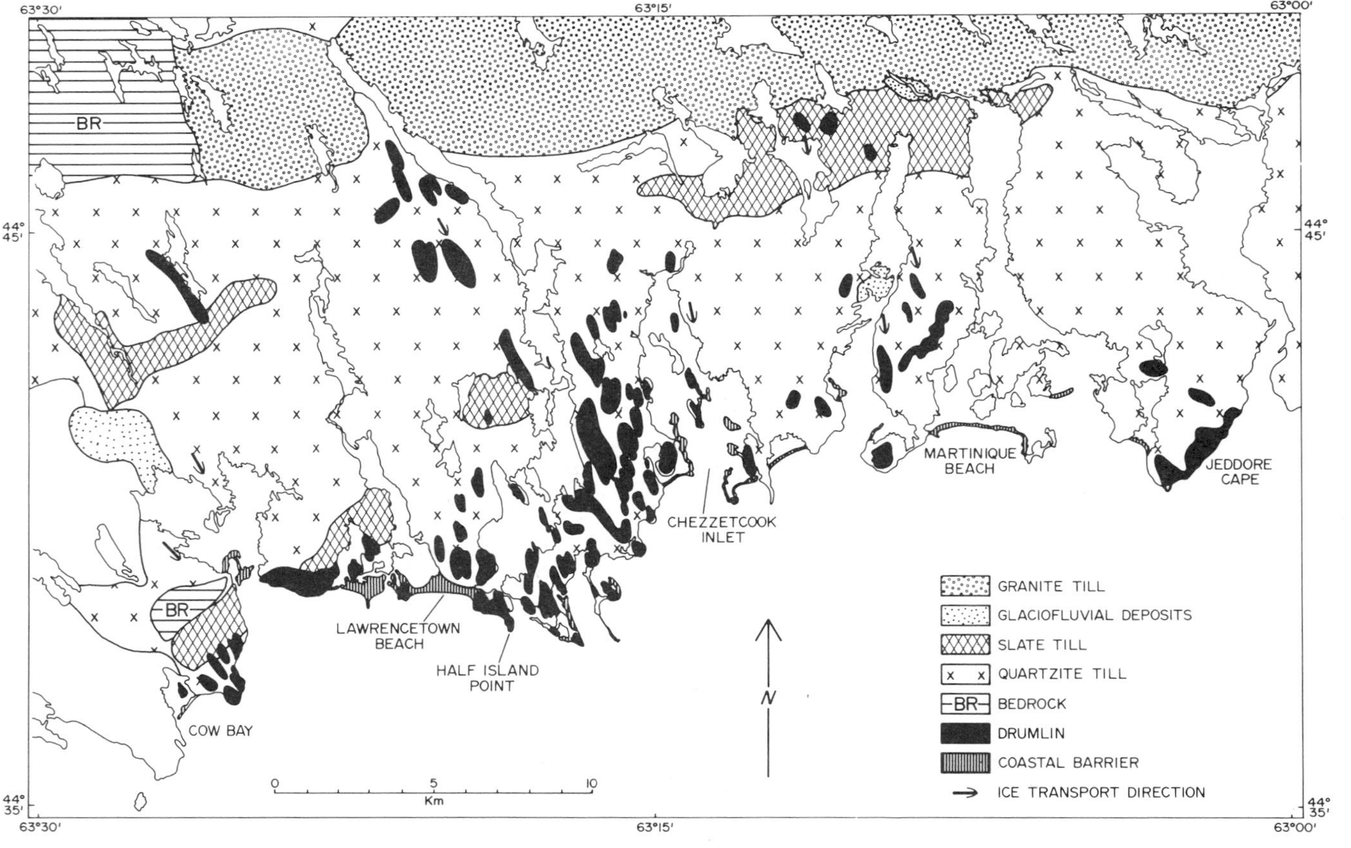

FIG. 3. Unconsolidated sediments on the Eastern Shore of Nova Scotia are primarily Wisconsinan tills. These tills consist of thin, locally derived lodgement and ablation tills such as Slate till and Quartzite till, together with thicker, transported drumlin tills. Barrier systems form where drumlins intersect the coastline. Drumlins, estuaries, and glacial striae all exhibit SE alignment indicating ice transport direction. [Modified from Stea and Fowler (1979).]

Harbour and Jeddore Cape and extending north to latitude 44° 46′N (Fig. 3), 79 individual drumlins occur, ranging in size from 200 m × 100 m to 3.4 km × 1 km.

V. Eastern Shore Processes and Morphology

A typical coastal compartment on the Eastern Shore consists of three major components: (a) headlands, (b) bay barriers and associated tidal inlets, and (c) estuaries or coastal bays. The boundary of each compartment is formed by a set of exposed headlands. Headlands are often a seaward continuation of the NW–SE trending ridge systems produced by glacial scouring. Eroding till cliffs and drumlins are subjected to rapid marine erosion in exposed headland locations. Sand and gravel are transported alongshore from headland sources to infill the mouths of submerged coastal valleys with barrier systems. The backbarrier valleys gradually fill with sediment accumulating in either estuarine or coastal bay environments. Tidal inlets through the barriers transport coastal sediments and saline water into the estuaries and bays.

Although the major components are readily identified in most coastal compartments, individual components and associated processes display a wide range of variability. Considerations of the temporal and spatial variability in each compartment have been investigated by radiocarbon dating, analysis of historical map and airphoto imagery, and the interpretation of coastal geomorphological features. These investigations form the basis for the subsequent development of a model to account for the genesis and evolution of the Eastern Shore.

A. Headlands

An analysis of unconsolidated sediment on the Eastern Shore indicates that drumlins are the major primary source for marine erosion. Other glacial sediments are too thin, and the Meguma Group lithology is primarily resistant quartzite. Eroding drumlin headlands occur (see Figs. 1 and 3) at Hartlen Point, Osborne Head, Conrads Head, Lawrencetown Head, Half Island Point, Graham Head, Wedge Island, Rudeys Head, Sellars Head, Story Head, Petpeswick Head, Flying Point, and Jeddore Head. Repetitive surveys of drumlin cliffs reveal relatively rapid rates of erosion up to 1 m/yr (Fig. 4), with occasional retreat of 3 m/yr during severe winter storm seasons (Boyd and Bowen, 1983).

Each drumlin headland along the Eastern Shore is in a different stage

FIG. 4. Cliff erosion along the Half Island Point drumlin. The wide range of grain sizes available can be seen in the exposed cliff face. The drumlin is being truncated at sea level (foreground), where it is overlain by a boulder retreat shoal in the process of formation.

of erosion. Drumlins on Petpeswick Head or inside Chezzetcook Inlet (see Fig. 12) have only recently been intersected by the transgressing shoreline. In contrast, drumlins such as Story Head and Wedge Island represent an advanced stage of erosion and have been almost totally consumed. The sites of former drumlins are marked by boulder retreat shoals, which result from the inability of wave processes to remove the largest boulder sizes present in the drumlins (Fig. 4). Boulder retreat shoals mark the site of former drumlins at Egg Islet, along the Rocky Run barrier and off Graham Head. Similar boulder retreat shoals are also apparent from sidescan sonar surveys of the inner shelf seaward of Osborne Head, Story Head, Petpeswick Head, Jeddore Head, and many other locations evident from the complex bathymetry of Fig. 1. The headlands and their offshore extensions separating coastal embayments also often serve as effective barriers to sediment exchange between embayments. This lack of exchange is indicated on sidescan sonar by the isolated nature of shoreface sand bodies off each embayment (Hall, 1985) and by the contrasting sediment texture and composition found in adjacent embayments (Boyd and Bowen, 1983). Therefore, during transgression on the Eastern Shore, sediment is confined within each embayment as it undergoes landward translation.

B. Bay Barriers and Tidal Inlets

Bay barriers on the Eastern Shore are constructed from sediment supplied by adjacent drumlin headlands by longshore transport, or from barriers that previously existed further seaward within the same embayment. Longshore supply is demonstrated by sediment size fining away from drumlin sources. Eastward fining occurs away from Petpeswick Head along Martinique Beach, westward fining away from Half Island Point along Lawrencetown Beach, and northeastward fining away from Osborne Head along Rainbow Haven Beach. The variable longshore transport direction on the Eastern Shore is a complex resultant of SW wave approach direction, nearshore bathymetry, and location of suitable sediment source. Further evidence of longshore sediment supply can be seen in the orientation of beach ridges such as those at Lawrencetown Beach (Fig. 5) indicating westward supply from Half Island Point, and the divergent transport away from Conrad Island and Red Island drumlins (see Figs. 11 and 12). Onshore transport from previously existing drumlins is shown by Steering Beach in the entrance to Musquodoboit Harbour and also by the sand shoals accreting to Red Island and Conrod Island in Chezzetcook Inlet (see Fig. 12).

Eastern Shore barriers are thus often composite features, supplied partly by contemporaneous drumlin erosion and partly from preexisting barriers further seaward. This was demonstrated by Sonnichsen (1984) for Half Island Point, the largest drumlin source on the Eastern Shore, and its sediment sink in the Lawrencetown barrier. After calculating the drumlin sediment budget based on drumlin volume and erosion rate, Sonnichsen concluded that Half Island Point was capable of supplying 1.76×10^4 m^3 of sediment to the barrier system each year. However, the total volume of the barrier and shoreface to a depth of 17 m (the shoreface base as measured on offshore seismic profiles) was 4.55×10^7 m^3. This would require 2560 yr of continuous supply, if the only source was the Half Island Point drumlin. However, 2560 yr is much longer than the 800 yr B.P. date of barrier formation according to Hoskin's (1983) radiocarbon dates. In addition, Half Island Point is only capable of providing a total of 1.3×10^7 m^3 or 25% of the total sediment needed to form the Lawrencetown barrier. It seems that for large barriers like Lawrencetown, a single drumlin source even as large as Half Island Point is insufficient to create a complete barrier system.

FIG. 5. The Lawrencetown coastal compartment is an example of barrier progradation. Here, sediment derived from the Half Island Point drumlin to the right (east) is transported westward into a beach-ridge plain, infilling the Lawrencetown Lake embayment (upper left). Beach-ridge alignment was controlled by a midbeach drumlin, now seen as a boulder retreat–shoal, and a cliffed drumlin now located behind the beach near its eastern end.

Eastern Shore barrier morphology displays wide variability, depending on whether the barrier has recently begun forming, is a well-established barrier or is in the process of retreating and being destroyed. Barriers that have recently begun forming occupy sheltered positions inside the mouths of embayments such as at Chezzetcook Inlet (see Fig. 12) and Musquodoboit Harbour. These new barriers stretch between terminal anchoring points in drumlins, bedrock outcrops or tidal channel levees. Obvious downdrift progradation of recurved spits and beach ridges from freshly eroded drumlin sources is often evident. Other features include large subaqueous bar systems migrating onshore.

Airphoto analysis shows Steering Beach, a series of intertidal shoals which existed at the entrance to Musquodoboit Harbour in 1945, migrated 550 m landward (20 m/yr) to establish a new linear barrier by 1974. Sediment comprising Steering Beach was almost exclusively derived from the preexisting 1945 shoal complex, which stabilized on one new bedrock anchor point at Indian Island and another on the main Musquodoboit Harbour tidal channel levee. Red Island, Gaetz Island, and Conrod Island are three sites of new barrier formation within Chezzetcook Inlet (Scott, 1980). Red Island grew from a small bedrock nucleus just over 100-m long in 1854 (see Fig. 11a) to an 1100-m-long barrier with recurved spits by 1974 (see Fig. 11b). Erosion of Conrod Island drumlin further to the west has provided sediment to join Conrod Island to Gaetz Island with a 300-m-long barrier displaying incipient beach ridge plain formation (see Fig. 12).

Established barriers exhibit a deeper open-ocean shoreface, which indicates onshore sediment supply has diminished. These barriers also often contain wide beach ridge plains and dune complexes such as those at Lawrencetown (Fig. 5) and Lawrencetown Inlet (Fig. 6). Tidal inlets de-

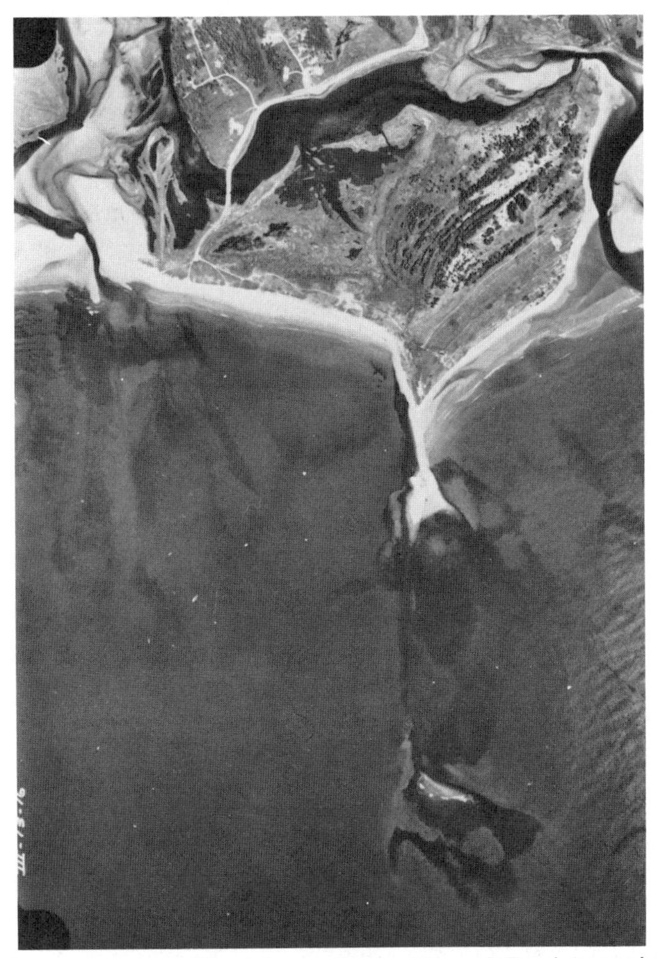

Fig. 6. Lawrencetown Inlet Beach (center right) and Conrads Beach (center left) are prograded beach ridge plains formed partly from sediment derived from the Egg Island–Fox Island depleted drumlin source lying immediately offshore.

velop through barriers that constrict bays and estuaries with a sufficiently large tidal prism. Ebb tidal deltas through established barriers such as those at Rainbow Haven Beach, Eastern Passage into Halifax Harbour, and Lawrencetown Inlet Beach (see right margin of Fig. 10b) develop where wave refraction provides partial shelter from incident wave energy, and where an abundant sand supply constricts a large tidal prism.

Lawrencetown exhibits many of the features characteristic of well-established barriers. The Lawrencetown compartment originally contained four drumlin sources. These included (Fig. 5) Lawrencetown Head to the west, Half Island Point to the east, a drumlin behind the present beach between Lawrencetown Lake and Porters Lake, and a drumlin that now exists as a boulder retreat shoal in the midbeach location. The original shoreline position is shown by the cliffed seaward margin of the drumlin between Lawrencetown and Porters lakes. Sediment from this source and Half Island Point further east combined to form the 16 eastern ridges of the beach ridge plain. Further west, erosion of the midbeach drumlin controlled the formation and orientation of eight recurved spits trending perpendicular to the present beach. At their time of formation, the spits probably paralleled a tidal inlet at the western end of the beach. The final phase of barrier building used sediment arriving from the Half Island Point source to bypass the midbeach drumlin and link to the Lawrencetown Head drumlin, closing the western tidal inlet. Hoskin (1983), using ridge morphology and carbon-14 dating, indicated that the initiation of ridge sedimentation at Lawrencetown Beach began around 800 yr B.P. and was completed by 350 yr B.P.

Barriers that are retreating or being destroyed on the Eastern Shore are thin and display cliffed foredunes (Fig. 7), washovers (Fig. 8) that develop into tidal inlets (Figs. 6 and 9a), and rapid shoreline erosion rates (Fig. 10). Aerial photo evidence indicates the thin eastern end of Martinique beach was progressively eroded prior to breaching and washover development in 1974 (Fig. 6). Sand fencing and revegetation were used to artificially close the washover channel in 1979. Similar overwash and breaching of Conrads Beach occurred in 1962. In this case the channel was not artificially closed and eventually developed into a permanent tidal inlet. Silver Sands Beach across the entrance to Cow Bay illustrates a system whose retreat has been artificially accelerated by aggregate mining. McIntosh (1916) described the beach system as consisting of a series of gravel ridges and two midbeach drumlins that backed an extensive sand beach. Relative stability continued until 1954. However, between 1956 and 1971 some 2 million tons of sand and gravel were removed from the beach. As early as 1960, the sand at the western end of the beach had been removed, the beaches that were tied to the shoals east of Hartlen Point had disappeared, and the inlet to Cow Bay had widened noticably. The main beach became much thinner and by 1964 averaged only half the

FIG. 7. Foredune scarp at the eastern end of Martinique Beach. This feature is characteristic of retreating barriers in stage 4.

FIG. 8. The Martinique Beach compartment is an early stage 4 coastal system. Note the thin barrier and multiple washover fans at the near (eastern) end of the beach.

(a)

(b)

width of 1954 (Huntley, 1976). The remainer of the sand not trucked away during the mining operation was removed by tidal processes to accumulate in the Cow Bay flood tidal delta. The result of mining combined with natural evolution produced 50–80 m of beach retreat between 1954 and 1971 and reduced Silver Sands barrier to a thin boulder beach.

Aerial photos have been used to date a sequence of barrier destruction at Rocky Run between Graham Head and Half Island Point (Fig. 10). Photographs from 1954 show two midbeach drumlins and a composite sand/gravel barrier with a small tidal inlet. By 1962 the beach had become extremely thin and the drumlins consumed. By 1982 the subaerial barrier was destroyed and transported landward into the Rocky Run estuary as a series of intertidal shoals. A similar history seems likely for the old barrier between Rudeys Head and Sellars Head. This embayment is characterized by the absence of an extensive landward estuary, and sediments from the destroyed barrier have infilled a small backbarrier bay with submerged washover sands.

C. Estuaries and Coastal Bays

Many of the submerged coastal valleys of the Eastern Shore display the characteristics of true estuaries, with marine water mixed with an appreciable freshwater input. This is the case at Musquodoboit Harbour, Porters Lake, Lawrencetown Lake, and Petpeswick Inlet. Other coastal valleys are closer to lagoons and bays that have been restricted by prograded barrier systems, as at Cow Bay, Cole Harbour, West Marsh, Three Fathom Harbour, Chezzetcook Inlet, and Oyster Pond. Estuaries and bays with tidal inlets and an available sand supply are gradually infilled. Where tidal inlets are recent and sand supply low, as at Porters Lake, only a saltwater wedge advances into the estuary. Where tidal inlets are well established, tidal channels extend landward for over 5 km and deposit estuarine and lagoonal sediments in salt marsh (Fig. 11), tidal channel levee, crevasse splay (Fig. 12), and flood tidal delta environments (Fig. 9b). Figure 9a shows the infilling of West Marsh by a 0.5-km^2 flood tidal delta since the tidal inlet developed at Conrods Beach in 1962. Figure 11 illustrates salt marsh infilling of Chezzetcook Inlet between 1945 and 1974.

FIG. 9. (a) This breach in Conrads Beach occurred in 1962 and by 1982 had transported a considerable volume of barrier sand into a flood tide delta system in West Marsh (cf. Fig. 9b). Conrads Beach is an example of an advanced stage 4 coastal system. (b) A tidal channel at Lawrencetown Inlet Beach (lower left) has been active throughout historical times and succeeded in infilling most of lower Lawrencetown Lake with a large flood tidal delta complex.

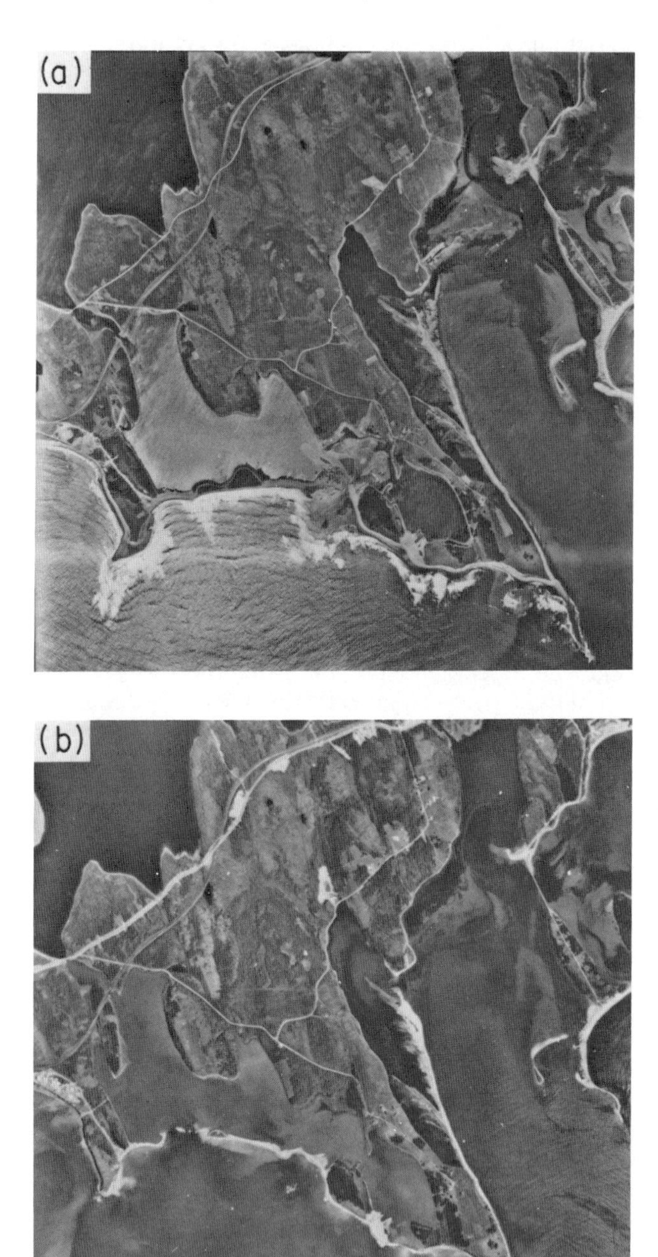

FIG. 10. Rocky Run barrier represents a historical example of Stage 5 barrier destruction. (a) A wide barrier (1945) extends from Half Island Point (left) to Graham Head (right) and includes two depleted midbeach drumlins. (b) By 1954 the beach is considerably thinner, the drumlins almost disappeared, and a breach has formed near Half Island Point.

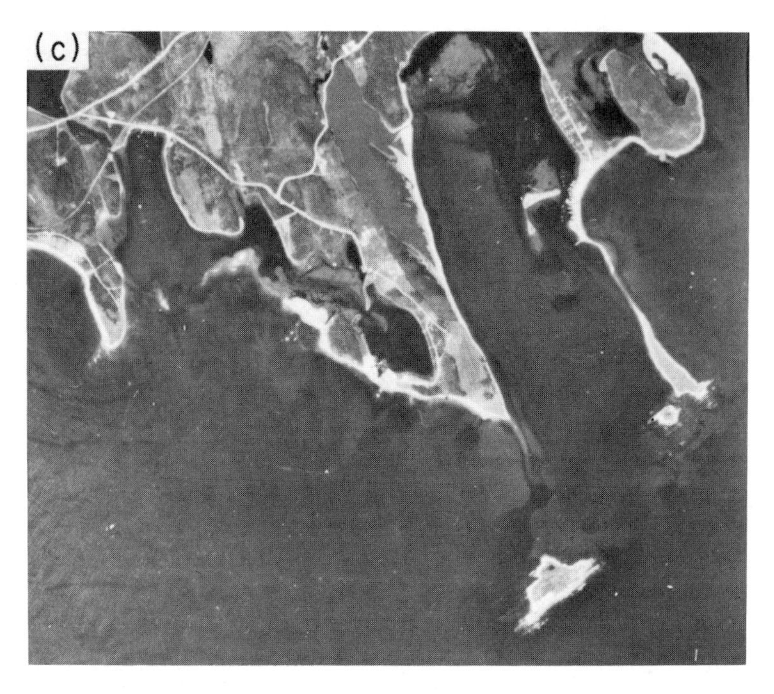

FIG. 10. (c) By 1974 the barrier has been destroyed and exists only in the form of intertidal and subtidal sand shoals. In the lower right-hand corner are Wedge Island (far right), an example of a depleted drumlin source, and Shut-In Island, an exhausted drumlin source, existing today as an offshore bedrock shoal. The eventual result of diminished sediment supply from drumlin sources in stage 4 is barrier destruction in stage 5, as illustrated by the Rocky Run barrier.

VI. Evolutionary Coastal Model

The range of evidence available from individual coastal embayments has enabled a general model to be constructed that explains the genesis and evolution of coastlines with similar processes and morphology to the Eastern Shore. This model provides a method for summarizing development of all 13 embayment–barrier compartments in this region, each of which exhibits a wide variety of physical characteristics and most of which are in different stages of evolution. In the model, a representative coastal compartment is constructed. It contains a composite set of physical features that are characteristic of several Eastern Shore compartments at that stage of evolution. Specifically (Fig. 13), stages 1 and 2 identify coastal genesis in early Holocene times and are not repeated. Stages 3 through 6 currently occur in a repetitive, cyclic form for each compartment. The time scales and detailed geomorphology of this evolutionary process depend on the transgression history as new estuaries and sediment sources are progressively encountered.

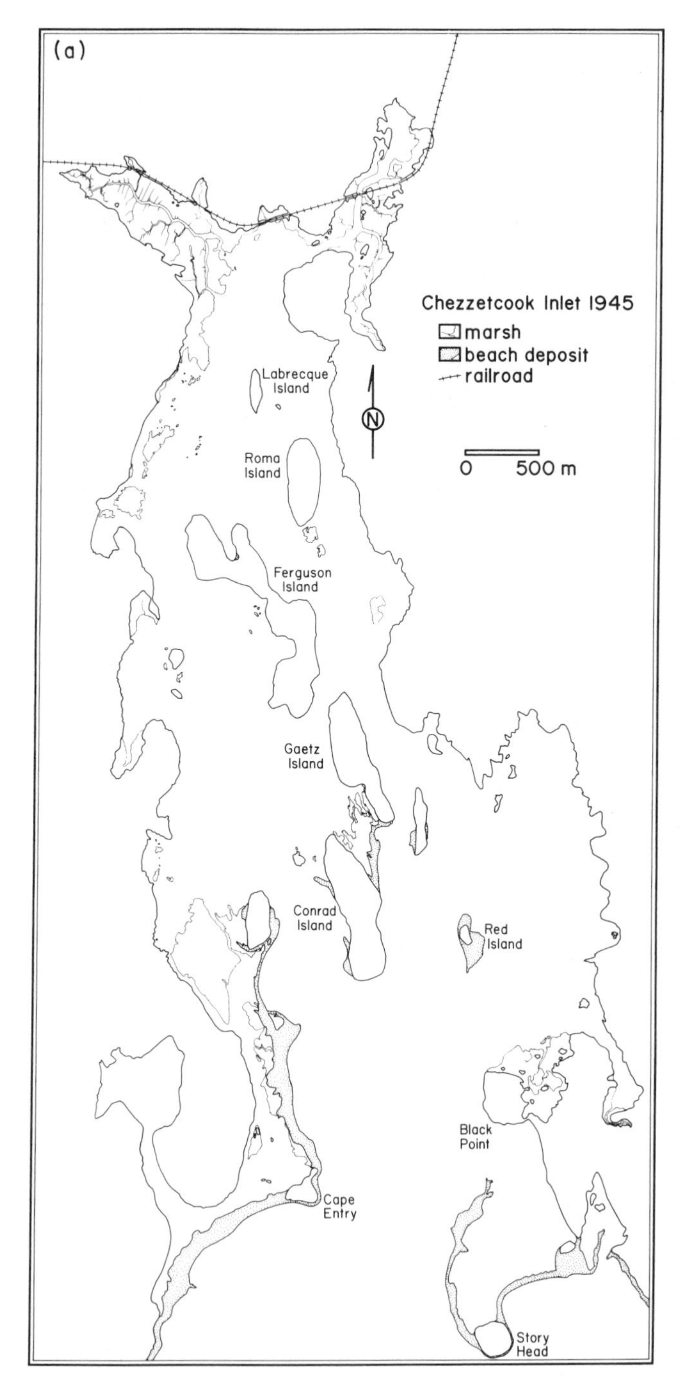

Fig. 11. The evolution of Chezzetcook Inlet between (a) 1945 and (b) 1974 is an example of stage 6 barrier reestablishment. Over a 29-yr period the inlet has infilled to the high tide level, allowing the development of extensive marsh and tidal flat environments. Sediment from old barrier systems near Cape Entry and Story Head have accreted new stage 6 barriers at Conrad Island and Red Island. [From Scott (1980), with permission.]

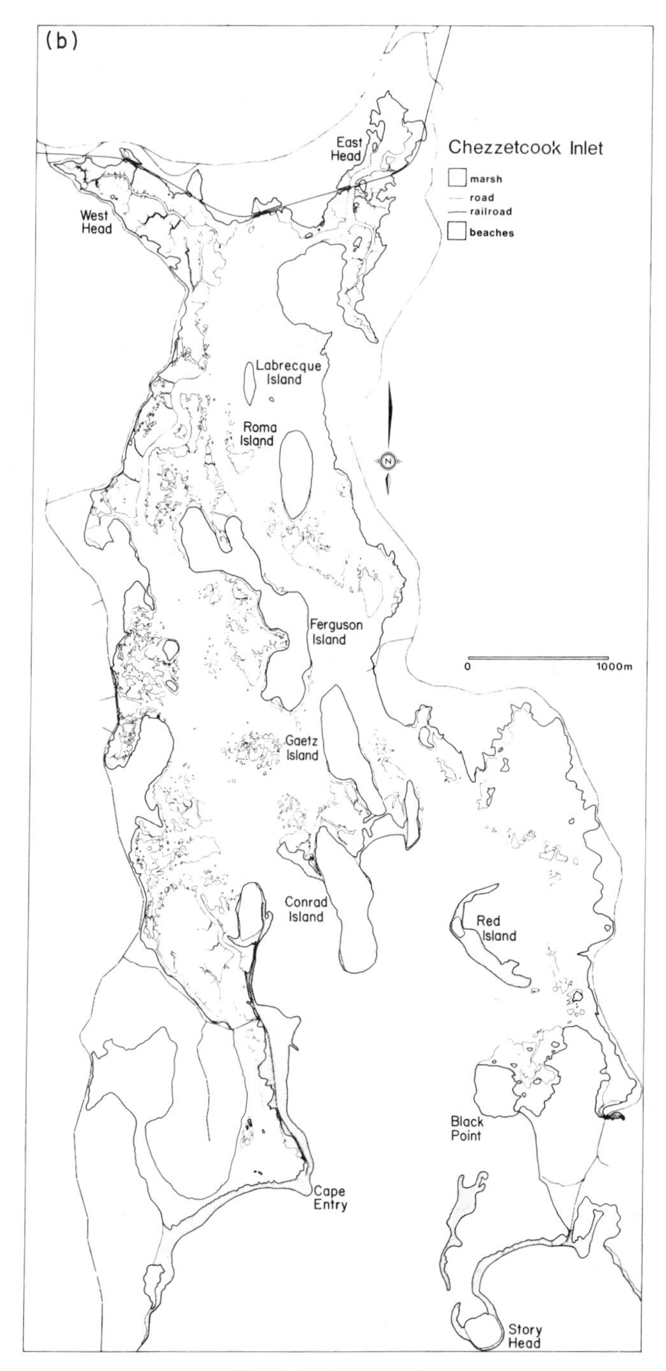

FIG. 11. (*Continued*)

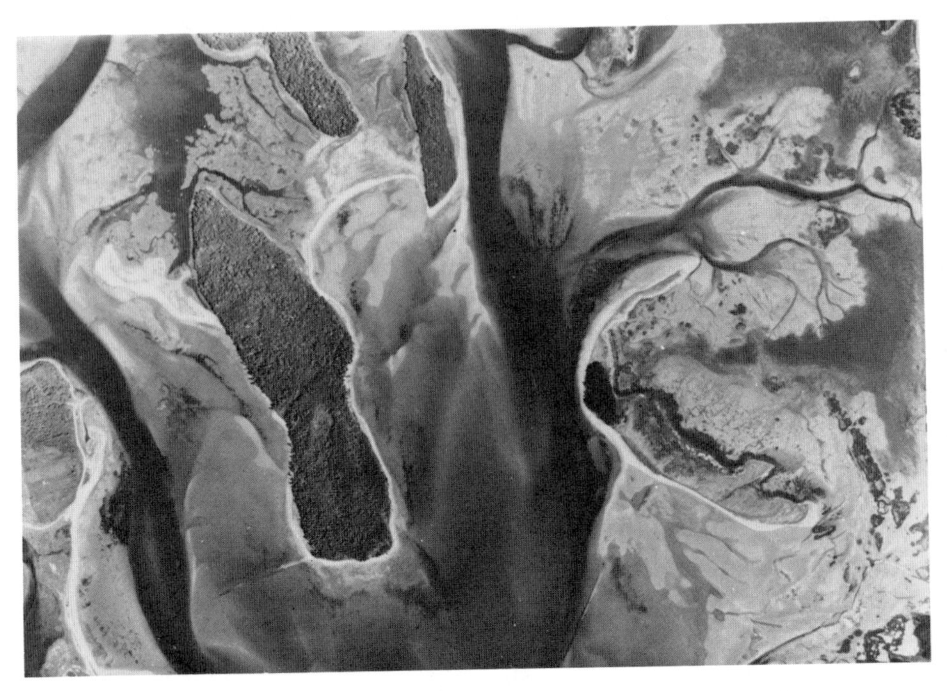

Fig. 12. Stage 6 barrier reestablishment inside the entrance to Chezzetcook Inlet. A beach-ridge plain has commenced formation at Conrads Island (center left), while Red Island (center right) has grown by over 1000 m between 1854 and 1974. Note the extensive development of remnant stage 5 intertidal sand shoals, moving landward to generate further accretion at the new barriers.

A. Stage 1—Continental Glacier and Ice Shelf

The Nova Scotia mainland and continental shelf experienced a complex glacial history during Pleistocene times. The majority of glacial stratigraphy postdates the last interglaciation (around 125,000 yr B.P.). The glacial sequence appears to be derived from a three-till–three-phase glaciation (Stea, 1982, 1983) composed of early, middle, and late Wisconsinan events. Wis-

---

Fig. 13. Coastal sedimentation along the eastern shore can be summarized in a six-stage evolutionary model. Sediments are initially supplied from Pleistocene glacial sources (stage 1). Following deglaciation, relative sea-level rise produces coastal transgression, transforming glacial valleys into embayments (stage 2). Subsequent evolution during stage 3 consists of marine reworking of glacial deposits into prograding barrier systems. As sediment supply diminishes, barriers retreat (stage 4), losing sand to washover, dune, and tidal inlet sinks until finally they are destroyed (stage 5). In stage 6, barriers encounter new sediment sources or headland anchor points and reestablish further landward.

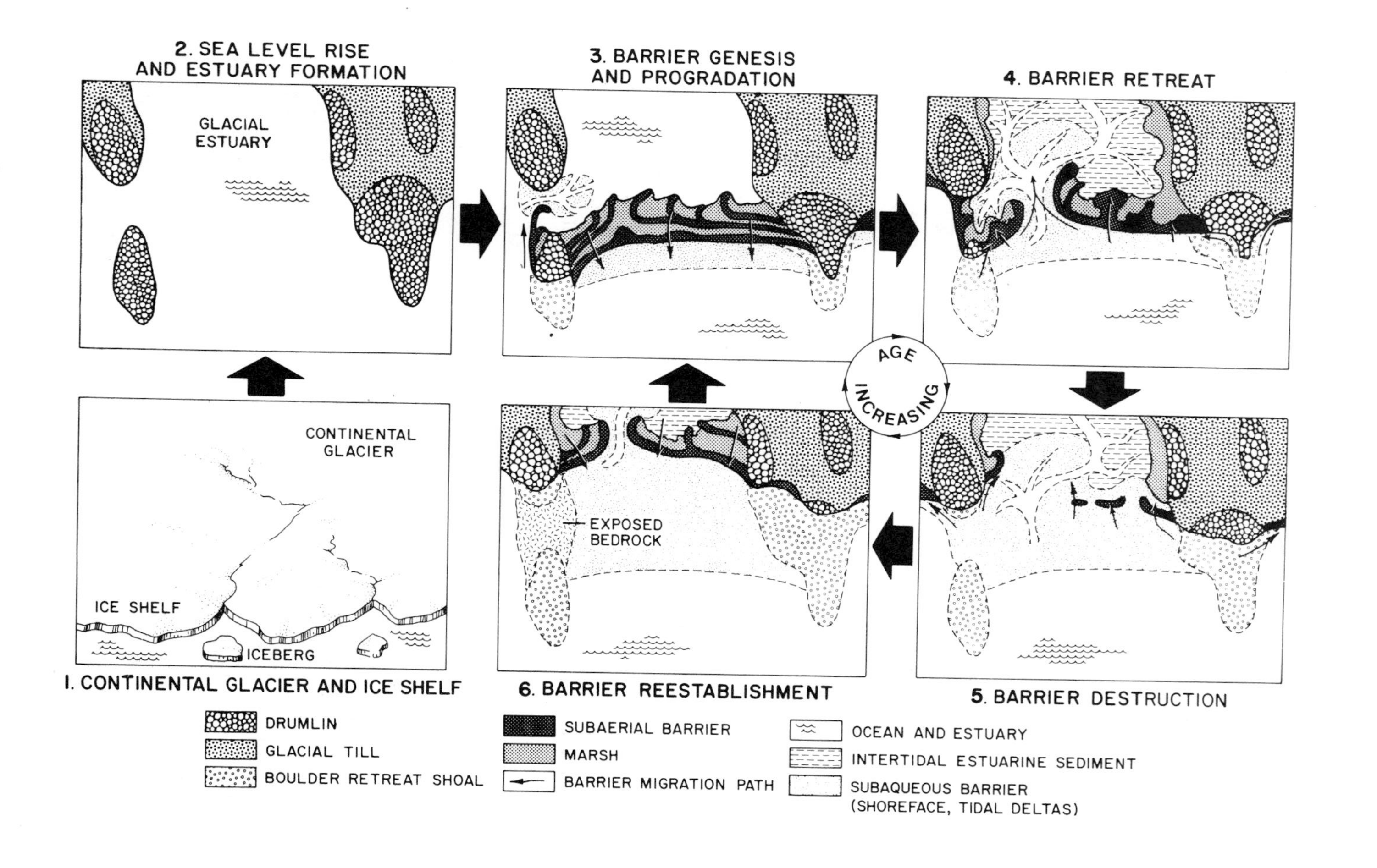
2. SEA LEVEL RISE AND ESTUARY FORMATION
GLACIAL ESTUARY
3. BARRIER GENESIS AND PROGRADATION
4. BARRIER RETREAT
CONTINENTAL GLACIER
ICE SHELF
ICEBERG
I. CONTINENTAL GLACIER AND ICE SHELF
AGE INCREASING
EXPOSED BEDROCK
6. BARRIER REESTABLISHMENT
5. BARRIER DESTRUCTION
DRUMLIN
GLACIAL TILL
BOULDER RETREAT SHOAL
SUBAERIAL BARRIER
MARSH
BARRIER MIGRATION PATH
OCEAN AND ESTUARY
INTERTIDAL ESTUARINE SEDIMENT
SUBAQUEOUS BARRIER (SHOREFACE, TIDAL DELTAS)

consinan ice advances were responsible (Fig. 13, stage 1) for Eastern Shore deposition of local slate till, quartzite till, and the overlying Lawrencetown Till (Stea 1980, and Fowler 1979). The ice-thickness estimates of Quinlan and Beaumont (1981) suggest a 250-m-thick icecap covered the Eastern Shore until around 16 ka BP, with deglaciation complete by around 12 ka BP. Wisconsinan ice-flow direction on the Eastern Shore maintained a consistent NW to SE orientation with glacial striae trends ranging from 121 to 174°E of N. Drumlinoid till accumulations and the linear bedrock scours now occupied by lakes and estuaries on the Eastern Shore also conform to a NW–SE orientation, suggesting formation and/or major modification by Wisconsinan ice.

B. Stage 2—RSL Rise and Estuary Formation

This second model stage represents a transition phase after the glacial cover of stage 1 and before the active coastal erosion and barrier formation of stage 3. Initially, in stage 2, RSL fell to at least −27 m. With eventual removal of the ice, the Eastern Shore would have been exposed as a glacially sculpted landscape consisting of linear scours and ridges formed in the Meguma bedrock or the remaining drumlinoid till deposits. The final ice ablation phase infilled glacial scours with a linear network of lakes aligned in the SE direction of ice flow (see Fig. 3). These lakes remain a common feature of the Eastern Shore landscape, with typical densities of 50 lakes per 100 km^2. The RSL rise along the Eastern Shore commenced around 7,000–12,000 yr B.P. At this time transgression of the inner Scotian shelf began, turning the most seaward lakes into marine to brackish estuaries and bays (Fig. 13, stage 2).

C. Stage 3—Barrier and Estuarine Sediment Accumulation

The combination of rising RSL and moderate to high wave power results in extensive reworking of glacial deposits intersected by a transgressing shoreline. Stage 3 describes a period in which drumlins and other till accumulations act as primary sediment sources for clay- to boulder-sized material (Fig. 4). Sediments follow a variety of dispersal paths, depending on their grain size. Waves are incompetent to transport boulders away from the seaward face of the drumlin, and boulder retreat shoals form. The mud component of the drumlin is transported (1) away from the sediment source areas and carried in suspension to midshelf basins (King, 1970; King and Fader, 1986) or (2) into lower energy depositional sites within estuaries.

Very fine sand- to cobble-sized material is moved by wave-generated longshore transport across estuary mouths to form barrier systems downdrift of the drumlins.

The extent of sand and gravel accumulation in estuary mouth barriers depends on the balance between the rate of sediment supply and the rate of RSL rise. If the sediment supply balance is positive, barrier systems may develop and also prograde seaward (Fig. 13, stage 3). The first barrier accumulations are spits adjacent to the drumlins. Further deposition results in longshore progradation of spits into beach ridges as at Lawrencetown Beach.

If an estuary has a sufficiently large tidal prism with a single outlet, a tidal inlet may be maintained throughout stage 3 barrier evolution. Under these conditions a significant proportion of sediment potentially available for barrier building will be transported to tidal inlets and accumulate in flood and ebb tidal deltas. This appears to be the case at Rainbow Haven barrier, where the large tidal prism of Cole Harbour has maintained a stable tidal inlet.

D. Stage 4—Barrier Retreat

During stage 4, the critical balance between sediment supply and RSL changes, resulting in the ending of barrier progradation and the onset of barrier retreat. The most obvious reason for this change lies in the reduction of sediment supply from the depleted drumlin and preexisting barrier sources. Typical drumlins along the Eastern Shore are 1 km long and 0.5 km wide. For cliff retreat rates of 1 m/yr, the lifespan of a drumlin and of its supply of sediment to the barrier is less than 1000 yr. The coastal retreat of the drumlin will also eventually limit beach progradation as these two adjacent shoreline segments become more aligned. Reduced sediment supply can also result from secondary causes such as diversion into tidal inlets or eolian bypassing. Barrier systems respond to negative changes in the sediment supply/RSL balance by entering a stage of coastal retreat (Fig. 13, stage 4). This stage is first characterized by barrier erosion as the shoreline successively occupies a more landward position. In some cases sediment eroded from the beach may initiate active dune growth through landward eolian transport early in stage 4, while the retreating foredune is marked by a scarp such as at the western end of Martinique Beach (Fig. 8). Further development into stage 4 is characterized by barrier breaching in the form of overwash and tidal inlet processes. High water levels resulting from storm surge and wave action (see, e.g., Boyd and Penland, 1981) enable transport of water and sediment through locations of lowest barrier elevation. Overwash channels in stage 4 may progress to become permanent tidal inlets in response to rising RSL and resulting increase in estuarine tidal prism.

E. Stage 5—Barrier Destruction

In the presence of a rapidly rising RSL, a coastal barrier system must maintain an active sediment supply or ultimately be destroyed. By stage 5, Eastern Shore barriers receive minimal sediment from their depleted sources. At a rate of RSL rise around 35 cm/100 yr, a typical barrier along the Eastern Shore with no active sediment supply would be submerged in less than 700 years. In reality, the effects of marine and subaerial erosion processes acting in combination with rising RSL accomplish the barrier destruction in a much shorter time. Each of the processes operating during stage 4 serves to remove sediment from the critical area of the beach face to dune systems, washovers, or backbarrier areas, including flood tidal deltas. At these locations the sediment is no longer available for beach rebuilding during poststorm recovery phases. The result of continuing sediment depletion and RSL rise is an increase in washover intensity, the formation of more tidal inlets, and finally the destruction of the barrier. During this phase, the sand and gravel that once formed the barrier are actively reworked into spits and intertidal to subtidal shoals (Figure 13, stage 5). Some beach material from the destroyed barrier systems, together with other finer grained material, accumulates in the estuarine tidal flat, distal flood tidal delta, and marsh environments. This material is not completely or permanently lost to the barrier system. As the shoreline eventually reaches further up the estuaries, previously deposited back-barrier material may once again contribute to barrier sediment supply as it is exhumed at the shoreface.

Flying Point at the eastern end of Martinique Beach and Wedge Island off Three Fathom Harbour are examples of depleted drumlin sources (Fig. 10c). A further and final phase of drumlin depletion takes the form of armored offshore islands such as Shut-In Island near Three Fathom Harbour or Egg Islet near Conrads Beach. Bedrock may be exposed at the seafloor or along the coast where original sedimentary deposits were thin or where bedrock highs crop out.

F. Stage 6—Barrier Reestablishment

The final evolutionary model stage is a period when the destroyed barrier remnants of stage 5 migrate up the estuary to reestablish new barrier systems in a more landward location (Fig. 13, stage 6). The movement of sediment is driven by ongoing RSL rise and storm processes. The formation of new barriers requires (a) a new sediment source and/or (b) a new set of headland anchor points. Stage 6 barrier generation differs from the original phase of barrier formation in that it has two additional sediment sources. These two sources are (1) the old stage 5 barrier remnants and (2) the back-barrier sands of the estuary and flood tidal delta. Both these sources of sediment are derived through shoreface retreat (Fischer, 1961; Swift, 1975) as earlier barrier and backbarrier sediments are reexposed on the retreating shoreface.

The procedure for new barrier formation in stage 6 is as follows. The intertidal shoals of stage 5 migrate landward until they encounter new headland anchor points. Usually these anchor points are either drumlins or bedrock outcrops. Their location within the estuary provides a lower-energy site for barrier accumulation than along the open coast. Initially spits prograde from the headland anchors until stabilization takes place. The large reservoir of sand remaining in the nearshore zone is gradually incorporated into the new barrier as the shoreface retreats. Concurrently, new sediment may also be arriving from adjacent drumlin sites. This sediment influx again leads to beach-ridge development and beach-ridge plain progradation as at the entrance to Chezzetcook Inlet (Fig. 12). The reformation of new barrier systems in Stage 6 illustrates the ongoing cyclic nature of the model.

VII. Discussion

Two separate components of RSL rise may be present on the Eastern Shore: (1) a eustatic rise, reported by Hicks (1978), and indicated by Emery (1980) and Gornitz *et al.* (1982) to be around 12 cm/100 yr along the eastern seaboard of the United States; and (2) the continuing response to glacial forebulge passage, which would account for the remaining 28 cm/100 yr identified on the Halifax tide-gage records by Grant (1970).

Coastal evolution on the Eastern Shore represents a stepwise form of Swift's (1975) concept of shoreface retreat. Stepwise shoreface retreat here results from the distribution of discrete local sediment supplies and the location of suitable headland anchor points. As the transgressing shoreline encounters either of these factors, a period of barrier building commences. Progradation of barriers results from sediment derived from eroding till, drumlin headlands, or preexisting barriers. When a period of sediment supply nears an end, RSL rise again becomes the dominant factor and shoreface retreat resumes. This results in alternating periods of transgression and regression within an overall framework of submergence.

Landward sediment transport during shoreface retreat on the Eastern

Shore consists of two separate mechanisms. First, eolian, overwash, and tidal inlet processes remove sediment from the barrier and deposit it in flood tidal delta, washover, and estuarine environments. Second, when landward translation of the shoreline causes these sediments to be exhumed by the retreating shoreface, they are reworked onshore to accumulate at a new barrier location. The only extensive surficial sand deposits that remain on the inner shelf after transgression are on the shoreface of modern bay barriers (Hall, 1985).

Johnson (1925) has described a pattern of shoreline development for a drumlin coast at Nantasket, Massachusetts, that is similar to that of the Eastern Shore. However, Johnson's (1925) example consists of a more random distribution of drumlins that is not constrained by a ridge and valley topography. The evolution of the Eastern Shore and comparisons with comparable regions such as Nantasket Beach indicate the critical importance of paleotopography in controlling transgressive sedimentation. Since transgressions typically migrate across topographically irregular surfaces, paleotopography can be expected to influence the sediment dispersal patterns of most sea-level transgressions. This is the case even in regions of subdued relief such as the Holocene Mississippi Delta, where transgressive sediment dispersal is controlled by the location of preexisting headland sediment sources and lagoonal sediment sinks (Boyd and Penland, 1984). The higher relief paleotopography of Nova Scotia's Eastern Shore is the result of glacial erosion patterns. Because continental glaciation would not usually be expected to produce linear valley and ridge landforms, the Eastern Shore topography appears to result from glacial modification of earlier fluvial drainage patterns. Deposition during transgression is controlled by alignment of the NW to SE valley and ridge systems. Submergence of the valleys provides the only major site for deposition. Tidal inlets are necessary to exchange the large tidal prisms that result from the progradation of barriers across the mouths of submerged valleys. Tidal inlets in turn control much of the estuarine and flood tidal delta sedimentation that characterizes the Eastern Shore coastline.

VIII. Conclusions

Sea-level transgression on the Eastern Shore of Nova Scotia is driven by an RSL rise of 35–40 cm/100 yr. The RSL rise appears to have a small eustatic component and a larger component resulting from glacial forebulge passage.

The style of coastal sedimentation on the Eastern Shore is strongly influenced by paleotopography. Primary sediment sources are found in

Pleistocene drumlin fields. Sediment deposition takes place in submerged valley systems of glacial and perhaps earlier fluvial origin.

Sediment dispersal patterns can be summarized by an evolutionary model with glacial origin. The model consists of cyclic barrier progradation followed by destruction and subsequent landward translation within an overall framework of submergence.

Acknowledgments

This research was supported by Canadian Government grants NSERC A8425 and EMR 202 (through the Atlantic Geosciences Centre at Bedford Institute). Additional support was provided by the Nova Scotia Department of Lands and Forests. R. K. Hall was supported by an Imperial Oil University Fellowship. Tom Duffett coordinated the field work and was assisted by Cecily Honig, Gary Sonnichsen, Tony LaPierre, Gary Frotten, and Paul Davidson. Development of concepts presented in this chapter was aided by input from Ralph Stea, David Scott, and Don Forbes. R. Boyd wishes to acknowledge the assistance of the Geology Department, University of Newcastle, in preparing this manuscript while there on study leave.

References

Boyd, R., and Bowen, A. J. (1983). "The Eastern Shore Beaches." Unpubl. Tech. Rep., Cent. Mar. Geol., Dalhousie Univ., Halifax, Nova Scotia.

Boyd, R., and Penland, S. (1981). Washover of deltaic barriers on the Louisiana coast. *Trans. Gulf Coast Assoc. Geol. Soc.* **31**, 243–248.

Boyd, R., and Penland, S. (1984). Shoreface translation and the Holocene stratigraphic record: Examples from Nova Scotia, the Mississippi Delta, and Eastern Australia. *Mar. Geol.* **60**, 391–412.

Canadian Hydrographic Service (1984). "Canadian Tide and Current Tables." Can. Gov., Dep. Fish. Oceans, Ottawa.

Davies, J. L. (1980). "Geographical Variation in Coastal Development," 2nd Ed. Hafner, New York.

Emery, K. O. (1980). Relative sea-levels from tide-gauge records. *Proc. Natl. Acad. Sci. USA* **77**, 6968–6972.

Fischer, A. G. (1961). Stratigraphic record of transgressing seas in the light of sedimentation on the Atlantic coast of New Jersey. *Bull. Am. Assoc. Pet. Geol.* **45**, 1656–1666.

Gornitz, V., Lebedeff, S., and Hansen, J. (1982). Global sea-level trend in the past century. *Science* **215**, 1611–1614.

Grant, D. R. (1970). Recent coastal submergence of the Maritime Provinces, Canada. *Can. J. Earth Sci.*, **7**, 676–689.

Grant, D. R., and King, L. H. (1984). A stratigraphic framework for the Quaternary history of the Atlantic Provinces. *In* "Quaternary Stratigraphy of Canada—A Canadian Contribution to IGCP Project 24" (R. J. Fulton, ed.), *Geol. Surv. Pap. (Geol. Surv. Can.)* No. 84-10.

Hall, R. K. (1985). "Inner shelf acoustic facies and surficial sediment distribution of the Eastern Shore, Nova Scotia." *Tech. Rep.* **8**. Cent. Mar. Geol., Dalhousie Univ., Halifax, Nova Scotia.

Hicks, S. D. (1978). An average geopotential sea-level series for the United States. *J. Geophys. Res.* **83,** 1377–1379.

Hoskin, S. (1983). Coastal sedimentation at Lawrencetown Beach, Eastern Shore, Nova Scotia. B. S. Honors Thesis, Geol. Dep., Dalhousie Univ., Halifax, Nova Scotia.

Huntley, D. A. (1976). Brief assessment of the replenishment rates and topographic changes in Cow Bay. Unpubl. rep., Dep. Oceanogr., Dalhousie Univ., Halifax, Nova Scotia.

Johnson, D. (1925). "The New England-Acadian Shoreline." Wiley, New York.

King, L. H. (1967a). Use of a conventional echo-sounder and textural analyses in delineating sedimentary facies: Scotian Shelf. *Can. J. Earth Sci.* **4,** 691–70.

King, L. H. (1967b). On the sediments and stratigraphy of the Scotian Shelf. *In* "Collected Papers in the Geology of the Atlantic Region" (E. R. W. Neale and H. Williams, eds.), *Geol. Assoc. Can. Spec. Pap.* **4,** 71–92.

King, L. H. (1969). Submarine and moraines and associated deposits of the Scotian Shelf. *Geol. Soc. Am. Bull.* **80,** 83–96.

King, L. H. (1970). Surficial geology of the Halifax–Sable Island map area. *Mar. Sci. Pap. (Ottawa)* **1.**

King, L. M., and Fader, G. B. (1986). Wisconsinan glaciation of the Atlantic continental shelf of Southeast Canada. *Geol. Surv. Can. Bull.* No. 363.

McIntosh, K. (1916). A study of Cow Bay beaches. *N. S. Inst. Sci., Proc. Trans.* **14,** Part 2, 109–119.

Piper, D. J. W., Mudie, P. J., Letson, J. R. J., Barnes, N. E., and Iuliucci, R. J. (1986). The marine geology of the Inner Scotian Shelf off the South Shore, Nova Scotia. *Geol. Surv. Pap. (Geol. Surv. Can.)* No. 85–19.

Prest, V. K. (1984). The Late Wisconsinan glacier complex. *In* "Quaternary Stratigraphy of Canada—A Canadian Contribution to IGCP Project 24" (R. J. Fulton, Ed.), *Geol. Surv. Pap. (Geol. Surv. Can.)* No. 84-10.

Quinlan, G., and Beaumont, C. (1981). A comparison of observed and theoretical post-glacial sea levels in Atlantic Canada. *Can. J. Earth Sci.* **18,** 1146–1163.

Schenk, P. S., Lane, T. E., and Jensen, L. R. (1980). Paleozoic history of Nova Scotia—a time trip to Africa (or South America?). *Geol. Assoc. Can. Field Trip Guideb.* **20.**

Scott, D. B. (1980). Morphological changes in an estuary: a historical and stratigraphical comparison. *In* "The Coastline of Canada" (S. B. McCann, ed.), *Geol. Surv. Pap. (Geol. Surv. Can.)* No. 80-10, p. 199–205.

Scott, D. B., and Medioli, F. S. (1982). Micropaleontological documentation for early Holocene fall of relative sea level on the Atlantic coast of Nova Scotia. *Geology* **10,** 278–281.

Sonnichsen, G. (1984). The relationship of coastal drumlins to barrier beach formation along the Eastern Shore of Nova Scotia. B. S. Honors Thesis, Geol. Dep., Dalhousie Univ., Halifax, Nova Scotia.

Stea, R. R. (1980). A study of a succession of tills exposed in a wave cut drumlin, Meisners Reef, Lunenburg County. *N. S. Dep. Mines Energy Rep.* **80-1,** 9–19.

Stea, R. R. (1982). The properties, correlation and interpretation of Pleistocene sediments in central Nova Scotia. M. S. Thesis, Dalhousie Univ., Halifax, Nova Scotia.

Stea, R. R. (1983). Surficial geology of the western part of Cumberland county, Nova Scotia. Current research, Part A. *Geol. Surv. Pap. (Geol. Surv. Can.)* No. 83-1A, 195–202.

Stea, R., and Fowler, J. (1979). Regional mapping and geochemical reconnaissance of Pleistocene till, Eastern Shore, Nova Scotia. *N. S. Dep. Mines Rep.* (D. Gregory, ed.), No. 78-1, pp. 5–14, map.

Swift, D. J. P. (1975). Barrier island genesis: evidence from the central Atlantic shelf, eastern U.S.A. *Sediment. Geol.* **14,** 1–43.

Wang, Y., and Piper, D. J. W. (1982). Dynamic geomorphology of the drumlin coast of southeast Cape Breton Island. *Marit. Sediments Atl. Geol.* **18,** (1),1–27.

CHAPTER 5

Holocene Evolution of the South–Central Coast of Iceland

DAG NUMMEDAL

Department of Geology and Geophysics
Louisiana State University
Baton Rouge, Louisiana

ALBERT C. HINE

Department of Marine Science
University of South Florida
St. Petersburg, Florida

JON C. BOOTHROYD

Department of Geology
University of Rhode Island
Kingston, Rhode Island

The south–central coast of Iceland consists of broad alluvial outwash plains, "sandurs." Landward, the sandurs are flanked by piedmont glaciers and high basalt cliffs (late Pleistocene sea cliffs); seaward they terminate in extensive barrier spits. The sandurs receive most of their sediment during glacier bursts (jökulhlaups) caused by subglacial volcanic activity or the failure of marginal ice dams retaining meltwater. Large volcanogenic bursts carry discharges as high as 100,000 m^3/s.

The wave energy along the Atlantic shore of these sandurs is among the highest in the world. Extratropical cyclones passing south of Iceland in the winter season reach their peak energy when the associated winds blow onto the south Icelandic coast from the southeast. Consequently, longshore sediment transport is generally directed to the west along the east–west shorelines. Convergence zones, characterized by rapid coastal progradation, are found within bights facing the southeast.

The combination of frequent glacier bursts, a nearly vegetation-free coastal zone, and extreme coastal storm dominance has produced in south Iceland a suite of sedimentary environments with few analogs elsewhere.

I. Introduction

The south–central coast of Iceland consists of broad alluvial outwash plains, "sandurs," extending seaward from the margin of piedmont glaciers or the escarpment of an interior plateau consisting of Neogene–Quaternary volcanics. Locally, late Holocene and Recent basaltic lava fields constitute the coastal lowlands.

The Holocene evolution of this coast has been controlled by a unique combination of glacial, fluvioglacial, volcanic, and marine processes. The objective of this chapter is to describe this spectrum of geological processes and to evaluate their role in the Holocene history of the region.

Glaciation has influenced this coastline by (a) eroding deeply incised valleys across the margin of the plateau, (b) eustatically lowering sea level to near the edge of the Icelandic continental shelf, (c) loading the island, followed by rapid postglacial isostatic readjustment and coastal uplift, and (d) producing large quantities of terrigenous clastics.

Fluvioglacial processes in Iceland include normal summer meltwater discharge and catastrophic flood events. The annual discharge pattern is characterized by high sediment transport during the summer by meltwater drainage from the ice margin in a series of braided streams. Catastrophic discharge events, glacier bursts (in Icelandic: *jökulhlaups*), are of two kinds. Some are limnoglacial, caused by a sudden drainage of ice-dammed meltwater lakes along the glacier margins. Others are caused by subglacial volcanic eruptions and melting over subglacial hot springs. Volcanic activity in historic time is also known to have generated debris avalanches and mudflows.

Finally, the large quantities of clastics, released by all the above mechanisms, are deposited along a coastline almost devoid of vegetation and exposed to the high wave energies of the subarctic North Atlantic.

Iceland's location on the Mid-Atlantic ridge, its high latitude, and its position relative to the Atlantic extratropical storm tracks cause this spectrum of sedimentary processes, which is unique on a global scale. The morphology and the suite of sedimentary environments found here have no close analogs elsewhere in the world.

II. Glaciation

A. General

The late Wisconsinan ice sheet covered nearly all of Iceland, except for small areas along the north–central and northwestern coasts (Fig. 1). A dry Wisconsinan climate along the northern coast probably accounts for the nonglaciated regions. Glaciers extending northward from the ice sheet followed relatively narrow valleys and scoured broad U-shaped valleys. The thickness of the northern valley glaciers probably was between 300 and 500 m, yet generally less than the associated relief of the valley (Milanovsky, 1982). In south Iceland, in contrast, late Wisconsinan glaciers covered the entire landscape. With beginning deterioration and thinning of the ice sheet in the latest Wisconsinan, the ice began to form extensive piedmont glaciers, probably similar to those along the present southern margin of Vatnajökull (Fig. 2). The glacial erosion surface beneath the outwash plains of south Iceland appears to have a relatively flat relief. Seismic refraction studies at the coast of Skeidararsandur suggest bedrock

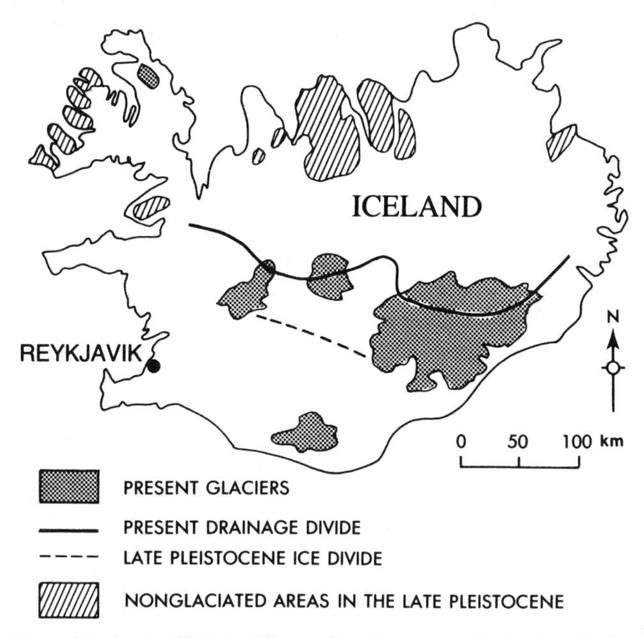

FIG. 1. Map of Iceland with late Wisconsinan ice cover. Note that the ice divide was located south of the present fluvial divide. [Modified from Einarsson (1961).]

FIG. 2. Landsat image of south Iceland. Note Vatnajökull with its major piedmont glaciers Sidujökull (1) and Skeidararjökull (2) and the extensive Skeidararsandur outwash plain (3).

at a depth of about 100 m (T. Einarsson, 1973 personal communication). The clastic wedge overlying the bedrock represents probably a number of Pleistocene and Holocene glacial–interglacial cycles.

B. ISOSTASY

The 500- to 600-m-high basalt cliffs at the landward edge of the Icelandic sandurs commonly have caves and beach gravels at their base, ample evidence of former sea cliffs. The beach gravels are part of an extensive set of terraces, some coastal and others fluvial, found all along the coast of Iceland. Carbon-14 dating suggests that the most landward coastal terrace gravels were deposited since 11,000 yr B.P., corresponding to the Alleröd interstadial (Einarsson, 1966; Milanovsky, 1982). The 11,000-year date is a maximum age. Dated Holocene basalt flows, which locally overlie the coastal gravel terraces, give a minimum terrace age of about 8,000 yr B.P. suggesting that the early Holocene transgression responsible for these inland beach gravels was of relatively short duration. It is inferred that this reflects rapid isostatic uplift after deglaciation. Only in the earliest Holocene did local sea level actually rise in Iceland.

This early phase was followed by a rapid regression. Since about 6,000 yr B.P. there appears to have been little relative sea-level change (Einarsson, 1961). This rapid isostatic rebound of Iceland, as compared to Scandinavia and Canada, which are still rising, is probably due to a lower mantle viscosity associated with Iceland's position on the Mid-Atlantic Ridge.

The marine terrace gravels also provide information about the magnitude of the postglacial rebound of Iceland. Postglacial marine terraces are extensive (Fig. 3). Their highest elevations are found in the interior valleys of the southwest, where some are as high as 110 m above present sea level. Maximum terrace elevations in the east are significantly less (Fig. 3). Eustatic sea level at the time of deposition of the marine gravels (8,000 to 11,000 yr B.P.) probably was between 20 and 35 m below the present

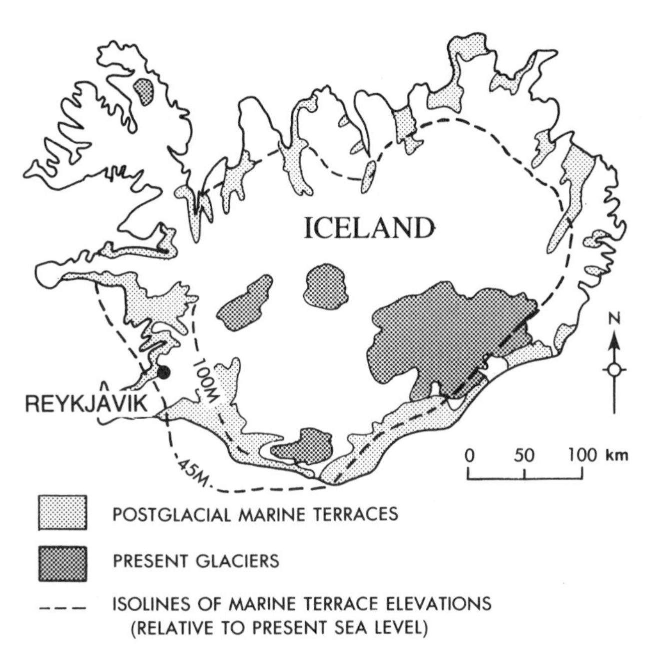

FIG. 3. Map of the distribution and elevation of late Pleistocene–Holocene marine terraces along the coast of Iceland. Note that the highest elevations are found in regions closest to the maximum thickness of the late Pleistocene ice sheet. [Modified from Einarsson (1961).]

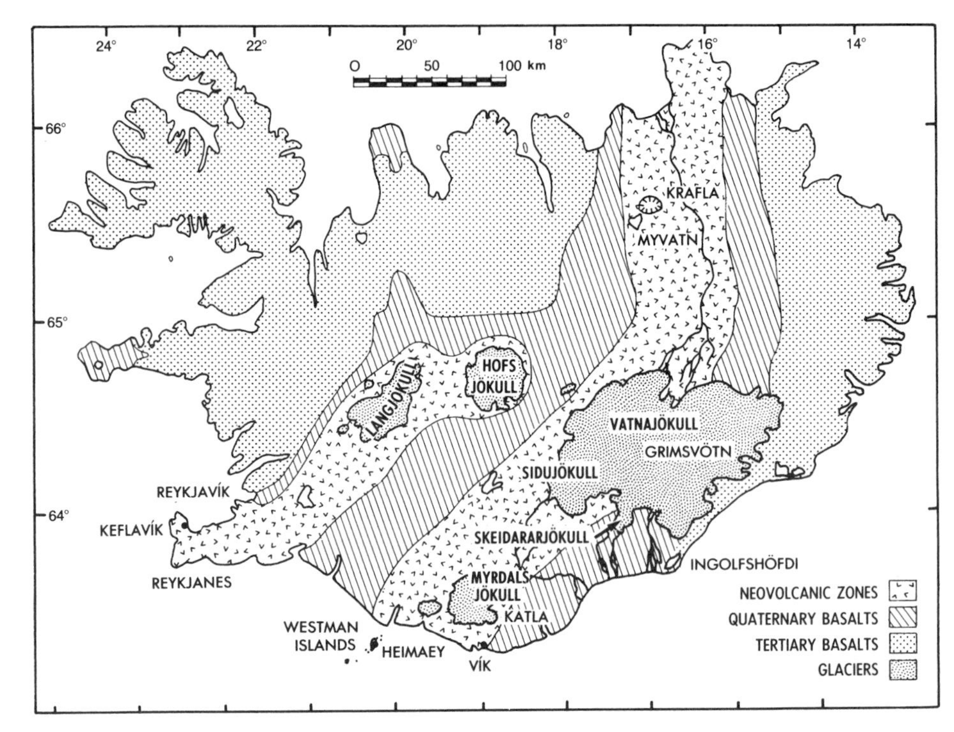

Fig. 4. Generalized geologic map of Iceland with Neovolcanic zones and major ice fields.

(Dillon and Oldale, 1978). Therefore, the recorded elevations suggest an uplift in the interior of western Iceland of 130–145 m. The magnitude of postglacial uplift along the south–central sandur coast may have been about 65–80 m.

The late Pleistocene ice divide extends from Sidujökull west to Langjökull (Figs. 4 and 1). This probably was the zone of maximum ice thickness. Because the amount of ice loading is proportional to its thickness, the postglacial uplift of the interior has been higher than that along the coast.

Little precise information is available about the recent tectonic vertical movement in Iceland. It is the opinion of some Icelandic geophysicists, however, that vertical crustal movement along the southern coast has been insignificant over the last few thousand years (S. Björnsson, personal communication).

C. The Historical Record

Excellent historical records were traditionally kept in Iceland. Old Norse sagas and the medieval Icelandic Annals contain detailed accounts of major geologic events during the last millenium, particularly glacial changes, glacier bursts (jökulhlaups), and volcanic eruptions. Analysis of much of these data (Nummedal *et al.*, 1974) yields the following picture.

Holocene glaciers in Iceland probably attained their minimum size about 4,000–5,000 yr B.P. This is indicated, in part, by wood fragments and peat of this age found in morainal debris along the present margin of Skeidararjökull (Fig. 4). At the time of early settlement in Iceland (tenth century A.D.) it appears that Vatnajökull was much smaller than it is at present, and probably split in two. Its Old Norse name of Klauvajökull ("split glacier") indicates the existence of a nonglaciated valley between its eastern and western parts. Probably this valley ran through the Grimsvötn depression (Fig. 4).

Deteriorating climate in late medieval time (1,200–1,400 A.D.; Fig. 5a,b) caused Vatnajökull to grow. Its middle valley was closed and its southern large piedmont glaciers prograded onto the outwash plains. Based on historically dated ice-margin positions, Skeidararjökull attained its maximum southern extent in 1749 A.D. (Figs. 6 and 7). Since that time there has been a steady retreat of its margin at a rate of about 20 m/yr (Figs. 5c and 6).

III. Fluvioglacial Processes

Sediment delivered to the ice-sheet margins is carried seaward by a series of braided streams constituting the vast Skeidararsandur (1,300 km^2) and other south-Icelandic outwash plains (Fig. 2). Proximally, the active streams have gradients of 6–17 m/km; longitudinal gravel bars are the dominant bedform. Distally, the rivers are sandy with a gradient of 2–3 m/km, containing a complex array of linguoid bars (Boothroyd and Nummedal, 1978). Although seasonal meltwater contributes to sediment transport, the dominant transport factor is the periodic glacier burst.

A. Origin of Glacier Bursts

Iceland is a subaerial exposure of part of the Mid-Atlantic Ridge. Active subaerial volcanism since the early Miocene (Fridleifsson, 1982) has produced extensive basalt flows, perhaps as much as 10,000 m thick (Einarsson, 1960; Walker, 1965). At present, the Mid-Atlantic Ridge enters Iceland at the Reykjanes peninsula and strikes northeast past Langjökull (Fig. 4). Offset about 150 km to the east is the second major neovolcanic zone. At its southern end this zone includes recently active volcanoes in the Westman Islands (Surtsey erupted in 1963; Heimaey in 1973). Trending

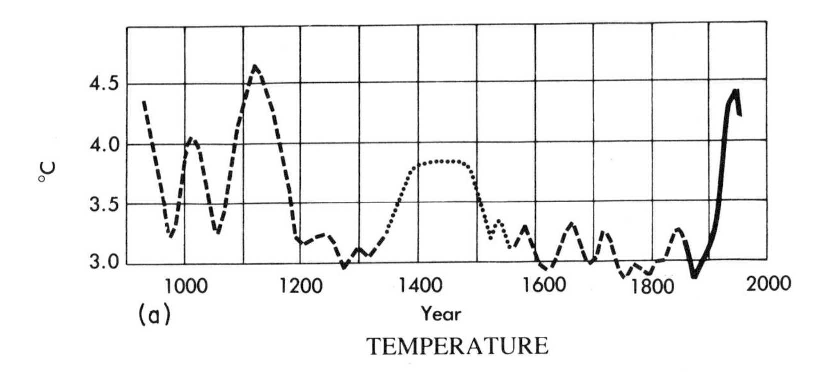

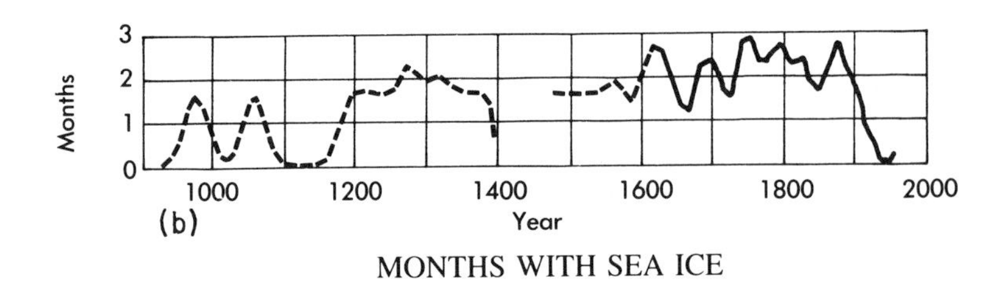

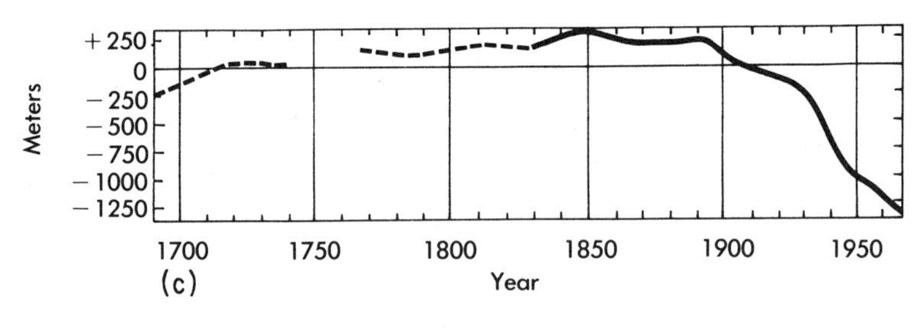

SOUTHERN MARGIN OF VATNAJÖKULL

FIG. 5. Fluctuations in Icelandic climate during the last millenium. (a) Temperature reduction after the twelfth century was accompanied by (b) an expansion in the sea ice cover and (c) an advance of the southern margin of Vatnajökull. [From Eythorsson and Sigtryggsson (1971), with permission.]

northeast, this zone passes beneath the large ice fields of Myrdalsjökull and Vatnajökull. Within this zone are the subglacial volcano Katla, at the southeastern margin of Myrdalsjökull, and a line of volcanoes and hot springs under Vatnajökull, including the Grimsvötn depression (Fig. 4). North of Vatnajökull this neovolcanic zone changes its trend to due north. Here it includes the Myvatn area with the recent, repeatedly active, Krafla

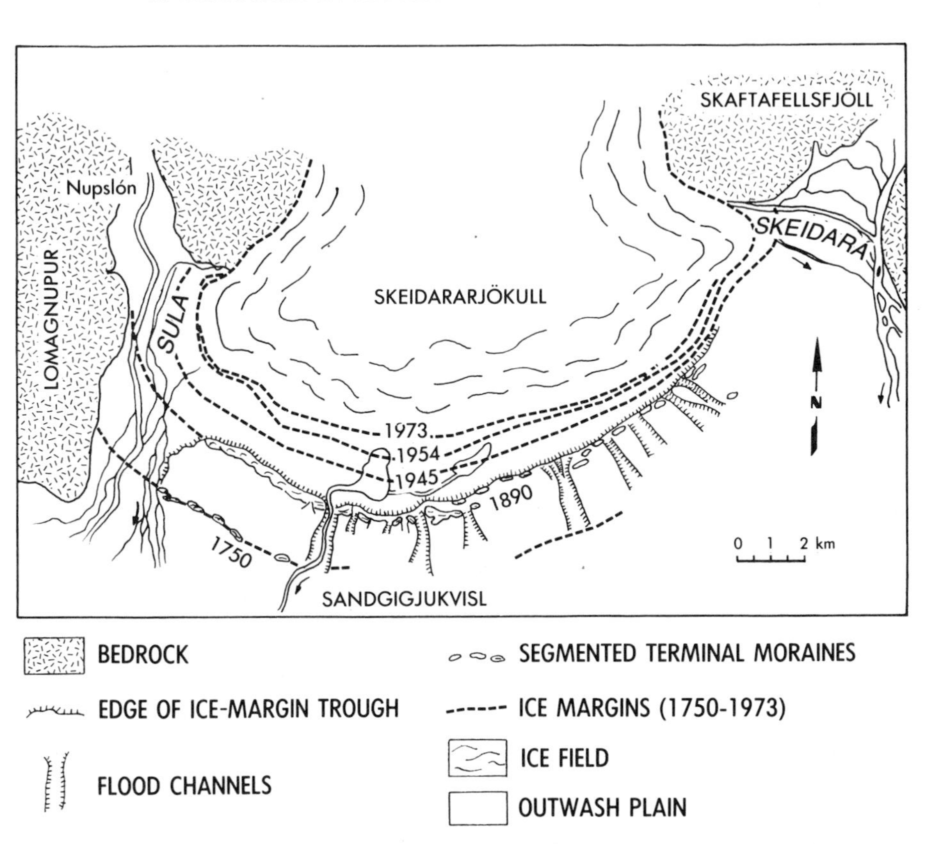

FIG. 6. Retreat of the margin of Skeidararjökull since 1749. The map is based on the locations of moraines, aerial photos, and previous field studies by Todtman (1960).

caldera. For further general information about the geology, geophysics, and crustal structure of Iceland, see Palmason and Saemundsson (1974), Saemundsson (1979), and Fridleifsson (1982).

The extensive outwash plains south of Myrdalsjökull and Vatnajökull are, in large part, the direct result of subglacial volcanism within this neovolcanic zone. Eruptions of subglacial volcanoes are accompanied by mudflows and large glacier bursts. Bursts from Katla, the largest recorded in Iceland, have attained an estimated discharge of 100,000 m^3/s at their peak. Such bursts, however, last but a few days (Thorarinsson, 1960). For the sake of comparison, the mean discharge of the Amazon River, near Obidos, Brazil, is about 170,000 m^3/s (Shen, 1971).

Bursts from Grimsvötn, discharging beneath the southern margin of Skeidararjokúll (Figs. 6 and 8), have attained peak discharges of 40,000–50,000 m^3/s and, because of longer duration than the Katla bursts, have had higher cumulative flow volume (Thorarinsson, 1960).

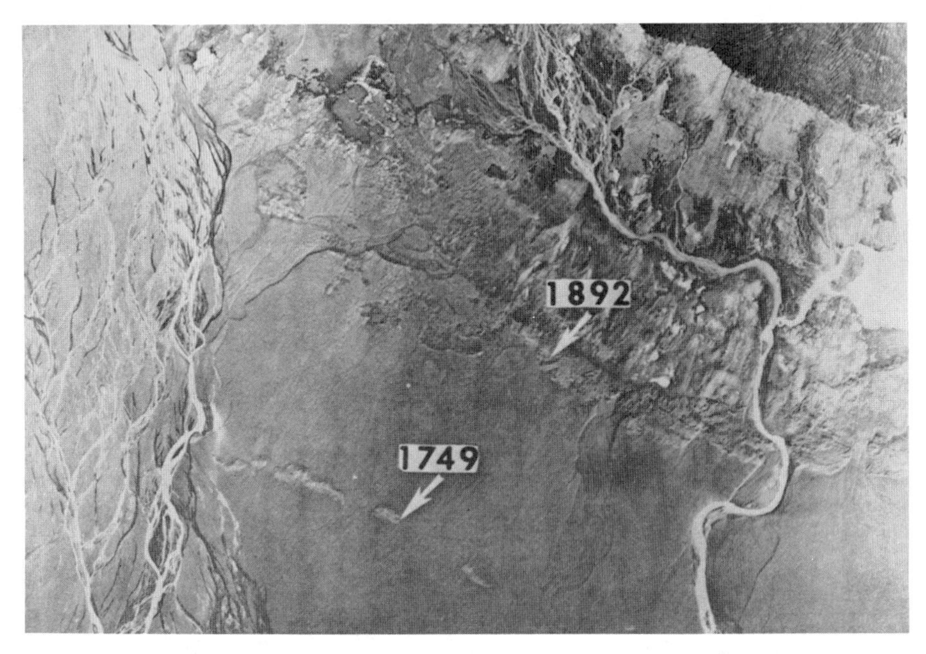

FIG. 7. Vertical air photo of northwest Skeidararsandur. A few remains of the 1749 moraine exist between Sula (left) and Sandgigjukvisl (right). Farther east the moraine is completely eroded. The huge 1892 moraine forms a distinct break between the sandur to the south (outwash) and a fluted basal till plain. (Photo courtesy of Landmaelingar Islands.)

Glacier bursts are also caused by the sudden draining of ice-dammed lakes (limnoglacial bursts), due to breaching of the ice dam, supraglacial flow, or subglacial tunneling. The largest bursts of this kind in Iceland come from Lake Graenalon, an 18-km^2 lake trapped against the Lomagnupur Mountains at the northwest margin of Skeidararjökull (Fig. 8). These bursts discharge through the river Sula.

B. FREQUENCY OF BURSTS

The earliest recorded bursts of Skeidará occurred in the middle of the fourteenth century. Bursts at that time apparently destroyed settlements located at what are today the active outwash plains of Skeidararsandur and its western neighbor Myrdalssandur (T. Einarsson, 1973 personal communication).

No historical accounts are found for bursts in Skeidará prior to the middle of the fourteenth century, suggesting that the bursts are related to the late medieval climatic deterioration and ice-field growth (Fig. 5). As

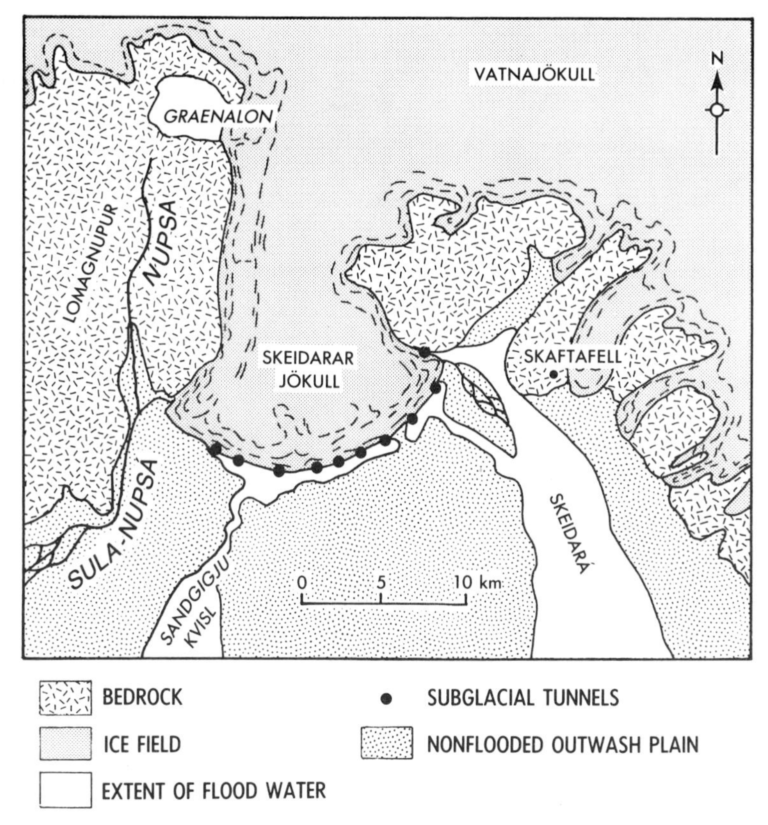

FIG. 8. Map of upper Skeidararsandur during the burst of July 18, 1954. Discharge occurred from 10 subglacial tunnels distributed along the entire glacier front. Note the ice-marginal drainage pattern and the complete inundation of the active braid plains of the rivers Skeidará and Sandgigjukvisl. Maximum discharge was estimated at 50,000 m^3/s. [Modified from Rist 1957).]

suggested above, there probably was a nonglaciated valley through the Grimsvötn depression before this time, permitting continuous discharge from the lake. Growth of Vatnajoküll and ice closure of the Grimsvötn outlet created the present situation, where meltwater above this subglacial hot spot accumulates until it attains a critical threshold for release beneath the ice cap.

An extensive list of bursts in the Skeidará River since 1684 was compiled by the late Dr. Sigurdur Thorarinsson in an unpublished manuscript. With the permission of Dr. Thorarinsson, this information is presented in Table I. Known occurrences of bursts in the river Sula are included. Where available, published information is referenced in the table.

TABLE I. RECORDED BURSTS (HLAUPS) IN SKEIDARÁ (FROM GRIMSVÖTN) AND SULÁ (FROM GRAENALON) IN HISTORICAL TIME

Suluhlaup	Skeidararhlaup	Remarks	Source
1201			Thorarinsson (1939, p. 22)
	1389	Eruption in Grimsvötn	Thorarinsson (1939, p. 22)
	1684/1685		S. Thorarinsson (unpublished)
	1702(?)		S. Thorarinsson (unpublished)
	1706		S. Thorarinsson (unpublished)
	1716		S. Thorarinsson (unpublished)
	1725/1726		S. Thorarinsson (unpublished)
	1774	Eruption in Grimsvötn	S. Thorarinsson (unpublished)
	1784 (April)		S. Thorarinsson (unpublished)
	1796 (November)		S. Thorarinsson (unpublished)
	1816 (June)	Major burst	S. Thorarinsson (unpublished)
	1838	Major burst	S. Thorarinsson (unpublished)
	1851		S. Thorarinsson (unpublished)
	1861	"The big burst"	S. Thorarinsson (unpublished)
	1867	(*a*)	S. Thorarinsson (unpublished)
	1873	(*a*)	S. Thorarinsson (unpublished)
	1883	(*a*)	S. Thorarinsson (unpublished)
	1892	(*a*)	S. Thorarinsson (unpublished)
	1897	(*a*)	S. Thorarinsson (unpublished)
1898 (November)			Thorarinsson (1939, p. 22)
	1903		S. Thorarinsson (unpublished)
	1913		S. Thorarinsson (unpublished)
1913			Eyolfur Hannesson
	1922 (September)		S. Thorarinsson (unpublished)
	1934 (April)	Eruption in Grimsvötn Total discharge: 15 km^3	Nielsen (1937, p. 71)
1935 (September)			Thorarinsson (1939, p. 226)
	1938 (May)	Total discharge: 10 km^3	Thorarinsson (1939, p. 226)
	1939 (June)	Minor burst	Rist (1957, p. 30)
1939 (July)			Thorarinsson (1939, p. 224)
	1941	Minor burst	Rist (1957, 1970)
	1945 (September)		Rist (1957, 1970)
	1948 (January)		Rist (1957, 1970)
	1954 (July)	Maximum discharge: 10,500 m^3/s Total discharge: 3.5 km^3	Rist (1957, 1970)

TABLE I. (*continued*)

Suluhlaup	Skeidararhlaup	Remarks	Source
	1960		S. Rist (1973 personal communication)
	1965		S. Rist (1973 personal communication)
	1971		S. Rist (1973 personal communication)
1973 (August 7, 8)[b]			Observed by authors

[a]Detailed description in Nummedal *et al.* (1974).
[b]Suluhlaups have occurred regularly every 2–3 yr since 1939 and annually since 1969 (Rist, 1973 personal communication).

The absence of tabulated bursts between 1389 and 1684 and again in the time interval between 1725 and 1774 is probably due to incomplete records rather than the absence of bursts (S. Thorarinsson, personal communication). The records indicate that bursts in Skeidará occurred about once a decade until the mid-1930s. Since that time the frequency has increased to a burst every 6 years. The regular timing of bursts suggests that their source, the Grimsvötn lake, builds in volume in response to a fairly constant heat flux until some critical volume of meltwater is attained, at which time a breakout is triggered. Some of the extreme bursts in the past, which also were associated with major ashfalls (e.g., the 1867 burst; Nummedal *et al.*, 1974), probably were triggered by subglacial volcanic eruptions.

Not only has the frequency of bursts in Skeidará increased the last decades; bursts in Sula have occurred every 2–3 years since 1939 and annually since 1969 (Rist, 1973, personal communication). In all probability, this increase is related to the decrease in ice thickness during the last decades. Although few reliable estimates of the variation in ice thickness exists, it is directly related to the position of the ice margin (Nye, 1959), which has undergone significant retreat during the last decades (Fig. 6). In fact, there were no recorded bursts in Sula until 1898. Prior to that time, the ice was too thick to permit drainage through the dam. Instead, Lake Graenalon maintained a steady water-surface elevation high enough to permit discharge into the Nupsá River through a col in the mountains south of the lake (Fig. 8).

In addition to these regular bursts in the Skeidará and Sula rivers, there have been multiple bursts from Myrdalsjökull farther west. A particularly large burst there was triggered by eruption of the subglacial vol-

(a)

(b)

cano Katla in 1918. Large mud and debris flows have been triggered by eruptions of a volcano under Orafajökull, to the east of Skeidararsandur, in 1662 and 1726.

C. Observations of Glacier Bursts

Glacier bursts are responsible for most of the sediment transportation on Skeidararsandur. Each burst is a short-lived "catastrophe" that significantly alters the morphology and sediment pattern on an outwash plain. The initial description of Icelandic glacier bursts by Thorarinsson (1939) contributed greatly to the correct interpretation of the morphologic imprints of many late Pleistocene catastrophic floods in the western United States (Pardee, 1942; Bretz, 1969; Baker, 1973; Baker and Nummedal, 1978).

Unfortunately, precise measurements of the hydraulic parameters of Icelandic glacier bursts have been precluded for logistical reasons. One early estimate of the discharge of a burst in Skeidará was based on the amount of water released from the Grimsvötn depression as measured by the volume of the subsequent ice caldera (Thorarinsson, 1939). Rist (1957) applied Manning's equation and field observations on flood-channel width and depth to determine discharge values for the 1954 burst in Skeidará. He estimated a peak discharge of 50,000 m^3/s and a total flow volume for the burst of 3.5 km^3, which is equivalent to an entire normal meltwater season's flow volume. The large 1954 burst in Skeidará was associated with flow from 10 subglacial tunnels beneath Skeidararjökull. Only the eastern discharge points actually drained into the river Skeidará; the others drained west into the Sandgigjukvisl river (Fig. 8). Large, partly channelized, ice-marginal troughs in front of Skeidararjökull date back to this and similar events (Fig. 6). Glacier-burst channels, not occupied and modified by subsequent moderate flow events, show channel-floor lineations indicative of upper flow regime conditions at the bed of 5- to 10-m-deep channels (Fig. 9a) and large obstacle scours around stranded icebergs that have subsequently melted (Fig. 9b).

Fig. 9. (a) Photo of a major channel occupied during the 1954 glacier burst. This channel is seen in Fig. 8 as the one that connects discharge tunnels 2 and 3 (from the east) with the main Skeidará floodplain. Note linear gravel ridges (arrow) and current "tails" indicative of upper flow regime at the bed of this 5- to 10-m-deep channel. Kettle holes are seen in cross-section at the edge of the high-level outwash plain (arrow). Flow direction was from left to right. (b) Scour crescent around a (subsequently melted) iceberg in abandoned drainage channel in the proximal facies of the Skeidará outwash plain. Note person for scale (arrow). The crescent is about 3 m deep.

The presence of multiple, often changing, discharge points at the front of Skeidararjökull during bursts causes rapid destruction of older, downstream, terminal moraines. Only minor remnants, the "Sandgigur," are left of the 1749–1750 terminal moraine (Figs. 6 and 7). To the east of the Sandgigjukvisl River this moraine is removed or buried. The present course of the Sandgigjukvisl River through the 1892 moraine was established by a burst in 1945 (B. Larusson, 1973, personal communication).

During the 1973 field season we were fortunate enough to observe a burst from Lake Graenalón in the river Sula on August 8 (Fig. 10). The discharge of this burst was computed as follows. Theoretical studies by Kennedy (1963) demonstrated that, to a first approximation, the wavelength of standing waves (L) is related to the mean current velocity (V) according to the equation

$$V^2 = Lg/2\pi \tag{1}$$

where g denotes the acceleration of gravity. The equation was field tested in Iceland and found to be quite reliable (Nummedal *et al.*, 1974). Theory, flume data, and limited field data also show that Froude numbers associated with upper flow regime standing waves over coarse beds usually fall in the range of 0.80–1.0 (Chow, 1959; Allen, 1970). Froude number (F) is defined as

$$F = V/\sqrt{gD} \tag{2}$$

where D is mean flow depth; the other parameters are defined above.

Using air photos to determine the wavelength of standing waves (L), one can calculate V [Eq. (1)]. Subsequently, the flow depth is determined from Eq. (2), assuming $F = 1$. The total width of all channels is determined from air photos. Integration of depth-velocity data for individual channel segments determines total stream discharge.

Data for the burst in Sula on August 8, 1973, yield a current velocity of about 4.2 m/s for the largest channels, down to velocities of 2.8 m/s for the bar surfaces. These velocities imply typical depths of about 1.8 m for the deep channels and 0.8 m over the bars. The total river discharge was calculated to be 1,500 m^3/s. Even though this burst was a very small one compared to the major bursts in Skeidará, it did demonstrate the flow pattern and sediment transport potential of glacier bursts. At the peak of the burst the discharge was increased by a factor of about 100 compared to normal meltwater flow in Sula in 1973.

Current velocities during past glacier bursts can be calculated using Shields' criterion for initiation of bed-material motion to compute the

Fig. 10. Oblique air photo from an altitude of 700 m of flow in Sula during the burst of August 8, 1973. The area shown is about 4 km from the ice front. Stranded icebergs (the kind that produced the scour in Fig. 9b) on the bar surfaces indicate waning flood stage. The discharge is estimated at 1500 m^3/s.

critical shear stress needed to move the boulders observed in the large sandur channels (e.g., Fig. 9a). The calculations provide only minimum current velocities because of size limitations on the supplied glacial detritus. Such calculations indicate that current velocities much greater than those observed during the 1973 burst in Sula have occurred on Skeidararsandur in the past.

Calculations of sediment load during bursts are even more tenuous than those of discharge. One estimate was made by Rist (1957) for the 1954 burst in Skeidará. He found that the total suspended sediment load during that burst was 28.9 million tons. Based on repeated, detailed flow measurements on Skeidará during the year 1973, Nummedal *et al.* (1974) estimated that the total sediment load (suspended plus bed load) for that entire meltwater season was 380,000 tons. Consequently, an average burst in Skeidará (the 1954 burst) may transport more sediment than 75 years of normal meltwater discharge.

IV. Coastal Processes

A. Historical Shoreline Change

Despite the large amount of coarse sediment that is carried seaward by major glacier bursts, the Skeidararsandur shoreline is an almost straight, east–west trending, barrier spit shoreline (Figs. 11 and 12). Farmers east of the sandur have observed the shoreline near the mouth of Skeidará during glacier bursts. They report that bursts do not cause permanent progradation. (S. Arason, personal communication). A transient progradation of about 200 m is the maximum observed; this is quickly eroded by the waves.

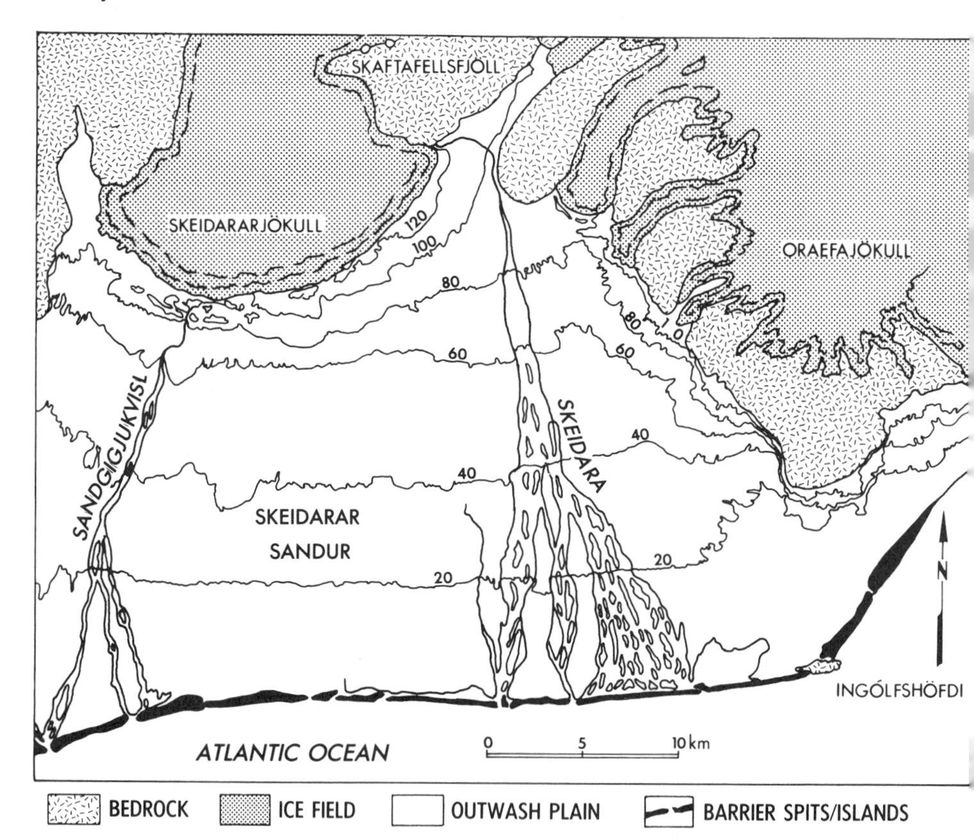

Fig. 11. Photogrammetric contour map of Skeidararsandur produced for this study by NAVOCEANO from vertical air photos flown in the winter of 1972–1973. The contour interval is 20 m. Note the shape of the coastal barriers, the convex topographic profile of the proximal sandur plain, and the planar nature of the distal sandur.

Fig. 12. Oblique aerial photograph (from about 1500 m) looking west along the Skeidararsandur shoreline. Note the narrow barrier spits (200–300 m wide) and the gentle bulge in shoreline orientation centered on the western distributary of Skeidará.

In the short term, the position of the shoreline west of Ingolfshöfdi (Fig. 4) is fairly stable. Historical maps (Fig. 13) indicate that this shoreline was stable over the past century. Distances and details in the interior of Iceland were not sufficiently known in earlier centuries to allow precise mapping. However, details along the coast, particularly the bearing of the coastline at any point, were remarkably accurately recorded (A. Bodvarsson, personal communication).

Vertical air photos of the Skeidararsandur shoreline taken in 1942 (courtesy of Landmaelingar Islands) and in 1972–1973 (courtesy of NAVOCEANO) were used to compile uncontrolled photomosaics from which two shoreline maps were constructed. Corroborating the historical evidence, these maps document little change in shoreline configuration over 30 years. Six bursts in Skeidará occurred in this period (Table I), yet no net progradation was recorded. The Skeidará distributaries, however, did change somewhat. Bursts are known to produce a number of transient distributaries, most of which are sealed by spit growth following the flood (Hine and Boothroyd, 1978). Also, the main river channels remain fairly stationary during years without bursts, but the distributary mouths migrate with the seasons. During low-discharge periods (winter), the distributary mouths migrate downdrift. During floods, the flow momentum realigns the distributary mouths to new updrift locations (Ward *et al.*, 1976).

Shipwreck maps indicate that the south Icelandic shoreline prograded more rapidly farther west. Six known shipwrecks have occurred along this coast since year 1900 (Fig. 14). The time of each wreck and its location and distance from the present high-tide swash are summarized in Table II.

The data demonstrate a westward increase in the rate of shoreline progradation. This increase is supported also by other historical data. The first party of settlers in Iceland, which arrived at Ingolfshöfdi in 874 A.D., found this cliff to be land-connected with the open ocean facing the south side, as it is today. Hjörleifshöfdi, a similar basalt mesa 120 km to the west, could also be reached by sea-going vessels in year 874 (S. Thorarinsson, 1973 personal communication) but is today about 2-km inland (Fig. 15).

A study conducted by the Icelandic National Harbors Authority (Hafnamálastofnun Rikisins, 1972) concluded that shoreline progradation along Myrdalssandur (east of Vik, Fig. 15) has been very rapid. Since the year 1900 this shoreline has prograded as much as 1 km. This is partly due to

Fig. 13. Map of Skeidararsandur and adjacent areas in 1852 (cartographer unknown). The trend of the coastline resembles the present, indicating minor migrations during the past century. The rivers Skeidará and Nupsá are located at their present positions. Map also shows that Lake Graenalon at this time discharged directly into Nupsá, and hence was not subject to glacier bursts.

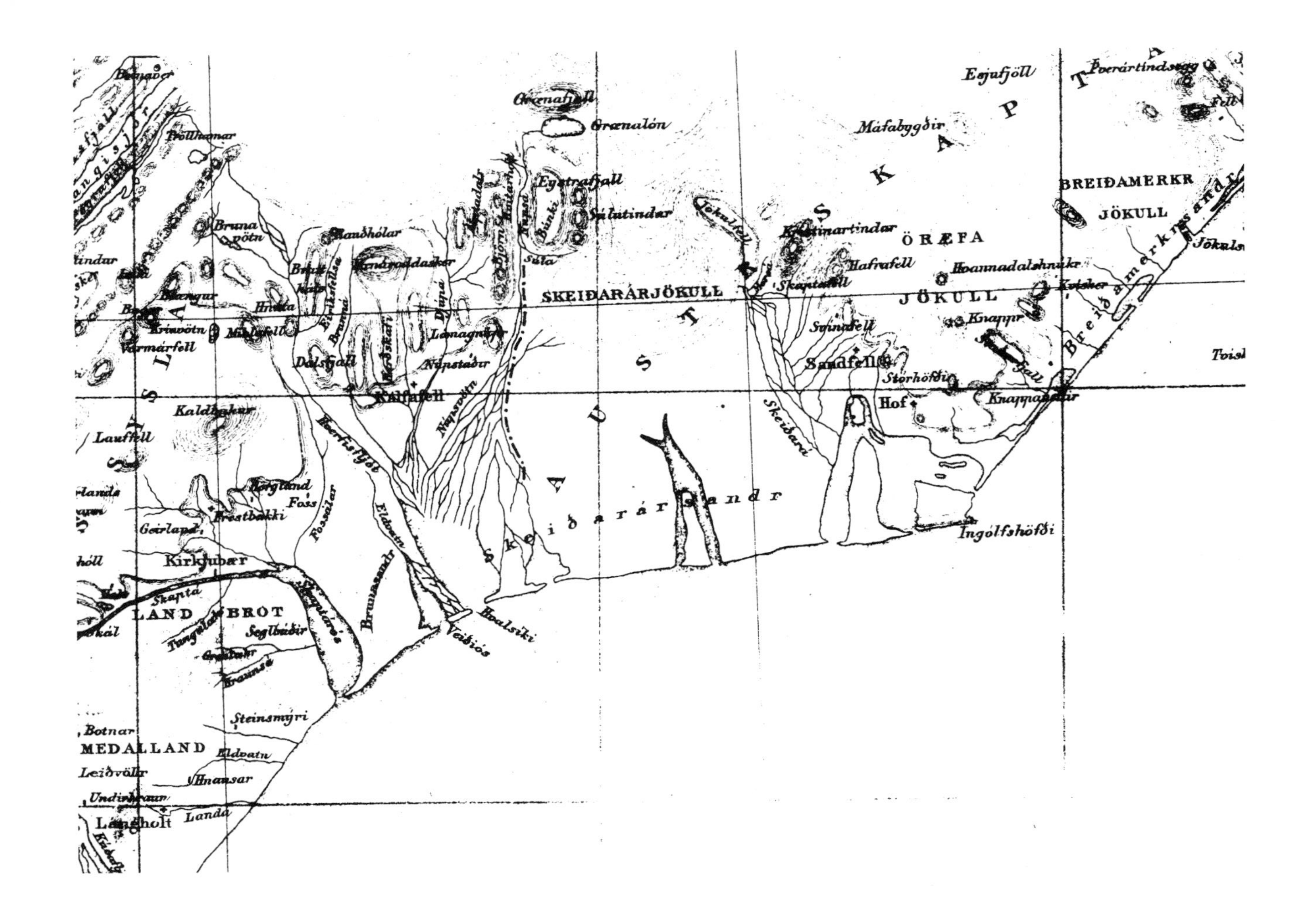
Eajufjöll
Máfabyggðir
Grænalón
BREIÐAMERKR
JÖKULL
ÖRÆFA
Egstrafjall
Súlutindar
Hafrafell
Hoannadalshnúkr
SKEIÐARÁRJÖKULL
JÖKULL
Knappr
Svínafell
Sandfell
Stórhöfði
Hof
Knappavöllr
Ingólfshöfði
Skeiðará
Skeiðarár sandr
Núpstaðr
Núpsvötn
Kálfafell
Dalsfjall
Bruna
pötn
Kaldbakur
Lauffell
Geirland
Prestbakki
Foss
Fossalar
Kirkjubær
Skapta
LAND BROT
Seglbúðir
Brunasandr
Eldvatn
Hoalsíki
Veiðiós
Steinsmýri
Botnar
MEDALLAND
Eldvatn
Leiðvöllr
Hnausar
Langholt
Landa

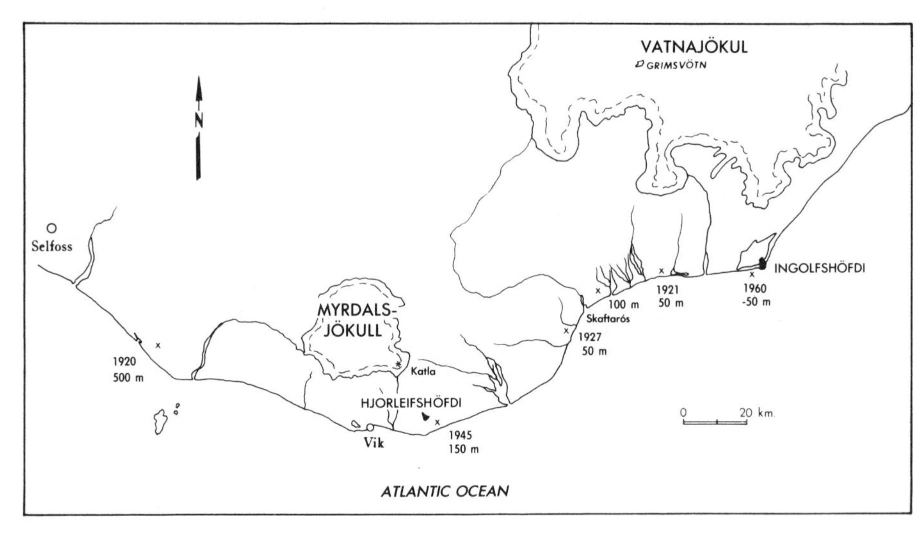

FIG. 14. Location of known shipwrecks on the southern coast of Iceland. The year of the wreck and its present location inland from the swash line are indicated. (Historical data courtesy of Mr. Sigurdur Larusson.)

TABLE II. KNOWN RECENT SHIPWRECKS ON THE SOUTH COAST OF ICELAND[a]

Ship	Time of shipwreck	Location	Distance west of Ingolfshöfdi (km)	Inland distance from high-tide swash (m), April 1973	Estimated annual shoreline progradation (m/yr)
British trawler *Lord Stanhop*	1960	Ingolfshöfdi	5	−50	−3.8
Trawler	1921–1922	Skaftafellsfjära	30	50	1.0
2000-Ton Norwegian cargo ship	1927	Skaftarós	58	100	2.2
Trawler	1927	Eldvatnsós	64	50	1.1
American cargo ship *Grimsby Town*	1945–1946	Hjörleifshöfdi	115	150	5.6
Trawler	1920	Landeyar	190	500	9.4

[a]Data on time and location of shipwrecks from S. Larusson (1973 personal communication). Distance of wreck from high-tide swash measured or estimated from aerial photographs. Inland distance for *Grimsby Town* estimated by Mr. Larusson.

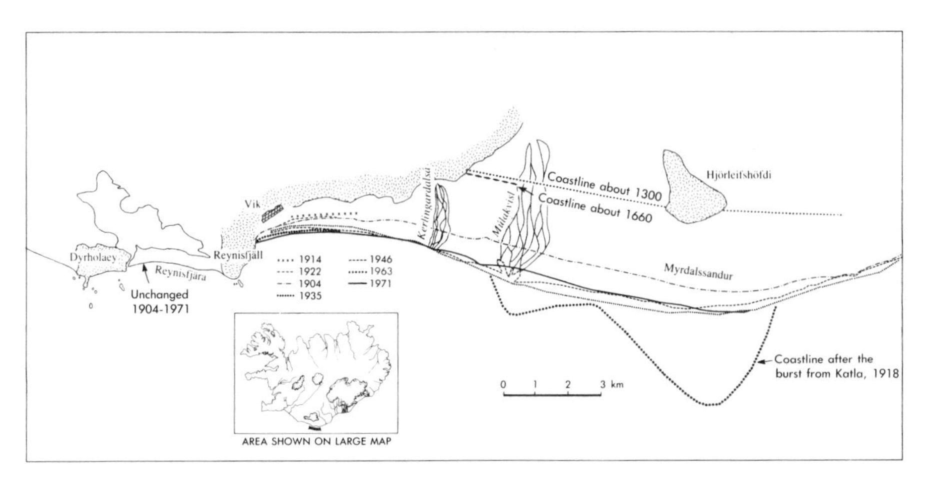

FIG. 15. Observed and inferred migrations of the shoreline near Vik in Myrdal since 1300 A.D. The progradation observed after the 1918 burst from Katla far exceeds what has ever been observed at Skeidará. Erosion of this fan delta by westward longshore currents has caused rapid shoreline progradation at Vik. [Data from Hafnamálastofnun Rikisins (1972).]

FIG. 16. Oblique aerial photo looking to the east across the basaltic coastal cliffs of Dyrholaey (foreground) and Reynisfjäll. The community of Vik (Fig. 15) is hidden behind Reynisfjäll in the upper left corner.

the exceptionally large glacier burst from the subglacial volcano Katla beneath Myrdalsjökull, which caused a transient progradation of the shoreline of about 3 km (Fig. 15). Sediment transported westward from southern Myrdalssandur by littoral currents is ultimately trapped east of the natural groin provided by the promontory of Reynisfjäll west of Vik (Fig. 16). Consequently, Vik, which had a good harbor a century ago, is today located more than 800-m inland.

B. Wave Climate

The most severe weather disturbances affecting Iceland and the adjacent North Atlantic Ocean are extratropical cyclones, commonly generated on the Atlantic Arctic and Polar fronts (Petterssen, 1969). The Atlantic Arctic frontal zone is the area of confluence of arctic and polar maritime air. The Atlantic Polar front represents the boundary zone of the polar continental and tropical maritime air. Both frontal areas are strongest in winter because of high latitudinal heat-induced pressure differences.

Cyclones generated on the Arctic Front near Greenland and Iceland generally move to the northeast into the Barents Sea. This cyclone travel path is illustrated by storm number 16 in December, 1972 (see Fig. 17a). Cyclones on the Atlantic Polar front generally develop off the east coast of the United States and move eastward along the Gulf stream, commonly attaining maximum intensity south of Iceland. This path is typified by storm number 10 in December 1972 (Fig. 17a).

The reduced intensity of latitudinal pressure differences in the summer forces a breakdown in the zonal current with consequent decrease in frontal cyclone genesis. The few cyclones that do occur generally migrate across, or to the north of, Iceland, in contrast to the more southerly winter tracks. A typical midsummer cyclone path is that followed by storm number 2 in July 1972 (Fig. 17b). A detailed analysis of monthly cyclone patterns across the north Atlantic for 1972 is presented in Nummedal *et al.* (1974).

The frequent and intense North Atlantic extratropical cyclones generate extremely high wave power off Iceland's southern coast. To quantify that power, and to estimate the magnitude of longshore sediment transport along the Skeidararsandur shoreline, we hindcast the wave conditions at 6-hr intervals for the entire month of December 1972. The Sverdrup–Munk–Bretschneider method (Coastal Engineering Research Center, 1973) was applied to data obtained from north Atlantic synoptic weather maps (obtained at U.S. Weather Bureau, National Climatic Center, Ashville, North Carolina). Selected for inclusion in the hindcast was any cyclone attaining a maximum surface wind speed at or above 15 m/s. Fetch areas were delineated by the following criteria: (1) wind direction relative to the south Icelandic shore, (2) the presence of fronts, (3) diverging isobars, and (4) coasts of adjacent land masses.

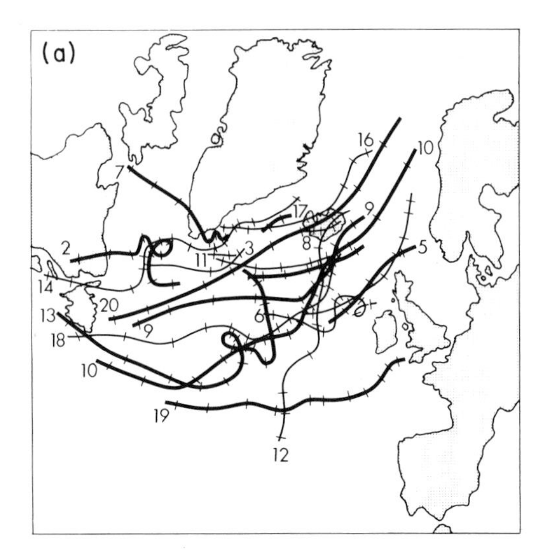

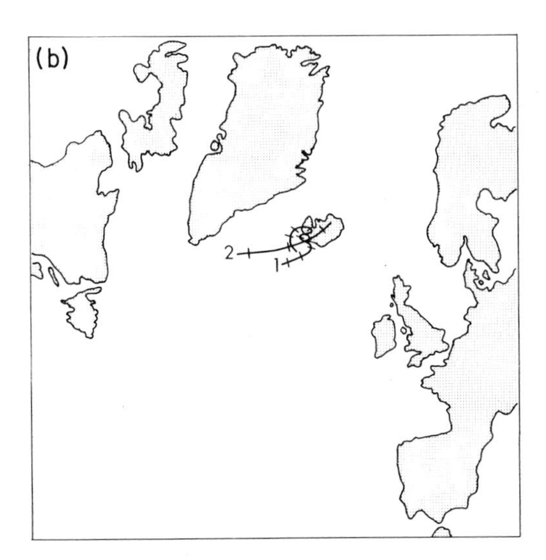

Fig. 17. (a) Cyclone tracks across the North Atlantic in December 1972. Thin lines indicate storms with maximum surface wind velocities less than 18 m/s. Thick lines denote storms with winds above 18 m/s. The hatch marks designate the position of the center of the low pressure system at successive 6 hr intervals. (Data from U.S. Weather Bureau, Northern Hemisphere Surface Charts.) (b) Cyclone tracks across the North Atlantic in July 1972. The symbols are the same as those in (a).

The hindcast wave climate for December, 1972 is expressed in terms of the distribution of deep-water wave power incident at the Skeidararsandur shoreline (Fig. 18). For methods used in the hindcasting, see Nummedal (1975) or Nummedal and Stephen (1978). Although the hindcast is limited to 1 month, there is good reason to believe that the findings may apply to winter conditions in general, because (a) the large number of north Atlantic cyclones during this month (19 storms) makes it, statistically, the most valid sampling period, and (b) a survey of weather data for south Iceland over the past 48 years (Vedráttan, 1924–1972) indicate that December 1972 was a typical winter month in terms of directions and intensities of observed coastal winds.

The dominant wave power off Skeidararsandur in December 1972 came from the ESE. A subpopulation of much more modest energy had a directional spectrum from the SW (Fig. 18). The reason for the observed

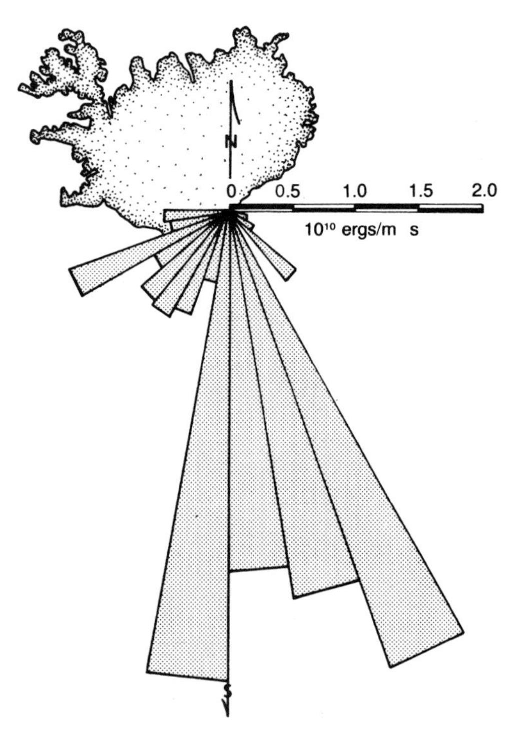

Fig. 18. Deep-water wave power distribution off southeast Iceland for December 1972. The numbers give the average wave power per unit width of wave crest for each 10° sector. The diagram is plotted as a wind rose—that is, the sectors indicate the direction from which the waves arrive.

power distribution is seen in the life history of typical storms. Typical winter storms in their growth stage, near Newfoundland, have an associated fetch azimuth, for waves affecting Iceland, between 180 and 250°, that is, from the southwest. Although this situation may prevail for many days, the associated wave power reaching Iceland is relatively low. This explains the southwesterly subpopulation in Fig. 18. As the cyclones move east they intensify, and the dominant fetch assumes an orientation between south and southeast. For example, storm number 9 in December 1972 (Fig. 17a) attained its maximum wave power on December 11 for a fetch azimuth of 135° (wind from the SE). Typically, the wave power associated with winds between south and southeast is an order of magnitude higher than that for southwesterly winds of the same storm system.

The observation that maximum wave power is associated with waves from the south-southeast suggests a net longshore sand transport toward the west along the east–west oriented shoreline of Skeidararsandur. East of Ingolfshöfdi, where the shoreline turns to the northeast (Fig. 11), one would expect sediment transport in that direction. To the west of Skaftaros (Fig. 14) one would again expect to find northeastward sediment transport due to the shoreline trend. Near Vik on the south-central coast (Fig. 15), longshore transport should be directed westward. These predicted transport directions, based on the hindcast winter wave climate, are entirely consistent with observed coastal morphology. The large magnitude of the wave power, when compared to other shores in the world (Davies, 1973; Nummedal and Stephen, 1978; Owens and Roberts, 1978), is consistent with the observations that the Skeidararsandur shoreline has remained straight and stable. Sediment from Skeidará is effectively dispersed westward.

C. Coastal Sediments and Morphology

Sediment distribution trends and morphology along the Skeidararsandur shoreline are all consistent with the westward sediment transport direction deduced from the wave climate study.

The barriers consist entirely of sand, in spite of the fluvial transport competence during glacier bursts. The sand is dominantly medium tc coarse, and consists of basalt and felsic volcaniclastic fragments. In general, the beaches fine to the west, from about 0.7 mm at East Skeidará River, decreasing to about 0.35 mm east of the Sandgigjukvisl River (Fig. 19). Near the mouths of West Skeidará and Sula the sediment population locally coarsens, suggesting that the rivers are local littoral sediment point sources. A general westward improvement in sorting (Fig. 19) is consistent with the overall inferred transport direction.

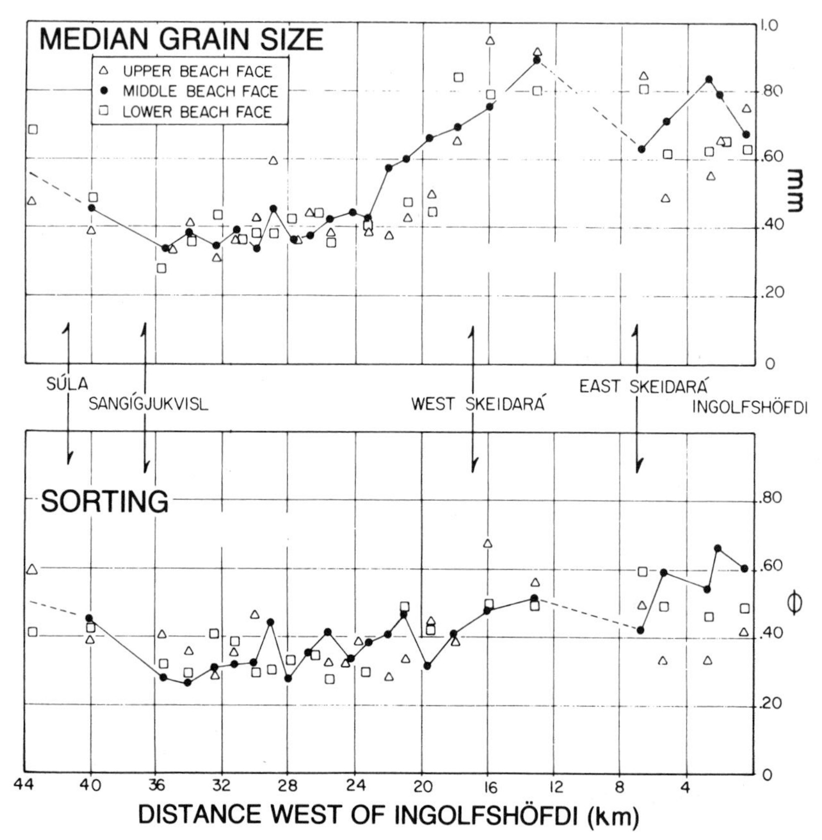

FIG. 19. Grain size parameters (mean size and sorting) for beaches along Skeidararsandur. Data were obtained during field work in the summer of 1973 and reflect the distribution during the fair-weather season. Note the westward decrease in size and improvement in sorting.

The Skeidararsandur shoreline is microtidal, with observed tides within the river distributaries ranging from 2 m at spring to about 1 m at neap tide (Ward *et al.*, 1976). This tidal range is much too low for the Icelandic wave climate to produce barrier morphologies characteristic of the "mixed-energy" regime of the South Carolina coast or the West Friesian Islands, regions where the tide range is not that much higher (Hayes and Kana, 1976; Nummedal and Fischer, 1978). High wave energy has produced a strongly wave-dominated morphology, with coastal barrier islands and spits averaging 5–10 km in length. The barriers increase in width westward, from 200 m near the Skeidará distributaries, to 1.4 km near Sandgigjukvisl. The Icelandic barrier islands are asymmetric in plan form, with one bulbous end and one narrow, tapering end (Figs. 11 and 20). Such barriers are common on a global scale and are often referred to as "drumstick-shaped" (Hayes and Kana, 1976). Wherever they have been examined, drumstick barriers are found to be oriented with the bulbous end updrift (Hayes and Kana, 1976; FitzGerald *et al.*, 1984). The drumstick shape of a mixed-energy barrier is normally the result of swash-bar accretion on the updrift beach. This process, however, is inoperative on the microtidal, high-wave-energy coast of Iceland. Distributaries here have no associated ebb-tidal deltas for swash-bar formation. Nevertheless, a related island geometry is developed because of the downdrift deflection of river distributaries and wind tidal-flat drainage channels (Hine and Boothroyd, 1978). Consequently, the orientation of the tapered Icelandic barriers is consistent with longshore sediment transport toward the northeast, to the east of Ingolfshöfdi, and westward transport along the shore of Skeidararsandur (Fig. 11).

The historical evidence for stability of the Skeidararsandur shoreline, combined with increased rates of progradation farther west (Fig. 14), is suggestive of a zone of longshore sediment convergence in the Skaftarós "embayment." Oblique air photos of the barriers at the eastern flank of this embayment (Fig. 20) clearly demonstrate their wider, more accretionary nature compared to barriers to the east (Fig. 12).

FIG. 20. Oblique air photo (from about 1500 m) of the coastal barriers near the distributaries of Sandgigjukvisl and Sula. Note the westward tapering of the barriers. Compare to barrier morphology at the mouth of Skeidará (Fig. 12).

V. Shoreface and Continental Shelf

Concurrent with our onshore investigations, a research vessel from Sjömaelingar Islands (Icelandic Hydrographic Survey) conducted bathymetric surveys and sediment sampling offshore. The survey took place in August and September 1973, extending offshore to a depth of 110 m (Fig. 21).

The bathymetry of the Skeidararsandur shoreface is extremely steep, as compared to shorefaces elsewhere. From the beach to about 75-m water depth the slope ranges from about 26 m/km (1 : 38) off the Skeidará River, to about 20 m/km (1 : 50) farther west. This westward decrease in shoreface slope indicates westward sediment transport also on the shoreface, in water depths beyond those affected by the wave-induced littoral currents. Beyond 75 m depth the shoreface merges with the regional south-Icelandic shelf, which has a gradient of about 6 m/km at the 100-m contour. Typical U.S. East Coast shorefaces have slopes at about 1 : 200, decreasing seaward to about 1 : 2000 where they merge with the inner continental shelf in water depths of 10–15 m (Niedoroda *et al.*, 1985). Beyond the 100-m contour, the Icelandic shelf is rough with numerous rock outcrops. Toward the east, along an alignment extending south from Ingolfshöfdi (Fig. 21), these rocks pierce the sea floor at depths as shallow as 50 m.

In the alongshore direction there is little morphologic variability in shoreface detail, the expected response to the strong littoral transport that operates along this coast. The upper shoreface, however, is characterized by longshore bars off the major rivers. The bars have a relief of about 5 m relative to the bottom of their associated landward troughs; they crest in about 7 m of water. Due to the intense wave climate, no observations were made on these bars during storms. Their presence only near the rivers suggests that high sediment supply is an essential factor in maintaining their long-term stability.

The shoreface sediments become finer offshore. Off the two major rivers, the Skeidará and the Sula, the 90% sand contour is about 3 km offshore, at the base of the shoreface in a water depth of 70 m (Fig. 22). Between the rivers, for example, at profile ICB-3, the 90% sand contour extends only out to 15-m water depth. On the inner shelf floor, in water depths between 75 and 100 m, the sediment consists of between 10 and 50% sand. The 50% sand contour falls generally between the 90 and 100 m depth contour. In contrast, sandy silts are typical at depths beyond 10 m on the southern California coast (Howard and Reineck, 1981).

Sand dispersal across the Skeidararsandur shoreface is undoubtedly controlled by downwelling, alongshelf storm flows in a manner similar

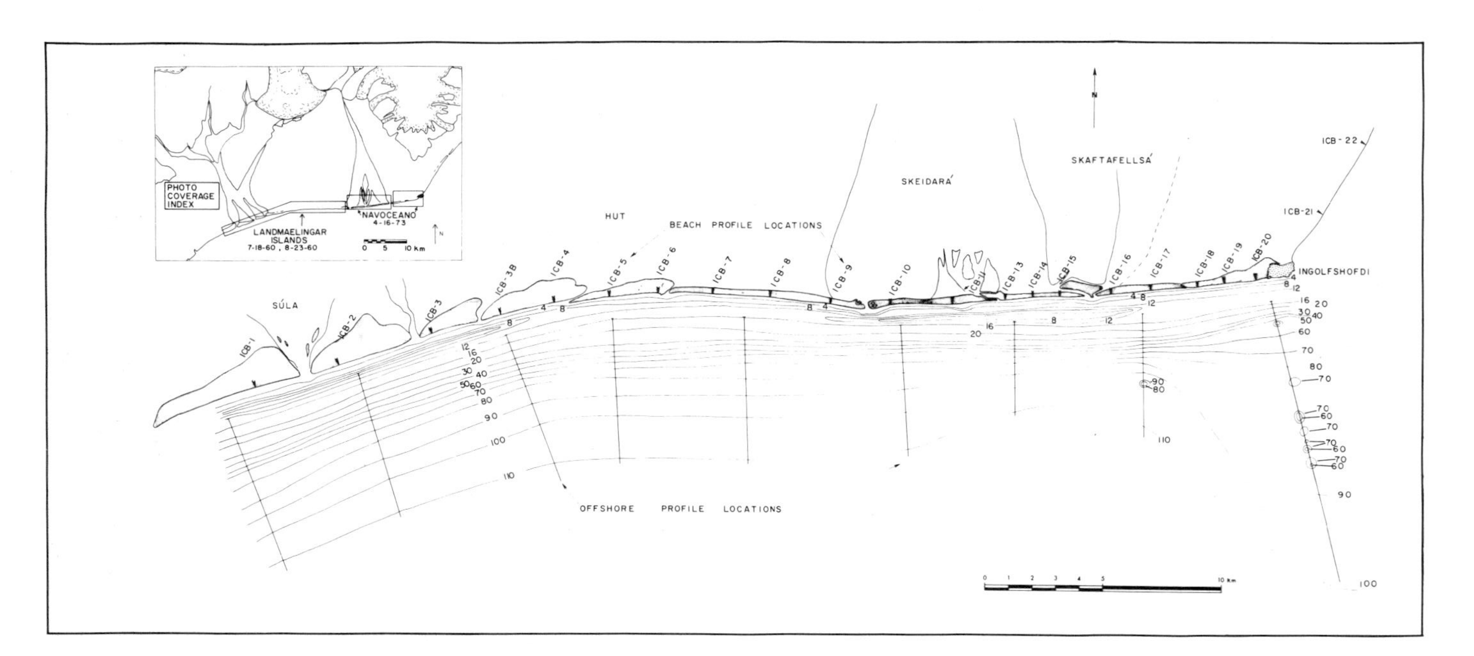

FIG. 21. Bathymetry off the Skeidararsandur coast. (Data based on echo-sounder surveys by the Icelandic Hydrographic Office in August and September of 1973.) Contours are in meters. Offshore survey profiles are indicated. Numbers onshore (ICB-3 etc.) refer to beach profiles.

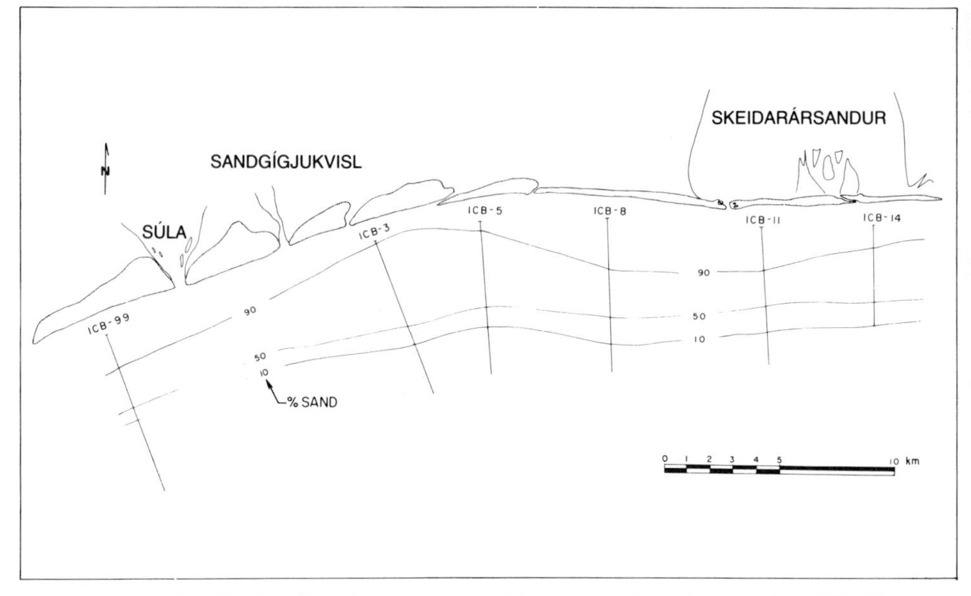

Fig. 22. Distribution of sand and mud (in weight percent) along the shoreface of Skeidararsandur. (Data were obtained by the Icelandic Hydrographic Office in August and September of 1973.)

to what we observe along the Long Island (Niedoroda *et al.*, 1985) and Texas (Snedden, 1985) shorefaces. The fact that the sandy shoreface off south Iceland extends to much deeper depths than at the two aforementioned locations is a clear response to stronger benthic storm currents. The shoreface is basically sandy down to its break in slope at 75 m, although it is muddier between the river mouths than in their immediate vicinity.

VI. Conclusions

The south–central coast of Iceland has developed its present characteristics in response to (1) the Late Pleistocene glaciation and the island's rapid isostatic adjustment to the ice removal, (2) the continued existence of interior ice fields across the neovolcanic zones producing frequent, and large, glacier bursts, and (3) the high wave energy of the subarctic North Atlantic.

Eustatic sea-level rise following the global retreat of the Wisconsinan ice sheets produced a transgression in Iceland that reached its maximum landward extent about 11,000 yr B.P. Rapid isostatic uplift of the Icelandic landmass caused a regression from 11,000 to 8,000 yr B.P. Beginning during this regressive phase and continuing into the present, large outwash plains (sandurs) have been prograding across the south Icelandic coast. The present progradational phase is the last in a series of Quaternary episodes of progradation that probably have produced a composite clastic wedge underlying the Icelandic sandurs.

At present, coastal-plain sedimentation is dominated by volcanogenic and limnoglacial bursts. About 92% of the long-term sediment transport across Skeidararsandur is due to glacier bursts; the remainder is accounted for by normal summer meltwater discharge. Individual glacier bursts may have discharges as high as 100,000 m^3/s, but they last for only a few days. Volcanogenic bursts from Grimsvötn in Vatnajökull occur about every sixth year, whereas the smaller limnoglacial bursts from Lake Graenalon now occur annually. This increased frequency of limnoglacial bursts is due to a thinning and retreat of Skeidararjökull due to climatic warming over the past century.

Wave energies along the south Icelandic coast are among the highest in the world. The associated strong westward sediment transport causes rather slow shoreline progradation near the Skeidará distributaries, in spite of the large influx of sediment during glacier bursts. Shoreline progradation is significantly faster farther west where the shoreline assumes a more southwesterly trend.

Progradation of Skeidararsandur during the Holocene, across the glacially scoured, basaltic floor of south Iceland, has produced a terrigenous clastic wedge that currently is about 100 m thick at the shoreline. Assuming that the present shoreface and fluvial outwash plains are typical of their Holocene precursors, this wedge is a coarsening-upward sequence on a scale of 100 m in thickness. Sandy muds, muddy sand, and fine sand are marine deposits. Clastics with grain sizes from medium sand up to cobbles and boulders characterize the overlying fluvial facies.

ACKNOWLEDGMENTS

This study was supported by the Naval Ordnance Laboratory by contract N60921-73-C-0258, Miles O. Hayes, principal investigator. Roger K. Johnson, Naval Systems Laboratory, Panama City, Florida, is acknowledged for his guidance throughout all aspects of the project. Cmdr. James D. McKnight, Icelandic Defense Forces, Keflavik, provided invaluable logistical support and advice while we were in Iceland. Excellent logistical support and field assistance were rendered by Bergur Larusson, Hjörtur Gudmundsson, Kristinn Gudbransson, Hörtur Haradsson, and Andra and Jon Heidberg. Discussions with the late Dr. Sigurdur Thorarinsson,

Dr. Trausta Einarsson, and Dr. Sveinbjörn Björnsson, University of Iceland, and Sigurjon Rist, Icelandic National Energy Authority, provided valuable information about the geology, geophysics, and hydrology of the Skeidararsandur region.

Gunnar Bergstensson, Icelandic Hydrographic Service, obtained data on the bathymetry and bottom sediment distribution off Skeidararsandur in August/September 1973 at the request of the authors. The great interest in this project shown by Adelsteinn Juliusson and Gisli Viggosson, Icelandic National Harbors Authority, is gratefully acknowledged. The late Dr. Robert K. Fahnestock, State University of New York at Fredonia, contributed numerous ideas both in the field and during the subsequent data analysis.

References

Allen, J. R. L. (1970). "Physical Processes of Sedimentation." Allen & Unwin, London.

Baker, V. R. (1973). Paleohydrology and sedimentology of Lake Missoula flooding in eastern Washington. *Geol. Soc. Am. Spec. Pap.* No. 144.

Baker, V. R., and Nummedal, D. (1978). "The Channeled Scabland; a Guide to the Geomorphology of the Columbia Basin, Washington." Planet. Geol. Program, NASA, Washington, D.C.

Boothroyd, J. C., and Nummedal, D. (1978). Proglacial braided outwash: a model for humid alluvial-fan deposits. *In* "Fluvial Sedimentology" (A. D. Miall, ed.), *Mem. Can. Soc. Pet. Geol.* No. 5, pp. 641–668.

Bretz, J. H. (1969). The Lake Missoula Floods and the Channeled Scabland. *J. Geol.* **77,** 505–543.

Coastal Engineering Research Center (1973). "Shore Protection Manual," 3 vols. U.S. Gov. Print. Off., Washington, D.C.

Chow, V. T. (1959). "Open-Channel Hydraulics." McGraw-Hill, New York.

Davies, J. L. (1973). "Geographical Variation in Coastal Development." Hafner, New York.

Dillon, W. D., and Oldale, R. N. (1978). Late Quaternary sea level curve: reinterpretation based on glacio-eustatic influence. *Geology* **6,** 56–60.

Einarsson, T. (1960). The plateau basalt areas in Iceland. *In* "On the Geology and Geophysics of Iceland" (S. Thorarinsson, ed.), Guide to Excursion No. A2, pp. 5–20. Int. Geol. Congr., 21st Sess., Reykjavik.

Einarsson, T. (1961). Das Meeresniveau an den Küsten Islands in post-glazialer Zeit. *Neue Jahrb. Geol. Palaeontol. Monatsh.* 6.

Einarsson, T. (1966). Late- and post-glacial rise in Iceland and subcrustal viscosity. *Jökull* **3,** 157–166.

Eythorsson, J., and Sigtryggson, H. (1971). The climate and weather of Iceland. *In* "The Zoology of Iceland" (E. Bertelsen *et al.*, eds), Vol. 1, Part 3. Munksgaard, Copenhagen.

FitzGerald, D. M., Penland, S., and Nummedal, D. (1984). Control of barrier island shape by inlet sediment bypassing: East Frisian Island, West Germany. *Mar. Geol.* **60,** 355–376.

Fridleifsson, I. B. (1982). The Icelandic research drilling project in relation to the geology of Iceland. *J. Geophys. Res.* **87,** 6363–6370.

Hafnamálastofnun Rikisins (1972). "Höfn vid Dyrholaey." Reykjavik.

Hayes, M. O., and Kana, T. W. (1976). "Terrigenous Clastic Depositional Environments," Tech. Rep. CRD-11. Dep. Geol., Univ. of South Carolina, Columbia.

Hine, A. C., and Boothroyd, J. C. (1978). Morphology, processes, and recent sedimentary history of a glacial-outwash plain shoreline, southern Iceland. *J. Sediment. Petrol.* **48,** 901–920.

Howard, J. D., and Reineck, H. E. (1981). Depositional facies of high energy beach to offshore sequence: comparison to low energy sequence. *Am. Assoc. Pet. Geol. Bull.* **65,** 807–830.

Kennedy, J. F. (1963). The mechanics of dunes and antidunes in erodable bed channels. *J. Fluid Mech.* **16,** 521–544.

Milanovsky, Y. Y. (1982). Geomorphology of Iceland. *In* "Iceland and Mid-Oceanic Ridge, Geomorphology and Tectonics," Chap. 1. Acad. Sci. USSR. (Engl. transl., Nat. Res. Counc., Reykjavik.)

Niedoroda, A. W., Swift, D. J. P., and Hopkins, T. S. (1985). The shoreface. *In* "Coastal Sedimentary Environments" (R. A. Davis, Jr., ed.), pp. 533–624. Springer-Verlag, Berlin and New York.

Nielsen, N. (1937). "Vatnajökull: Kampen mellom Ild og Is." Copenhagen.

Nummedal, D. (1975). Wave climate and littoral sediment transportation on the southeast coast of Iceland. *Proc. Int. Congr. Sedimentol., Nice, Fr.* Theme 6, 127–136.

Nummedal, D., and Fischer, I. A. (1978). Process–response models for depositional shorelines, the German and the Georgia Bights. *Proc. Coastal Eng. Conf., 16th, Am. Soc. Civ. Eng.* **2,** 1215–1231.

Nummedal, D., and Stephen, M. F. (1978). Wave climate and littoral sediment transport, northeast Gulf of Alaska. *J. Sediment. Petrol.* **48,** 359–371.

Nummedal, D., Hine, A. C., Ward, L. G., Hayes, M. O., Boothroyd, J. C., Stephen, M. F., and Hubbard, D. K. (1974). "Recent Migrations of the Skeidararsandur Shoreline, Southeast Iceland," Final Rep. for Contract No. N60921-73-C-0258. Nav. Ordnance Lab., Panama City, Florida.

Nye, J. F. (1959). The motion of ice sheets and glaciers. *J. Glaciol.* **3,** 493.

Owens, E. H., and Roberts, H. H. (1978). Variations of wave-energy levels and coastal sedimentation, eastern Nicaragua. *Proc. Coastal Eng. Conf., 16th, Am. Soc. Civ. Eng.* **2,** 1195–1214.

Palmason, G., and Saemundsson, K. (1974). Iceland in relation to the Mid-Atlantic ridge. *Annu. Rev. Earth Planet. Sci.* **2,** 25–50.

Pardee, J. T. (1942). Unusual currents in glacial Lake Missoula, Montana. *Geol. Soc. Am. Bull.* **53,** 1569–1600.

Petterssen, S. (1969). "Introduction to Meteorology," 3rd Ed. McGraw-Hill, New York.

Rist, S. (1957). Skeidararhlaup, 1954. *Jökull* **2,** 30–36.

Rist, S. (1970). Annall um jökulhlaup. *Jökull* **20,** 88.

Saemundsson, K. (1979). Outline of the geology of Iceland. *Jökull* **29,** 7–28.

Shen, H. W. (1971). "River Mechanics." Shen, Fort Collins, Colorado.

Snedden, J. W. (1985). Origin and sedimentary characteristics of discrete sand beds in modern sediments of the central Texas continental shelf. Ph.D. Thesis, Louisiana State Univ., Baton Rouge.

Thorarinsson, S. (1939). The ice-dammed lakes of Iceland with particular reference to their values as indicators of glacier oscillations. *Geogr. Ann.* **21,** 216–242.

Thorarinsson, S. (1960). The postglacial volcanism. *In* "On the Geology and Geophysics of Iceland" (S. Thorarinsson, ed.), Guide to Excursion No. A2, pp. 33–35. Int. Geol. Congr., 21st Sess., Reykjavik.

Todtman, E. M. (1960). Gletscherforschungen auf Island (Vatnajökull); Universität Hamburg, Abhandlungen aus dem Gebiet der Auslandskunde, Band 65, Reihe C. *Naturwissenschaften* **19.**

Vedráttan (1924–1972). "Mánadaryfirlit samid a vedurstofunni." Monthly weather summary from the Meteorological Institute of Iceland.

Walker, G. P. L. (1965). Evidence of crustal drift from Icelandic geology. *Philos. Trans. R. Soc. London, Ser. A.* **248,** 199–204.

Ward, L. G., Stephen, M. F., and Nummedal, D. (1976). Hydraulics and morphology of glacial outwash distributaries, Skeidararsandur, Iceland. *J. Sediment. Petrol.* **46,** 770–777.

CHAPTER 6

An Inventory of Coastal Environments and Classification of Maine's Glaciated Shoreline

JOSEPH T. KELLEY

Maine Geological Survey
Augusta, Maine
and
Department of Geological Sciences and Oceanography Program
University of Maine
Orono, Maine

A detailed examination of the variation in physiographic parameters and geological environments along the glaciated coast of Maine indicates that coastal variability is primarily controlled by regional changes in bedrock composition and structure, Quaternary sediment abundance and composition, and differential rates of relative sea level rise. The Q-mode factor analyses suggest that three coastal-zone cross sections of "end member" composition—mudflat, marsh, and ledge profiles—are the principal elements comprising an ideal Maine estuary. Regional variation in the relative importance of the three profiles results in a four-compartment classification of Maine's glaciated, estuarine coastline.

I. Introduction

Although it is less than 400 km from New Hampshire to the Canadian border, the tidally influenced shoreline of Maine is longer than any other state except Florida or Alaska, and stretches for almost 5600 km (Ringold and Clark, 1980). Despite the beauty and growing commercial importance of the coast, it has received scant attention from coastal geologists. Most areas of Maine's shoreline are still unstudied, and there has been no statewide synthesis of surficial–terrestrial and shallow-marine geology. The purpose of this chapter is to summarize the late-/postglacial history of coastal Maine, to evaluate the areal importance of physiographic parameters, and to classify environmental settings along the coast.

II. Coastal Climate

Maine possesses a humid, north temperate climate with average annual temperatures along the coast ranging from 5 to 10°C. Winter conditions are severe and commonly result in frozen beaches, marshes, flats, and shallow estuarine waters for several months per year (Fefer and Schettig, 1980). Precipitation falls throughout the year, but river discharge is bimodal with peaks in the fall and spring (U.S. Geological Survey, 1981). Winds vary seasonally in direction and are generally from the north–northwest in fall and winter and from the south–southwest in summer. Although tropical storms have been very rare, winter storms with winds greater than 100 km/hr out of the northeast occur three to five times per year and often last for several days.

Although wave height decreases greatly in a landward direction from exposed headlands and beaches, little is known of waves in Maine beyond hindcasting calculations (U.S. Army Corps of Engineers, 1957). Tides along the Maine coast are semidiurnal and increase in average range from 2 m in the southwest to greater than 6.5 m in the northeast (Fefer and Schettig, 1980). Tidal range also increases from the mouth to the head of large embayments like Penobscot Bay by as much as 1.2 m (Fig. 1).

III. Late Quaternary History

Glaciers advanced across Maine to their maximum extent near Georges Bank at the entrance to the Gulf of Maine about 20,000 yr B.P. (Thompson, 1979). Subsequent retreat brought the ice margin to near the present coast by about 13,200 yr B.P. (Stuiver and Borns, 1975). Since the weight of the ice had isostatically depressed the crust of northern New England beneath sea level, continued retreat of the ice from the coast left numerous, small moraines in close stratigraphic contact with marine deposits (Smith, 1982).

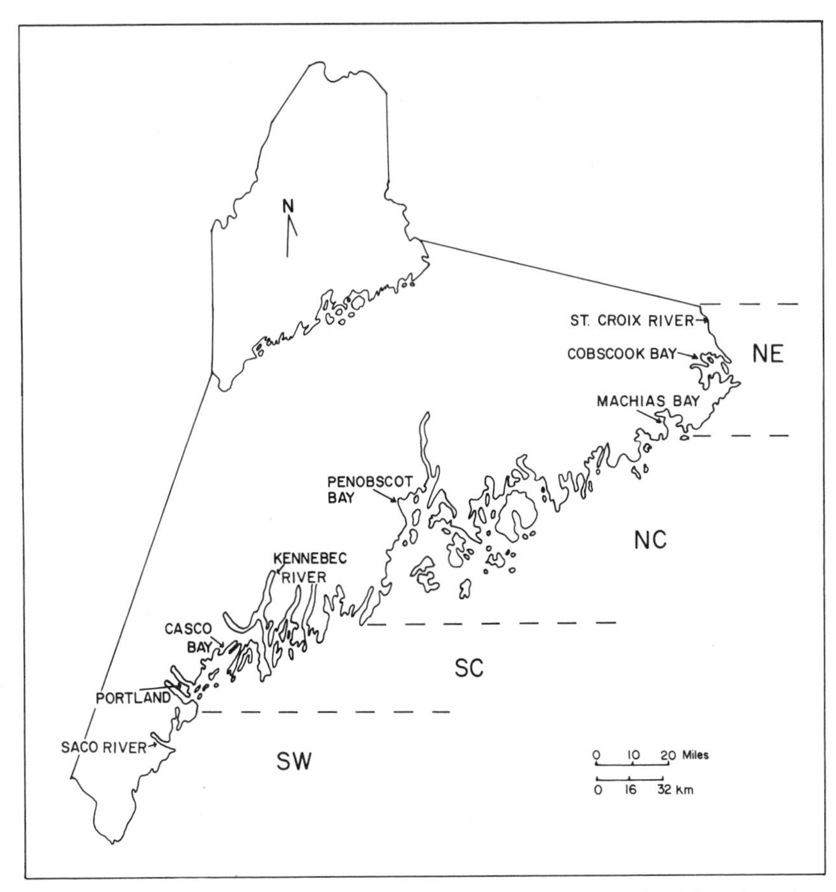

FIG. 1. Location of the study area. The coastal compartments (SW, SC, NC, and NE) are discussed later in the text.

A marine inundation accompanied disappearance of the ice and left many outwash deltas and a blanket of glaciomarine sediment (Presumpscot Formation) on land to mark the extent of maximum submergence (Bloom, 1963; Thompson, 1982). While deltaic sands are most abundant in southern Maine, Pleistocene marine sediment is ubiquitous in the present, coastal region (Thompson and Borns, 1985).

Radiocarbon dating of fossils within the marine sediment suggests that drowning of the coast was brief and ended by about 12,100 yr B.P. (Stuiver and Borns, 1975). Isostatic rebound of the ice-free landscape following submergence led to withdrawal of the sea from Maine and maximum emergence of the coastal region by about 8,500 yr B.P. (Schnitker, 1974).

Though few raised beaches or wave-cut benches mark the rapid regression of the sea, fluvially eroded till down to a present depth of 65 m indicates the apparent maximum extent of emergence (Schnitker, 1974; Kelley *et al.*, 1986; Belkrup *et al.*, 1986). Drowning of the coast has occurred since then. Both tide gage and releveling surveys on land suggest that subsidence of the coast today is not uniform (Anderson *et al.*, 1984), but is greatest in northeastern Maine where seismic energy release is also prevalent (Fig. 2).

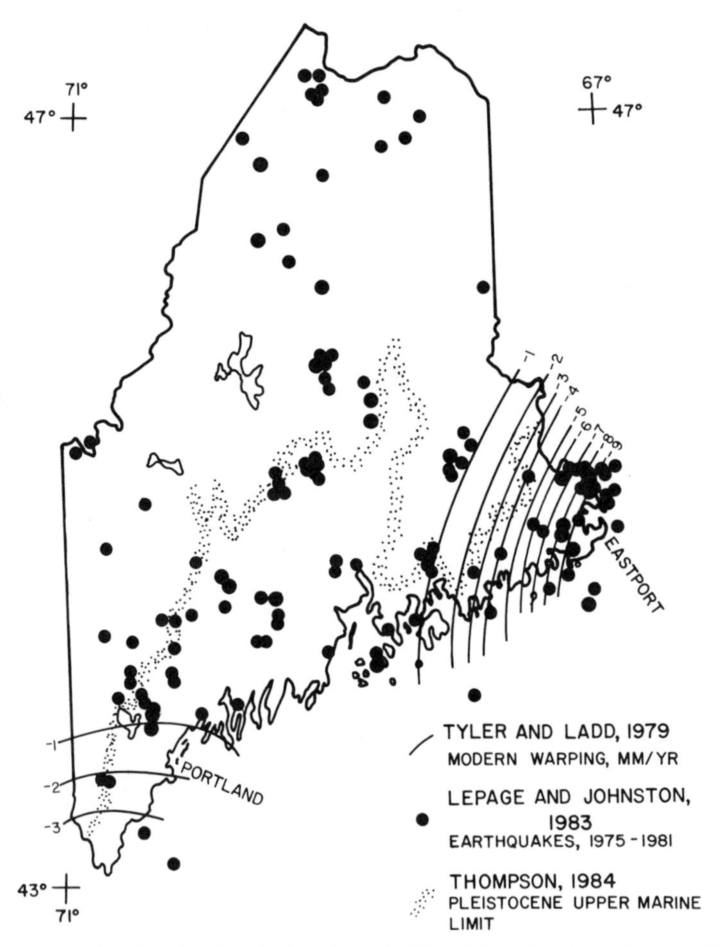

FIG. 2. Map of Maine showing the location of 1975–1982 earthquake epicenters (Lepage and Johnston, 1983), the rate of crustal subsidence (Tyler and Ladd, 1979), and the extent of postglacial marine submergence (the Presumpscot Formation; Thompson, 1984).

Fig. 3. An eroding bluff in the NC coastal compartment (Jonesport, Maine). Striated bedrock outcrops outside the field of view. It is overlain by (A) till and (B) Presumpscot Formation silty sands and clays. The sequence is capped by (C) possible salt marsh and (D) freshwater peats and (E) antropogenic fill. The bluff is retreating at a rate of about 1 m/yr.

An ideal Pleistocene stratigraphic column in coastal Maine thus contains till overlying polished, striated bedrock. The till is overlain by outwash sands and/or marine sediment, and occasionally capped by peat deposits (Fig. 3). In most locations in coastal Maine, the unconsolidated bluffs of Pleistocene sediment are eroding rapidly as contemporary sea level rises (Thompson, 1979).

IV. Bedrock Geology

The bedrock of coastal Maine is composed of complexly deformed and metamorphosed Paleozoic sedimentary and volcanic rocks, and igneous plutonic masses (Osbery *et al.*, 1985). The distribution of these rock types, their differential susceptibility to erosion, and their gross geometry led Jackson (1837) and later workers (Anonymous, 1983) to subdivide coastal Maine into four compartments (Fig. 1).

Southwestern (SW) Maine is underlain by northeast striking metasedimentary rocks punctuated by several plutonic bodies. In conjunction with the extensive glacial sand and mud deposits of southern Maine, the resultant physiography is one of rocky (plutonic) capes with intervening sand beaches fronting marshes (Fig. 1).

The south–central compartment (SC) extends from Cape Elizabeth to the western margin of Penobscot Bay and is also composed largely of high-grade metasedimentary rocks. These rocks strike more northerly than those to the south, however, and relief of the preglacial, strike-aligned valleys was accentuated by glacial scouring. This section of the coast is one of northeast-oriented peninsulas with intervening deep, narrow estuaries (Fig. 1).

The north–central (NC) compartment, which extends to slightly north of Machias Bay, is comprised of a subordinate proportion of low-grade metasedimentary rocks intruded by abundant granite plutons. The physiography is one of broad estuaries containing numerous resistant, granitic islands (Fig. 1).

The northeast compartment (NE) is underlain by low-grade metavolcanic rocks along the Atlantic coast, and metasedimentary rocks on the Passamaquoddy–Cobscook Bay shoreline. Because of their northeast strike, numerous northeast-trending faults, and relative resistance to erosion, the volcanic rocks comprise a nearly straight, high-cliffed coast, while the less-resistant low-grade metamorphic rocks form a highly indented, well-protected estuarine region (Fig. 1).

V. Methods

In an effort to synthesize existing physical and geological information in coastal Maine, and to classify environmental settings, a "census" of the coast was performed. By means of a linear probe and planimeter, each mile of tidally influenced shoreline was marked on 7.5-ft topographic maps (111 maps). Because islands less than a mile in circumference were neglected, only 3340 (of 3478) miles were located. From these, 334 mile-long sample units (MSUs) were designated on the basis of 334 random numbers chosen between 0 and 3340 (Beyer, 1971). The average distance between MSUs was 10 miles, and the modal spacing was 3 miles. The greatest distance between MSUs was 48 miles, while 70 MSUs were adjacent.

At five equally spaced locations along each MSU a traverse on a map was made normal to the coast (Fig. 4). Measurements made on each traverse are shown in Table I. The elevation was determined 92 m (300 ft) landward of mean high water and the depth 215 m (700 ft) seaward of mean low water. Exposure is the number of degrees surrounding the mean high-water mark with more than 0.8 km (0.5 miles) of unrestricted fetch.

Information on the coastal environments was reduced from the more complex Coastal Marine Maps (Timson, 1977), while data on the composition of the upland were derived from Surficial Geologic Quadrangle Maps (Maine Geological Survey, various authors). Measurements from the various environments were in length of environment along traverse. Width is the sum of the environmental length measurements. Examples of how specific determinations were made are presented in Fig. 4. Because the U.S. Government was the source of most of the raw data, measurements were generally made in English units.

VI. Results

A. Composition of Coastal Environments

To facilitate statewide comparison of coastal environments, observations have been grouped into the four compartments defined by bedrock (Fig. 1). Though these subdivisions are generally comparable to one another in straight-line length, the many coves in south and north–central regions greatly increase their tidally influenced length compared to the more linear northeast and southwest sections (Fig. 5).

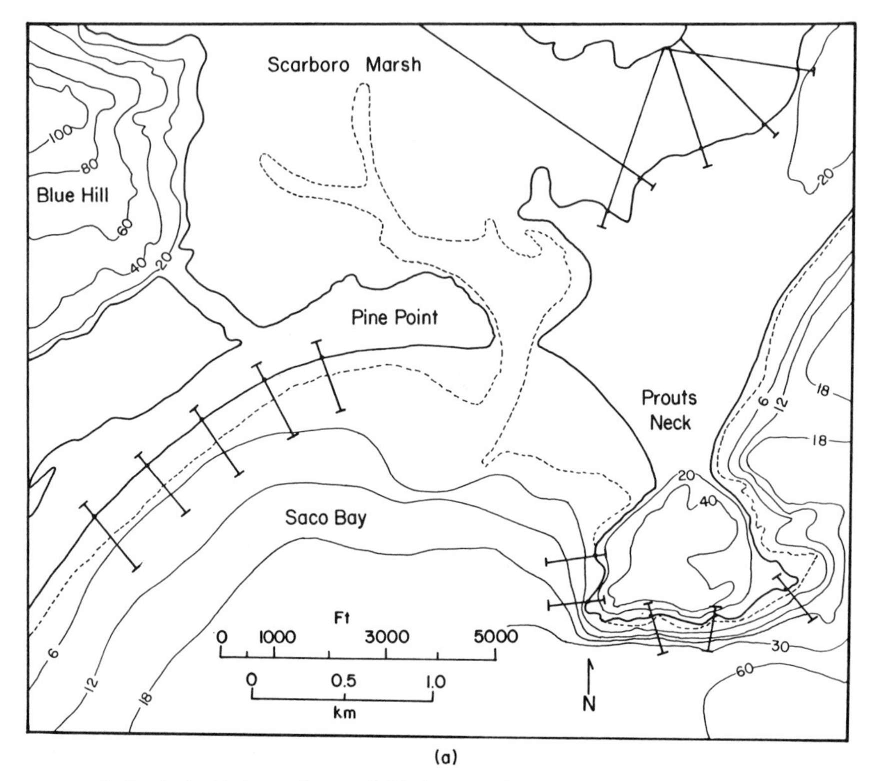

Fig. 4. Geological information available in coastal Maine. (a) Three MSUs comprised of five equally spaced traverses per mile are shown from the Prouts Neck area. Topographic and bathymetric data from the U.S.G.S. Prouts-Neck Quadrangle and Coast and Geodetic Survey Chart 231 permit measurement of height of the upland, depth offshore, coastal relief, and slope, as well as width of the intertidal zone, (b) Data modified from the more detailed Coastal Marine Maps (Timson, 1977) and Surficial Geologic Map (Thompson and Borns, 1985) provide information on environmental setting and upland composition.

In addition to differences in length, there are significant differences in the composition of the four compartments. Southwestern Maine is dominated by marshes with subordinate flat environments, while in the south–central region these relative abundances are reversed (Fig. 5). Overall, flats and marshes comprise more than 75% of the MSUs evaluated in the southern portion of the state. In contrast, both northern compartments are dominated by flats with subordinate ledge, and flat plus ledge environments comprise more than 75% of the MSUs in the northern half of the state.

Underlying these gross differences in environmental composition are more subtle, coastwide compositional variations within specific environ-

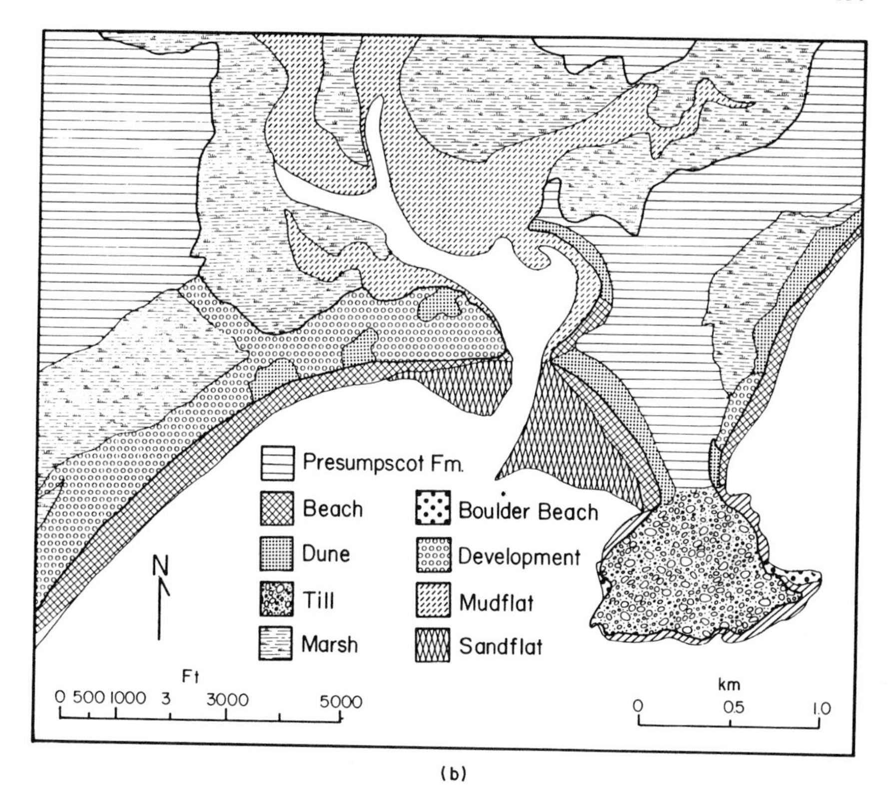

FIG. 4. (*Continued*)

ments. For example, not only do marshes become progressively less abundant proceeding from the southwest to northeast, there are also changes in the type of salt marshes along the coast and in their habit (Jacobsen *et al.*, in press). More than 99% of the marshes in the southwest are *Spartina patens* (high marshes) dominated, and exist along streams protected by barrier spits. By the geomorphic criteria of Frey and Basan (1985), these marshes are in an "old age" stage of evolution. There has been little change in the morphology or drainage of large southwest Maine marshes within historic times (Farrell, 1972), and they possess relatively thick peat deposits (>4 m; Bloom, 1963).

In the northern part of the state, marshes more commonly exist as narrow, fringing deposits on the margins of estuaries and at the head of riverine embayments. The proportion of *Spartina patens* decreases along with this change of habit such that 49% of the marsh in the northeast compartment is dominated by *Spartina alterniflora* (low marsh) (Table II) and the marshes are in a youthful stage of maturity (Frey and Basan,

TABLE I. PARAMETERS EVALUATED AND SOURCE OF DATA FOR THE "CENSUS" OF COASTAL MAINE

	1. U.S.G.S. Topographic maps (7.5')	2. National Ocean Survey charts	3. Calculated from 1 and 2	4. Maine Geological Survey coastal marine maps (7.5')	5. Calculated from 3 and 4	6. Main Geological Survey surficial geology maps (7.5')	7. Grouped as flat environment	8. Grouped as beach environment	9. Grouped as marsh environment
Elevation	X								
Depth	X	X							
Orientation	X								
Exposure	X								
Tidal range	X	X							
Relief			X						
Width				X					
Slope					X				
Upland composition						X			
Mud flat				X			X		
Coarse flat				X			X		
Ledge				X					
Dune				X				X	
Sand beach				X				X	
Sand and gravel beach				X				X	
Gravel beach				X				X	
High marsh				X					X
Low marsh				X					X
Lagoon				X					X

1985). The youthfulness of these marshes is also indicated by the thin nature of some northeast peat deposits (<2 m; J. T. Kelley, unpublished field notes).

Paralleling the decrease in high marsh is a southwest to northeast decline in the proportion of coarse-grained flats [intertidal flats comprised of sediment mostly coarser than sand (Anonymous, 1983)] with respect to mudflats. In the southwest section, the relatively small tidal range and extensive marshes preclude large mudflats; coarse-grained flats fronting small coves in the rocky capes comprise 33% of the flat environments. In the south–central and north–central compartments, mudflats increase to 90% of flat acreage due to the increasing tidal range and decrease in marsh. In the northeast section mudflats are most abundant and make up 97% of the

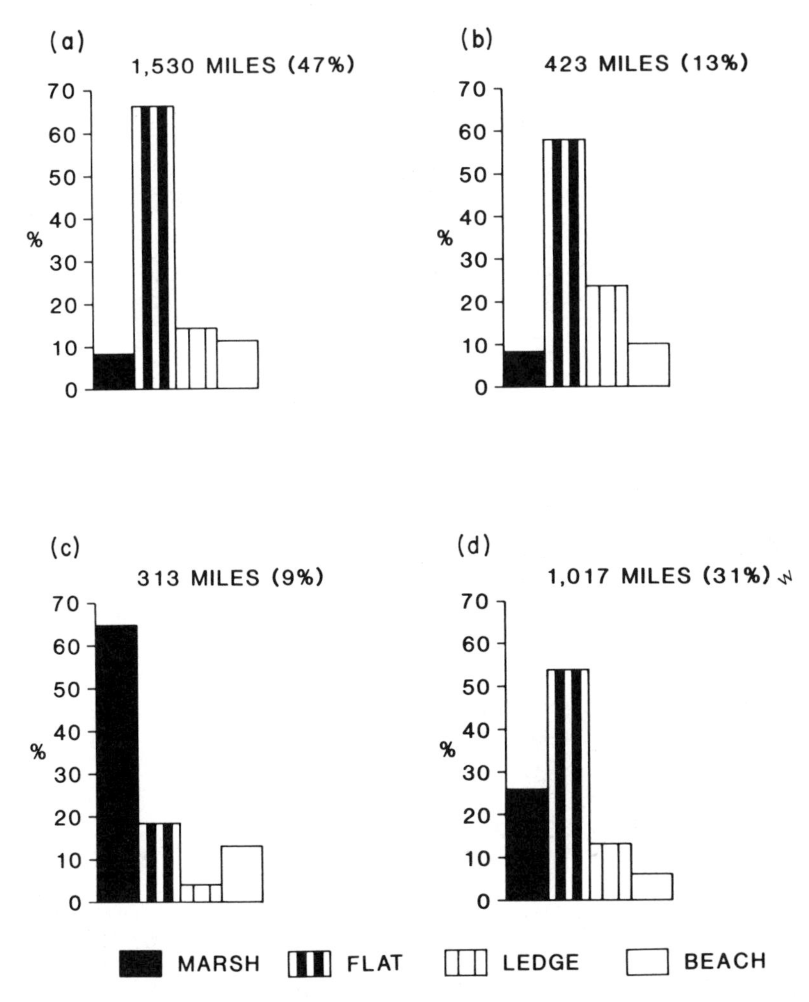

FIG. 5. Length of tidally influenced shoreline in the four coastal compartments: (a) NC, (b) NE, (c) SW, and (d) SC and percent of each compartment's MSUs dominated by marsh, flat, ledge, or beach environments.

flats evaluated, although ice-rafted boulders commonly litter the mudflat surfaces.

Though they represent a comparatively minor element of each coastal compartment, beaches display the greatest change in appearance along the coast (Fig. 6). In the southwest compartment, sandy barrier spits make up 90% of the beach environments while coarser grained, littoral deposits are rare. The dominance of sandy barrier spits continues into the south-central section, where beaches are concentrated near the mouth of Maine's

TABLE II. MEAN VALUES OF THE PARAMETERS EVALUATED IN THE FOUR COMPARTMENTS

						Beach[a]				Marsh[a]		Flat[a]			
Elevation (El) (ft)	Depth (DP) (ft)	Relief (RL) (ft)	Width (WD) (ft)	Slope (SP) (%)	Exposure (EX) (deg)	Sand (SB)	Sand and gravel (SGB)	Gravel (GB)	Boulder (BB)	High (HM)	Low (LM)	Mud (MF)	Course (CF)	Ledge (LD)	Tide (m)
SW 18	6	24	718	1.4	36.5	11.6	0.2	0.4	0.7	63.5	0.8	12.1	5.8	4.2	2–2.7
SC 41	13	55	595	3.4	44.7	4.2	0.4	0.7	0.8	22	3.9	48.6	5	13	2.7–2.8
NC 38	22	62	564	4.0	44.7	2.8	2.2	2.9	2.6	8	0.3	59.2	6.4	12.4	2.8–4.3
NE 47	21.5	69	584	4.4	58.5	1.0	6.7	0.6	1.4	4.5	4.2	56.2	1.7	23.8	4.3–6.5

[a] In percent of total acreage in compartment.

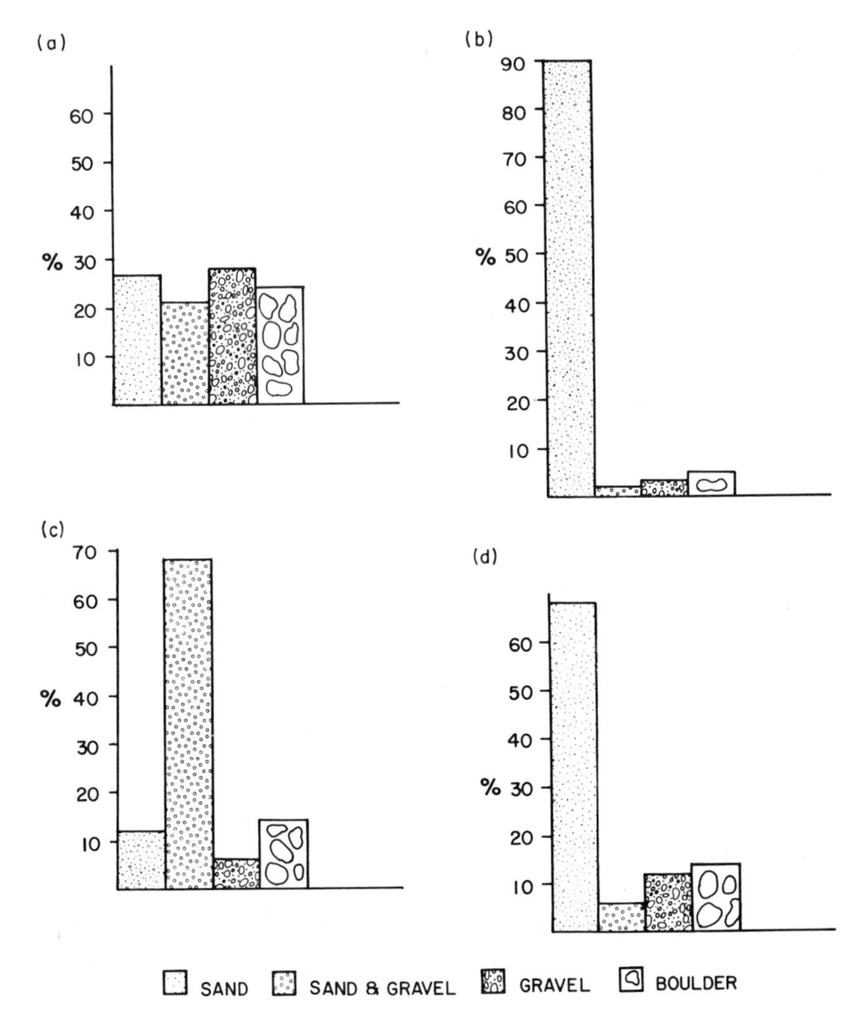

FIG. 6. Textural composition of Maine's beaches. The beaches are grouped by percent of coastal compartment: (a) NC, (b) SW, (c) NE, and (d) SC, where 100% is the sum of all traverses across beaches in a particular section.

largest river, the Kennebec. In the north–central compartment few barrier spits exist and the various coarse-grained pocket beaches become equally important (Fig. 6). It is of note that Maine's only carbonate beaches exist in this compartment (Raymond and Stetson, 1932; Leonard and Cameron, 1981). Finally, in the northeast compartment, sand and gravel beaches occupy the upper portion of the intertidal region. These beaches are commonly in a low-energy setting, and are derived from nearby, eroding bluffs of till or sediment from the Presumpscot Formation.

B. Physical and Geological Framework

To better understand variation in the composition of coastal environmental settings and to classify these, it is necessary to evaluate the physical or geological framework within which the environments exist. Figure 7 presents rose diagrams of the orientations or aspects of the traverses from each MSU. The orientations measured were grouped into 30° intervals and coded by the dominant, or most extensive, environment per traverse. In the southwest, the marsh-dominated traverses largely face the south and northeast (normal to streams entering the coastal region). Traverses dominated by other environments trend mostly to the northeast and southeast (perpendicular to the regional strike of the coastline). Although a chi-square test indicates the orientation of the traverses is nonrandom (few traverses face the west or northwest), the strongest preferred orientation is shown by the sand beaches, which are aligned to the predominant, southeast waves (Nelson and Fink, 1980).

In the three remaining coastal compartments, the orientation of bedrock controls the aspect of environmental settings. In the south–central section, the north–northeast strike of the metasedimentary rock peninsulas results in coastal settings facing the west–northwest and east–southeast. In contrast, the north–central compartment, with its broader, estuarine shoreline and many rounded islands, possesses environmental settings facing all directions equally except the northwest. The bedrock and majority of the traverses in the northeast section face the southeast (Atlantic shoreline) or northeast (St. Croix River shoreline), although flat-dominated settings face many directions within the irregular, Cobscook Bay (Fig. 7).

The relief (difference in elevation across the coastline) and slope (relief divided by width of intertidal zone) (Fig. 4) also show significant, systematic changes along the coast (Figs. 8 and 9). In the southwest, low elevation of the upland plus shallow depths offshore (Table II) result in a low-relief compartment (Fig. 8). Since wide marshes are common here, the average slope across the coastal zone is gentle. Although beach and ledge-dominated MSUs have somewhat greater relief and slope, the preponderance of marshes and flats in the southwest results in unimodal slope and relief histograms distributed around very low average values (Fig. 8 and 9).

In the mudflat-dominated south–central region, relief and slope histograms are also unimodal, but are normally distributed about a mean value more than double that in the southwest. This is because of the much greater elevation and depth in the region, and reduced intertidal width (Table II). As in the other compartments, there is a regular increase in the relief and slope associated with marsh, flat, beach, and ledge environments, respectively.

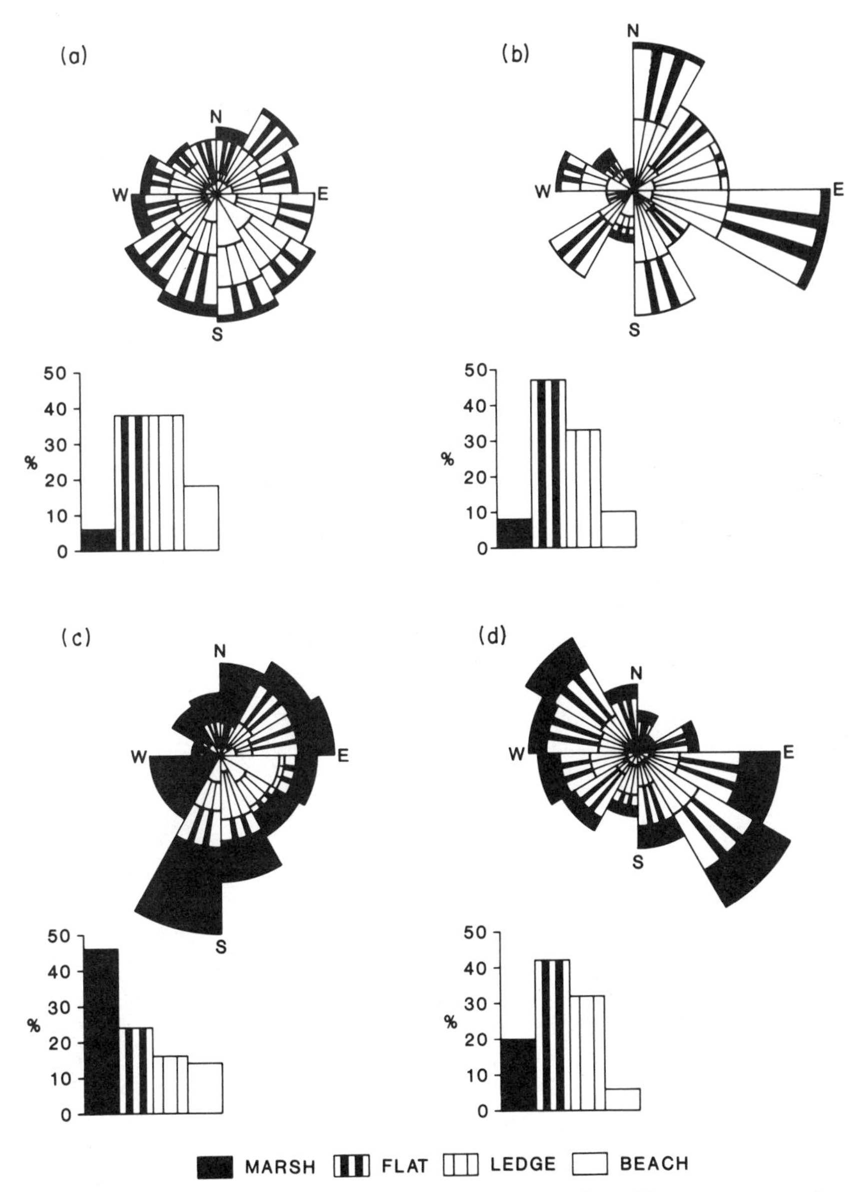

FIG. 7. The azimuth of each traverse made across the MSUs. The percentages refer to the number of traverses per compartment dominated by a particular environment: (a) NC, (b) NE, (c) SW, and (d) SC. Thus, the data for each compartment sum to 100%, although there were many more traverses in the south-central and north-central sections than in the other two compartments.

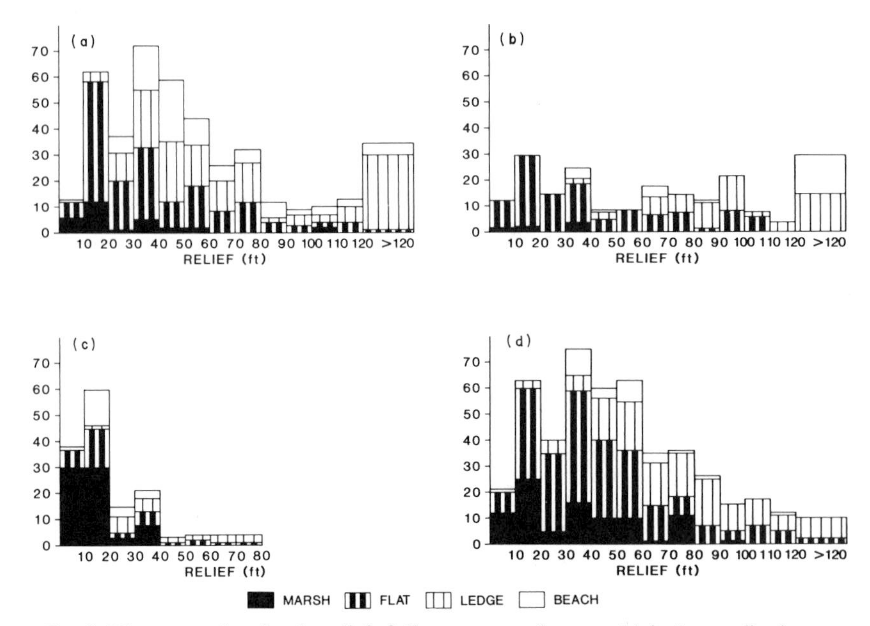

FIG. 8. Histograms showing the relief of all traverses made across Maine's coastline between points 92 m landward of mean high water and 215 m seaward of mean low water. Each traverse is coded to the most abundant environment encountered on a traverse. Although there are great differences in relief between the compartments, the relief associated with specific, dominant environments remains generally constant along the coast. (a) NC, $\overline{X}$ = 62; (b) NE, $\overline{X}$ = 69; (c) SW, $\overline{X}$ = 24; and SC, $\overline{X}$ = 55.

In the north–central compartment, greater elevation and depth associated with the increased abundance of ledge and coarse-grained beaches result in bimodal distribution of relief and slope. The broad range of values associated with beach environments reflects the near-equal abundance of beaches of different textures (steep, high-relief boulder beaches and gentle, low-relief sand beaches).

Average relief and slope continue to increase toward the northeast, where polymodal relief and slope histograms occur. While the upland here possesses the greatest average elevation along the coast, the depth of water is less than in the north–central section and intertidal width greater (Table II). It is of note that within the heterogeneous northeast compartment, those marshes in relatively high-relief settings appear to represent former freshwater bogs recently intersected by high tide. A vertical, seaward-facing scarp observed on many of these marshes suggests that they are eroding.

As with relief and slope (Figs. 8 and 9), the exposure of the MSUs also

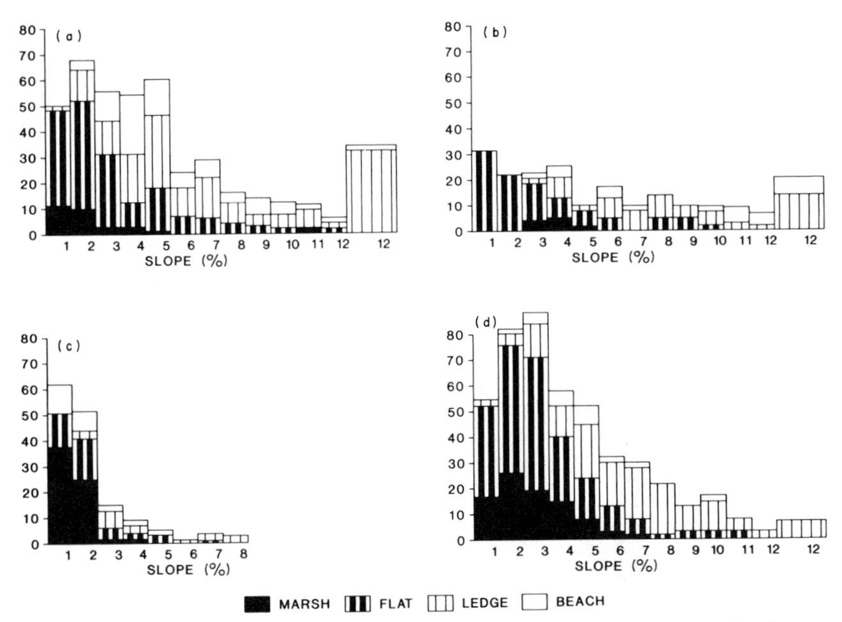

FIG. 9. Histograms showing the slope of all traverses made across Maine's coastline between points 92 m landward of mean high water and 215 m seaward of mean low water. Each traverse is coded to the most abundant environment encountered on a traverse as in Figs. 7 and 8: (a) NC, (b) NE, (c) SW, and (d) SC.

increases from southwest to northeast (Table II). This is not readily depicted graphically because of the wide range in the data. While marshes and flats often terminate in narrow creeks with no exposure (by the operational definition), ledge and beach MSUs commonly possess 180° or more exposure to waves. Thus, as with relief and slope, the southwest to northeast increase in exposure is associated with the changing abundance of the various environments.

In Fig. 10 a summary of the surficial geology of the upland is presented. Although important in its influence on the environmental composition of the coast, the surficial geology is difficult to assess. In some areas till may underlie Presumpscot Formation sediment, as discussed above, but only the Presumpscot Formation is indicated on surficial maps. In other areas the drift is too thin to identify, and till plus rock is recorded. Although till is generally coarser grained than Presumpscot Formation sediment, it is notoriously variable in texture, such that boulder, sand, and clay tills exist (Thompson, 1979). Similarly, the Presumpscot Formation is generally a "marine clay" (Bloom, 1963), but commonly possesses sand laminae and boulder dropstones.

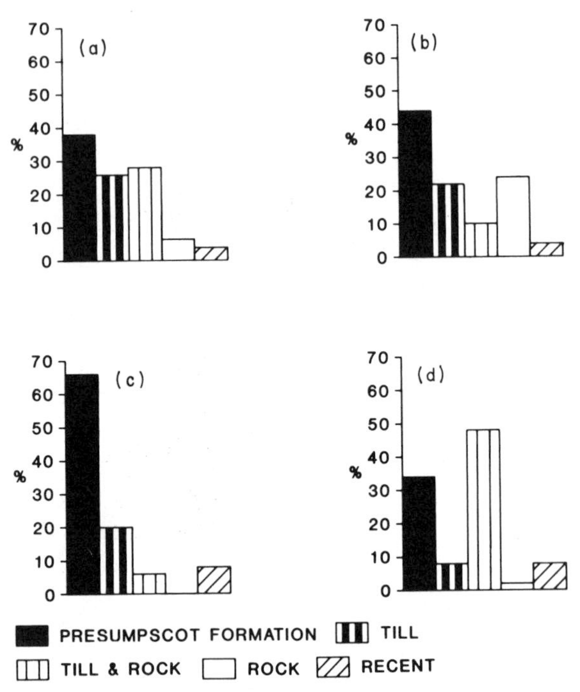

Fig. 10. Histograms of the surificial compositions landward of the MSUs from each coastal compartment: (a) NC, (b) NE, (c) SW, and (d) SC.

Despite these ambiguities, broad trends are discernable in the surficial data. The Presumpscot Formation is most abundant in the southwest (Fig. 10). Most marshes and flats here, as well as elsewhere on the coast, are found adjacent to the Presumpscot Formation. Till and till plus rock are most important in the central coastal sections, and most coarse-grained beaches are in close proximity to eroding till and coarser grained Presumpscot sediment. The proportion of exposed ledge lacking surficial sediment increases from near zero in the southwest to more than 20% of the MSUs in the northeast.

VII. Discussion

A. Coastal Classification

To reduce the complexity of data and organize observations from all traverses into a simple, coastwide classification, measurements from Table I for all MSUs were subjected to Q-mode factor analyses (SAS, 1982).

This technique reduces a matrix of n variables (Table I measurements) by m observations (MSUs) to one of n by f (factors) with a minimum loss of information (Hayden and Dolan, 1979). The few factors derived in this fashion possess "end member" compositions (in terms of the original MSUs) that may be summed in various combinations to produce a large proportion (>80%) of the original samples.

For this report, 13 variables (Table I) and 100 randomly chosen MSUs (the largest number permitted at one time by the SAS system) were repeatedly analyzed. Three factors consistently emerged from this and accounted for about 83% of the total variance. An examination of the properties of samples with high loadings on each factor led to construction of an average cross section representing each factor (Fig. 11).

Factor one, which accounts for 48.5% of the total variance, is most highly loaded on MSUs with extensive mudflats, and so is referred to as the mudflat factor (Fig. 11). Moderate average slope (2.1%) and relief (38 ft) and the occurance of minor ledge, marsh, and beach environments also

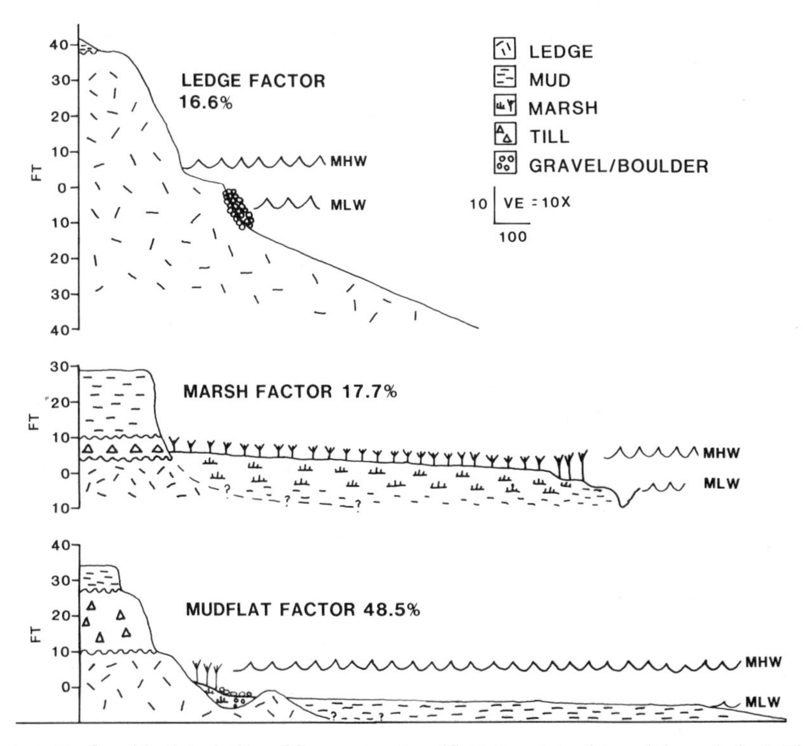

FIG. 11. Graphical depiction of factors one (mudflat), two (marsh), and three (ledge). The cross sections of the factors are drawn to scale and represent the average values of the 10 traverses most highly loaded on each factor.

tended to be associated with the ideal mudflat cross section. Factor two, the marsh factor, accounts for 17.7% of total variance and possesses a gentle average slope (1.7% average) and relief (29-ft average) and minor amounts of *Spartina alterniflora* in addition to the predominant *Spartina patens* high marsh. Factor three, the ledge factor, accounts for 16.6% of total variance, and has a steep average slope (7.1%) and relief (80.8 ft). This factor is more complex than the others, and the ledge is also associated with some coarse-grained beach environments. With the selection of more factors, a greater proportion of the total variance is accounted for (100 factors would account for 100% of the variance, or reproduce the original matrix). These additional factors are necessary to account for minor environments (coarse-grained flats, boulder beaches) or unusual MSUs, however, and are not of regional importance.

The results of the factor analyses suggest that although four coastal compartments may be defined on the basis of bedrock-controlled physiography (Fig. 1), they are not unique in composition. A ledge-dominated MSU from the northeast, for example, more closely resembles ledge environments in other compartments than marsh or flat-dominated MSUs from the northeast. A more simple classification of the coast is one of a series of estuaries comprised of varying combinations of the three ideal cross sections (Figs. 11 and 12). In this scheme the marsh cross section is most abundantly found at the riverine head of an embayment, while the mudflat and ledge cross sections occur successively seaward (Fig. 12). The four physiographic compartments thus represent the same environmental settings mixed in different proportions due to different types of bedrock, thicknesses of Quaternary sediment, and varying rates of relative sea-level rise.

In the southwest, relatively less resistant metasedimentary rocks have been preferentially eroded leaving topographic lows between a few high capes composed of igneous rocks Osberg *et al.*, 1985). The initially low relief of the area has been further subdued by burial beneath a thick section of outwash sands and marine clay (Bloom, 1963). The ongoing transgression has reworked the abundant glacial sand into tombolos and spits attached to rocky capes (Kelley *et al.*, 1986). The intervening low areas, protected by the barrier spits, have filled with a thick section of marsh sediment derived from locally abundant Presumpscot Formation material (Fig. 10) (Bloom, 1963). The southwest compartment is thus dominated by marsh cross sections (factor two) punctuated by occasional ledge and flat profiles at the capes.

The south–central section, perhaps because of deeper glacial scouring along its N–S oriented valleys, possesses double the relief of the southwest and is covered by a thinner, rockier Quaternary section (Fig. 10). The

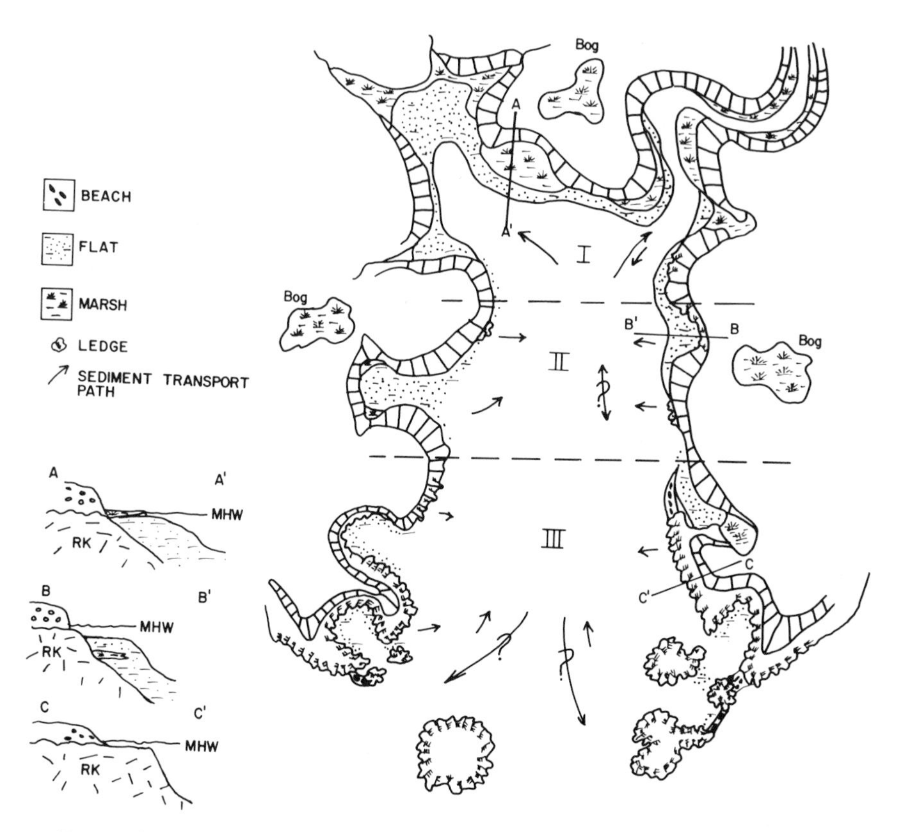

FIG. 12. Geographic organization of the three factor cross sections into a model of an ideal Maine estuary. At the landward end (A–A′) where riverine contributions are greatest and wave exposure minimal, rapid sediment accumulation permits marshes to grow out over prograding mudflats. In the central estuary where wave exposure is greater, marshes formed at a lower stand of sea level are eroding or have been removed, and sediment supplied to extensive flat profiles (B–B′) is episodically derived from nearby, eroding bluffs. At the mouth of the estuary the ledge profile (C–C′) is most abundant. In protected areas where former flats were extensive, a coarse-grained lag may be reworked into a new spit, and sediment may again begin to accumulate. Variation in the abundance of the three ideal profiles may describe the various estuaries of Maine with little loss of information.

elongate nature of the bedrock peninsulas has resulted in very long, protected estuarine segments with predominantly mudflat (factor one) cross sections. Marsh profiles (factor two) are of secondary importance here and are common only at the head of estuaries draining areas underlain by the Presumpscot Formation. Ledge-dominated cross sections (factor three) with beaches exist only at the tips of peninsulas and on islands.

The greater bedrock relief and local paucity of muddy rivers prevent extensive marshes from forming anywhere in the north–central compartment. Here, mudflat profiles derived from locally eroding outcrops of till, and many ledge cross sections on exposed shorelines are the dominant coastal components. This compartment possesses the largest number of coarse-grained beaches and flats that are not classified by the three-factor analyses.

The northeast compartment is dominated by ledge profiles along its Atlantic shoreline and flat profiles on its lengthy, Cobscook Bay coast. Many of the ledge cross sections possess boulder beaches, while the flats commonly have sand and gravel beaches on their landward margins. Marshes exist in this steep, rapidly sinking compartment (Fig. 2) mostly as eroding fringes along sheltered, estuarine coves, and a "pure" marsh cross section (factor two) is present only in one location.

B. Coastal Evolution

Sea level is rising everywhere in coastal Maine, and locally, the land is sinking still more rapidly (Fig. 2). The short-term response of the coast to this drowning varies with the typical coastal profile. Ledge profiles generally experience slow drowning with little erosion or deposition of sediment. Wave energy is too great to permit accumulation of any material except boulders near rocky cliffs, while Pleistocene bluffs are already so worn back by waves they are only eroded during extreme storms. Locally, where sand supply is great, spits and tombolos form landward of capes and eventually migrate or drown in place.

Landward from the "rockbound," open coast, mudflat profiles predominate. Landslides and storm erosion of bluffs occur commonly here (Thompson, 1979, Fig. 3) and episodically supply sediment to adjacent flats. Sediment accumulation or erosion is thus very variable on Maine flats, as is the texture of sediment (Anderson *et al.*, 1981). Flats generally appear to become finer grained and less erosional moving in a landward direction.

Far removed from open ocean waves, and in the estuarine region where rivers deliver fine-grained sediment, marshes are most extensive. Bluff erosion does not occur where protective marshes exist, though the most seaward marshes within estuaries tend to be themselves eroding. Where marshes are fronted by barrier spits, peat is commonly found outcropping on beaches, and time-series maps show a net historical shoreline retreat (Nelson, 1979).

Long-term evolution of the coast would ideally result in marsh peats sandwiched between mudflat deposits, with coarse-grained sediments and finally deep-water clays capping the sequence. Such a section would only be preserved in a very sheltered location enjoying an abundance of sediment and experiencing rapid sea-level rise. The scarcity of recent sediment over bedrock offshore of some estuaries (Kelley *et al.*, 1986; Belknap *et al.*, 1986) suggests few shallow-water deposits survive the transition to deep water intact.

VIII. Summary

The coast of Maine is most succinctly described as a series of estuaries that possess varying combinations of three ideal profiles: mudflat, marsh, and ledge. The parameters that determine the relative importance of the profiles in a specific area are bedrock lithology and structure, Quaternary sediment composition and thickness, and the local rate of relative sea-level rise.

Bedrock lithology and structure exercise a primary control on coastal morphology by determining the geometry and orientation of estuarine basins. The coast may be divided into four compartments on the basis of estuarine geometry:

(1) *Southwest (SW):* a section of low-relief estuarine embayments bounded by topographically high bedrock capes.
(2) *South–central (SC):* a compartment of deep, narrow, elongate estuaries that are parallel to the strike of the adjacent bedrock peninsulas.
(3) *North–central (NC):* a compartment of generally broad, deep estuaries possessing an abundance of rounded, granitic islands.
(4) *Northeast (NE):* a compartment with a very high-relief, outer cliffed component and an irregular, high-tidal-range protected estuary.

Quaternary glaciation has interacted with the bedrock by preferentially eroding less resistant rocks or deeply scouring basins aligned parallel to the direction of ice flow, and depositing various sediments of great textural diversity. In the southwest compartment the major impact of glaciation was to subdue the already low relief by mantling the terrane with glaciofluvial and glaciomarine sediments. The coarse components of these sediments have become barrier spits and tombolos, which protect marsh deposits whose inorganic sediments were largely derived from the Presumpscot Formation. In the south–central compartment, glaciers appear to have accentuated coastal relief by deeply scouring valleys parallel to ice flow. Erosion of till and Presumpscot Formation bluffs along estuarine margins has contributed sufficient sediment to form mudflats where bedrock peninsulas offer protection from ocean waves, but marshes exist mostly along river channels entering the deep estuaries. A few sand beaches near the mouth of the Kennebec River appear to have formed by the reworking of nearby Pleistocene deltaic deposits (Belknap *et al.*, 1986). Extensive, easily eroded rocks in the north–central section were also preferentially eroded by ice leaving broad estuaries with islands comprised of resistant granite. While bluff erosion of Pleistocene sediment has led to the formation of numerous coarse-grained beaches and flats, the exposed nature of these estuaries permits marshes to exist only in the most protected, landward coves. The extremely resistant, volcanic rocks of the northeast coast have been neither deeply eroded by ice nor mantled by extensive glacial deposits. The less resistant rocks of the Cobscook Bay area, however, were deeply scoured by ice. Bluff erosion of marginal till and Presumpscot Formation material here has begun to fill the basins with flat deposits, but marshes are scarce.

Though the long-term rise of sea level has led to the present location of the coast, contemporary sea-level change has only locally modified physiography. In the southwest, barrier spits and marshes have undoubtedly moved landward within historical times, but no modern morphologic changes are attributable solely to sea-level rise. In the south–central and north–central compartments, changes in the rate of sea-level rise through the Holocene may have resulted in differing degrees of preservation of Pleistocene sediment, but continuing erosion of bluffs is the only present effect of sea-level change. Only in the northeast, where the rate of sea-level rise is greatest and appears correlated with seismicity, is it possible to associate the paucity of salt marshes as a direct consequence of submergence.

Acknowledgments

I acknowledge financial support for this project from the Maine Geological Survey, Maine Sea Grant, and the Maine State Planning Office. Thanks are extended to Dr. Hal Borns of the Institute for Quaternary Studies, University of Maine, for introducing the author to many sections of the coast during field trips, and to Dr. Dan Belknap and Craig Shipp of the University of Maine for thoughtful reviews of the original manuscript. Drafting assistance from Alice Kelley is also very gratefully acknowledged.

References

Anderson, F. E., Black, L., Watling, L., Mook, W., and Mayer, L. (1981). A temporal and spatial study of mudflat erosion and deposition. *J. Sediment. Petrol.* **51,** 729–736.

Anderson, W., Kelley, J., Thompson, W., Borns, H., Sanger, D., Smith, D., Tyler, D., Anderson, R., Bridges, A., Crossen, K., Ladd, J., Andersen, B., and Lee, F. (1984). Crustal warping in coastal Maine. *Geology* **12,** 677–680.

Anonymous (1983). "The Geology of Maine's Coastline." Maine State Plann. Off., Augusta.
Belknap, D. F., Shipp, R. C., and Kelley, J. T. (1986). Depositional setting and quaternary stratigraphy of the Sheepscot Estuary, Maine: a Preliminary Report, *Geog. Phys. Quat.* **40,** 55–70.
Beyer, W. (1971). "Basic Statistical Tables." Chem. Rubber Publ. Co., Cleveland, Ohio.
Bloom, A. (1963). Late Pleistocene fluctuations of sealevel and postglacial crustal rebound in coastal Maine. *Am. J. Sci.* **261,** 862–879.
Farrell, S. (1972). Present coastal processes, recorded changes, and the Post-Pleistocene geologic record of Saco Bay, Maine. Ph.D. Thesis, Univ. of Massachusetts, Amherst.
Fefer, S. I., and Schettig, P. (1980). "An Ecological Characterization of Coastal Maine," 6 vols. U.S. Fish Wildl. Serv., Newton Corner, Massachusetts.
Frey, R., and Basan, P. (1985). Coastal salt marshes. *In* "Coastal Sedimentary Environments." (R. A. Davis, ed.), pp. 101–169. Springer-Verlag, Berlin and New York.
Hayden, B., and Dolan, R. (1979). Barrier islands, lagoons, marshes. *J. Sediment. Petrol.* **49,** 1061–1072.
Jackson, C. (1837). "The Geology of the State of Maine," 3 vols. Smith & Robison Publ. Co., Augusta, Maine.
Jacobson, H. A., Jacobson, G. L., and Kelley, J. T. (1987). "Distribution and Abundance of Tidal Marshes Along the Coast of Maine, Estuaries." In press.
Kelley, J. T., Kelley, A. R., Belknap, D. F., and Shipp, R. C. (1986). Variability in the evolution of two adjacent bedrock-framed estuaries in Maine. *In* "Estuarine Variability" (D. Wolfe, ed.), pp. 21–42. Academic Press, Orlando, Florida.
Leonard, J., and Cameron, B. (1981). Origin of a high latitude carbonate beach: Mt. Desert Island, Me. *Northeast. Geol.* **3,** 178–183.
Lepage, C., and Johnston, R. (1983). "Earthquakes in Maine, 1975–1982." Maine Geol. Surv. Open File Rep., Augusta.
Nelson, B., and Fink, K. (1980). "Geological and Botanical Features of Sand Beach Systems in Maine," Maine Sea Grant Publ. 14.
Nelson, R. (1979). Shoreline changes and physiography of Maine's sandy coastal beaches. M. S. Thesis, Univ. of Maine, Orono.
Osberg, P. H., Hussey, A. M., and Boone, G. M. (1985). "Bedrock Geologic Map of Maine," Maine Geological Survey, Augusta, Maine.
Raymond, P., and Stetson, H. (1932). A calcareous beach on the Maine coast. *J. Sediment. Petrol.* **2,** 51–62.
Ringold, P., and Clark, J. (1980). "The Coastal Almanac." Freeman, San Francisco, California.
SAS Institute (1982). "Users Guide to Statistics." SAS Inst., Cary, North Carolina.
Schnitker, D. (1974). Postglacial emergence of the Gulf of Maine. *Geol. Soc. Am. Bull.* **85,** 491–494.
Smith, G. (1982). End moraines and the pattern of last ice retreat from central and south coastal Maine. *In* "Late Wisconsin Glaciation of New England" (G. Larson and B. Stone, eds.), Vol. 10, pp. 195–210. Kendall-Hunt, Dubuque, Iowa.
Stuiver, M., and Borns, H. (1975). Late Quaternary marine invasion in Maine: Its chronology and associated crustal movement. *Geol. Soc. Am. Bull.* **86,** 99–104.
Thompson, W. (1979). "Surficial Geology Handbook for Coastal Maine." Maine Geol. Surv., Augusta.
Thompson, W. (1982). Recession of the Late Wisconsin Ice Sheet in coastal Maine. *In* "Late Wisconsin Glaciation of New England" (G. Larson and B. Stone, eds.), Vol. 10, pp. 211–228. Kendall-Hunt, Dubuque, Iowa.
Thompson, W. (1984). "The Marine Limit in Maine." Maine Geol. Surv. Open File Rep., Augusta.

Thompson, W., and Borns, H. (1985). "Surficial Geologic Map of Maine." Maine Geol. Surv., Augusta.

Timson, B. (1977). "Coastal Marine Geologic Maps." Maine Geol. Surv. Open File Rep. 77-1, Augusta.

Tyler, D., and Ladd, J. (1979). Vertical crustal movement in Maine. *In* "New England Seismotectonic Study Activities in Maine " (W. Thompson, ed.), pp. 99–154. Maine Geol. Surv., Augusta.

U.S. Army Corps of Engineers (1957). "Saco, Maine Beach Erosion Control Study," House Doc. No. 2, 85th Congr. U.S. Gov. Print. Off., Washington, D.C.

U.S. Geological Survey (1981). "Water Resource Data for Maine," Water Data Rep. ME-81-1. U.S. Geol. Surv., Augusta, Maine.

CHAPTER 7

Quaternary Stratigraphy of Representative Maine Estuaries: Initial Examination by High-Resolution Seismic Reflection Profiling

DANIEL F. BELKNAP

Department of Geological Sciences and Oceanography Program
University of Maine
Orono, Maine

JOSEPH T. KELLEY

Maine Geological Survey
Augusta, Maine
and
Department of Geological Sciences and Oceanography Program
University of Maine
Orono, Maine

R. CRAIG SHIPP

Oceanography Program
University of Maine
Orono, Maine

High-resolution seismic profiling is used to define submerged late Quaternary stratigraphic units on the coast of Maine. Wisconsinan glaciation sculpted bedrock valleys and later deposited till, glaciomarine sediments, and glaciofluvial sediments. The bedrock structure and variable sediment thicknesses and types produce a series of four different geomorphic zones along the coast. In order to examine differences in bedrock geology, glacial history, geomorphology, and Holocene coastal processes as controls on

Quaternary evolution of the Maine coast, three representative estuaries were chosen for investigation from the two central zones. A major control in this evolution is sea-level change, the result of glacioisostatic and glacioeustatic fluctuations. Relative sea level rose to at least 130 m above present 12,000 yr B.P. and fell to at least 60 m below present between 10,000 and 8,000 yr B.P., causing rapid coastal transgressions and regressions. An idealized stratigraphic section produced by these factors starts on bedrock, beginning with till overlain by glaciomarine mud and coarse deltaic sediments, further overlain by glaciofluvial outwash. This sequence is cut by an unconformity produced during the regression and subsequent transgression. Overlying this unconformity are Holocene sand, mud, and peat deposits in estuarine and inner shelf facies. Major factors in estuarine evolution are bedrock structure, direction of glacial scour, and nature and thickness of the glacial deposits. Secondary factors include littoral and terrestrial erosive processes during regression and subsequent transgression, and Holocene coastal processes of tides and waves. Maine estuaries show the effects of glaciation as a pervasive control. They differ from coastal plain estuaries of the U.S. Atlantic coast in the degree of structural control and types of sediment available. They are similar in many respects to Atlantic Canadian estuaries, except for their more extreme relative sea-level excursions since glaciation.

I. Introduction

The coast of Maine is a glacially sculpted bedrock terrain, mantled by till, stratified drift, and glaciomarine sediment of variable thicknesses and discontinuous distribution (Thompson and Borns, 1985). Bedrock lithology and structure control the morphology of the coast and location and thickness of Quaternary sediments. The bedrock has varying resistance to erosion, so the glacial ice created hills and valleys controlled by rock type and oriented in relation to bedrock structure. During retreat, the ice grounded first on the highs, while the valleys contained calving embayments. This inherited topography and localization of glacial sediments controlled fluvial and estuarine drainage and orientation of coastal systems to wave approach (Kelley, Chapter 6, this volume). The present setting is a low- to moderate-relief, rolling topography: a fjard coast. The purpose of this chapter is to present the preliminary results from a 1000-km seismic reflection survey in four Maine estuaries during 1983. The stratigraphic interpretations of high-resolution seismic data coupled with cores and surficial mapping allow evaluation of a model of Holocene evolution of

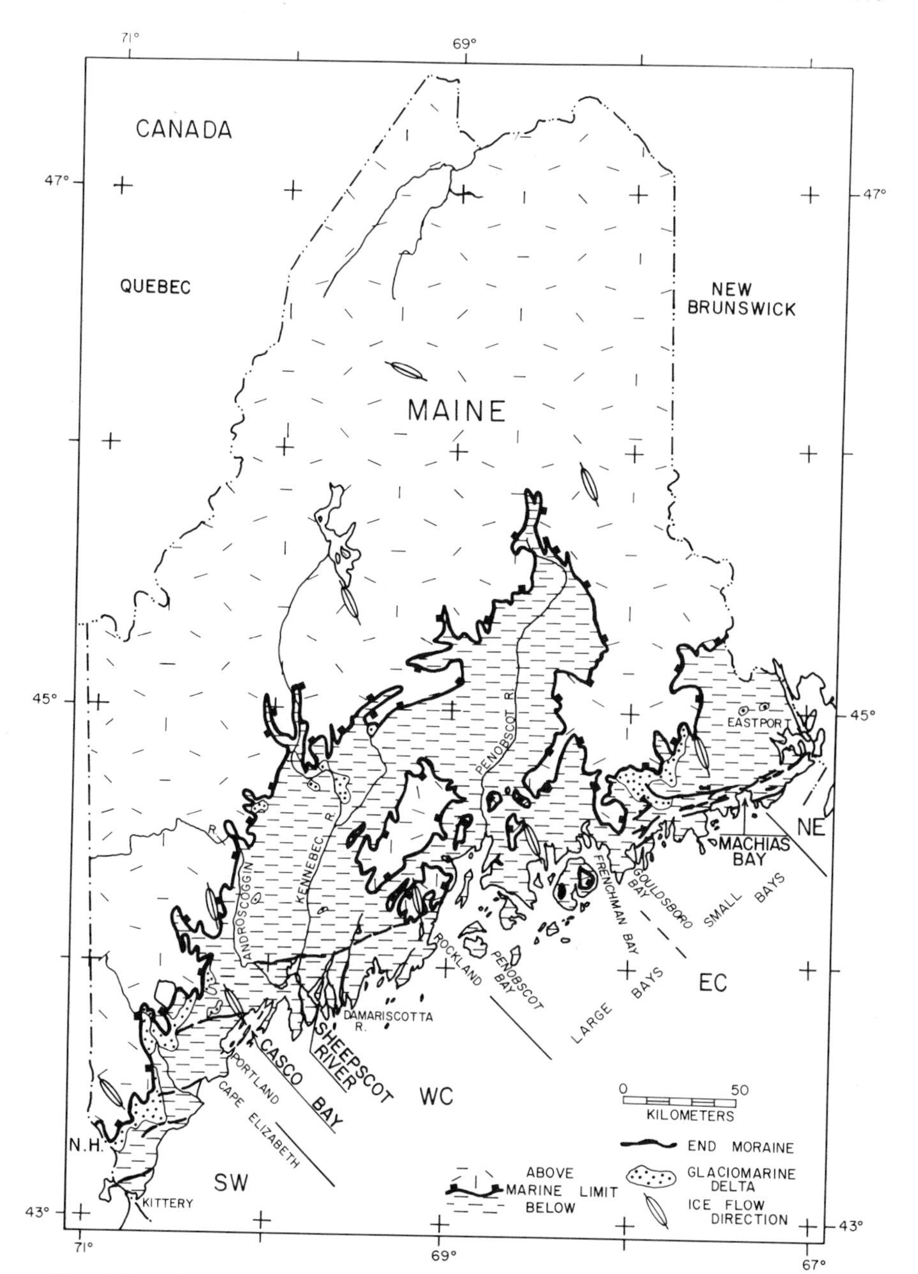

FIG. 1. Location map of Maine. Areas above and below the marine limit are shown in contrasting patterns, but marine sediments are found primarily in valleys (modified from Thompson and Borns, 1985). Locations mentioned in the text and the coastal geomorphic classification are also shown.

Maine's estuaries that couples bedrock framework and Quaternary sedimentary processes for a better understanding of this glaciated coast.

Early studies of the Quaternary stratigraphy of the region include Hitchcock (1873), Stone (1899), Leavitt and Perkins (1934, 1935), and Goldthwait (1949, 1951). Bloom (1960, 1963) formally named the Presumpscot Formation, a fine-grained glaciomarine unit that in coastal Maine is now exposed well above sea level due to glacioisostatic rebound. Borns (1973) and Stuiver and Borns (1975) clarified the deglacial stratigraphy of the Presumpscot with additional carbon-14 dating of fossil molluscs, wood, and marine algae. Field and airphoto mapping of geomorphic forms and natural and artificial exposures have continued to delimit glacial and deglacial features, as well as relative timing of events (Borns and Hagar, 1965; Borns and Calkin, 1977; Smith, 1982, 1985; Crossen, 1983; Thompson, 1979, 1982), culminating in a surficial geologic map of Maine (Thompson and Borns, 1985).

Although it has long been recognized that the Gulf of Maine was glaciated (Burbank, 1929; Murray, 1947), the offshore glacial to postglacial stratigraphy has been slower to develop. Numerous seismic, surface-sample, and coring transects have laid a framework for glacial stratigraphy of the Gulf of Maine (Knott and Hoskins, 1968; Schlee and Pratt, 1970; Oldale and Uchupi, 1970; Emery and Uchupi, 1972; Tucholke and Hollister, 1973; Oldale *et al.*, 1973), but nearshore studies have been limited and of a reconnaissance nature. Ostericher (1965) pioneered seismic profiling in coastal Maine, in Penobscot Bay. This area is presently being reexamined by us and by the U.S. Geological Survey (U.S.G.S.) (Knebel, 1986; Knebel and Scanlon, 1985). The nearshore shelf in southern Maine was extensively profiled by Folger and co-workers (1975), to assess sand and gravel resources. Schnitker (1972, 1974a) used 3.5-kHz seismic data in Montsweag Bay and Sheepscot Bay to define geomorphic criteria for a postglacial sea-level lowstand at 65 m below present. Approximately 1000 km of 3.5-kHz seismic profiling has allowed a preliminary assessment in the nearshore and estuarine segments of four parts of the Maine coast (Belknap *et al.*, 1985; Kelley *et al.*, 1984, 1986; Shipp *et al.*, 1984, 1985): Machias Bay, Gouldsboro Bay, the Damariscotta and Sheepscot region, and Casco Bay (Fig. 1).

II. Geologic Setting

A. Geomorphology

The bedrock of coastal Maine is of Paleozoic age. In southwestern Maine the rocks are primarily high-grade metasediments that have been tightly folded and faulted. In eastern Maine the rocks are less deformed chlorite-grade metavolcanics and metasediments. Numerous igneous bodies intrude the coastal regions. Later brittle fracturing of these rocks has further increased their structural complexity (Osberg *et al.*, 1984). Erosion of the Paleozoic basement has produced the present dominant bedrock control on coastal geomorphology. The latest modification of this coast by Pleistocene glaciation and Holocene sea-level rise have led to the present fjard setting (low- to moderate-relief glaciated terrain).

The coast has been divided into four coastal geographic compartments by Jackson (1837) and Kelley and Timson (1983). Figure 1 shows these compartments as well as the important glaciomarine sedimentary facies found in the coastal zone.

The southwest coast (SW), from Kittery to Cape Elizabeth, consists of arcuate barriers and protected marshes, pocket beaches, and rocky headlands. Spring tide range is 3.0 m. Abundant thick outwash deposits derived from the White Mountains have supplied sediments for this sandy coast through coastal erosive processes during Holocene sea-level rise. Wave and tidal processes redistribute this sediment.

The west–central coast (WC), from Cape Elizabeth to Rockland, is structurally controlled, consisting of a drowned topography of long, narrow estuaries and peninsulas shaped by variations in resistance to fluvial, glacial, and marine erosion. In addition, only thin till and outwash were deposited on the peninsulas, while the valleys were left only partly filled after the retreat of the Laurentide ice sheet. In this coastal compartment with 3-m spring tides sedimentary processes are dominated by tidal currents.

The east–central coastal compartment (EC), from Rockland to Machias, is controlled by large granitic intrusions. The resultant geomorphic form is of large bays interrupted by numerous islands and headlands. The bays are more equant in length-to-width ratio and more open to the Gulf of Maine than their southern counterparts. Shipp *et al.*, (1985) have further subdivided this compartment into a western large bay compartment (Penobscot Bay to Frenchman Bay) and an eastern small bay compartment (Gouldsboro Bay to Machias Bay). The western subcompartment has less glacial debris on land, but thick glaciomarine mud (Presumpscot Formation) on the bay floors. To the east, because of proximity to the Pineo Ridge moraine and related glaciomarine deltas and eskers, the small bay subcompartment has thicker valley fills and glacial drift cover on the inland parts of the peninsulas. Tidal currents are important in both the west–central and east–central segments, with spring range increasing to 4.4 m in the east, but the more open exposure of the east–central coast allows greater wave effects as well.

The northeastern (NE) coastal compartment is a straight, fault-controlled, cliffed coast cut in predominantly metavolcanic rocks. There are no large bays and few sediments to be found in the coastal zone; it is almost completely wave-dominated. North of the exposed Atlantic coast are the macrotidal (6–7 m) Passamaquoddy Bay and Cobscook Bay systems. These extensive digitate embayments have a complex distribution of estuarine and marsh environments, affected by tidal, biogeological, and, to some degree, wave processes.

B. Quaternary Evolution

The effects of Quaternary evolution on the coast are secondary only to the underlying bedrock control. Although many earlier glacial events must have scoured the region (Flint, 1971), there is clear evidence only for the latest Wisconsinan glaciation and subsequent deglaciation. Borns (1973) and Thompson (1982) have summarized this phase of Maine's geologic history. The influence of preglacial fluvial erosion is unknown; it is masked by later events. Glacial erosion sculpted the variably resistant bedrock into its present form during multiple advancing phases. The weight of the kilometers-thick ice also caused glacioisostatic depression of the crust. As the latest glacial stage waned, recessional moraines were deposited as far south as Georges Bank (Pratt and Schlee, 1969; King, 1969; Flint, 1971) and elsewhere in the Gulf of Maine as the terminus retreated north. In much of coastal Maine there is evidence for a marine-based ice sheet (Hughes, 1981; Mayewski *et al.*, 1981; Thompson, 1979; Borns, 1973; Smith, 1982, 1985; Crossen, 1983), with numerous glaciomarine outwash deltas and washboard moraines. The latter were deposited under an ice shelf at the glacial grounding line. Other glacial deposits include generally thin lodgment till, kames, drumlin and crag-and-tail ice streamlined deposits, and eskers. Seaward of the grounding line a glaciomarine rock-flour mud, the Presumpscot Formation (Bloom, 1960), was deposited. This cold-water deposit yields molluscs, microfossils, wood, and seaweed, which provide paleoenvironmental data. Time control is from numerous carbon-14 dates (Stuiver and Borns, 1975; Smith, 1985). During deglaciation (from ~12,500 to 11,500 yr B.P.) there was a retreating glacial front, a transgressing sea with calving embayments and estuaries, and island-bay complexes somewhat analogous to the present setting. This sea reached up the present Penobscot and Kennebec River valleys over 100 km from the present coast to at least the present elevation of 130 m (Fig. 1), while the coast was still isostatically depressed. This deglacial phase was interrupted by stillstands or minor readvances, which produced the Kennebunk moraine and the Pineo Ridge moraine and delta system. The topset–foreset contact of glaciomarine deltas approximately track the elevation of maximum sea level (Thompson *et al.*, 1983). The maximum position of the transgressive phase (the marine limit) is shown in Fig. 1.

By 11,000 yr B.P. the coast was rebounding rapidly due to glacioisostatic unloading. This caused a rapid fall of sea level, with only a few reported cases of shorelines preserved on delta fronts (Borns, 1973) and other areas (Willis, 1903). The stratigraphic record of this regression is undocumented. During the receding phase the Presumpscot Formation and other glacial deposits below the marine limit were probably eroded first by marine processes and then by subaerial processes progressively as sea level dropped and the shoreline advanced into the present Gulf of Maine. The lowest position of the sea is controversial (Fig. 2). Bloom (1960, 1963) discussed scenarios for approximately 10 m (minimum) of emergence ~6000–7000

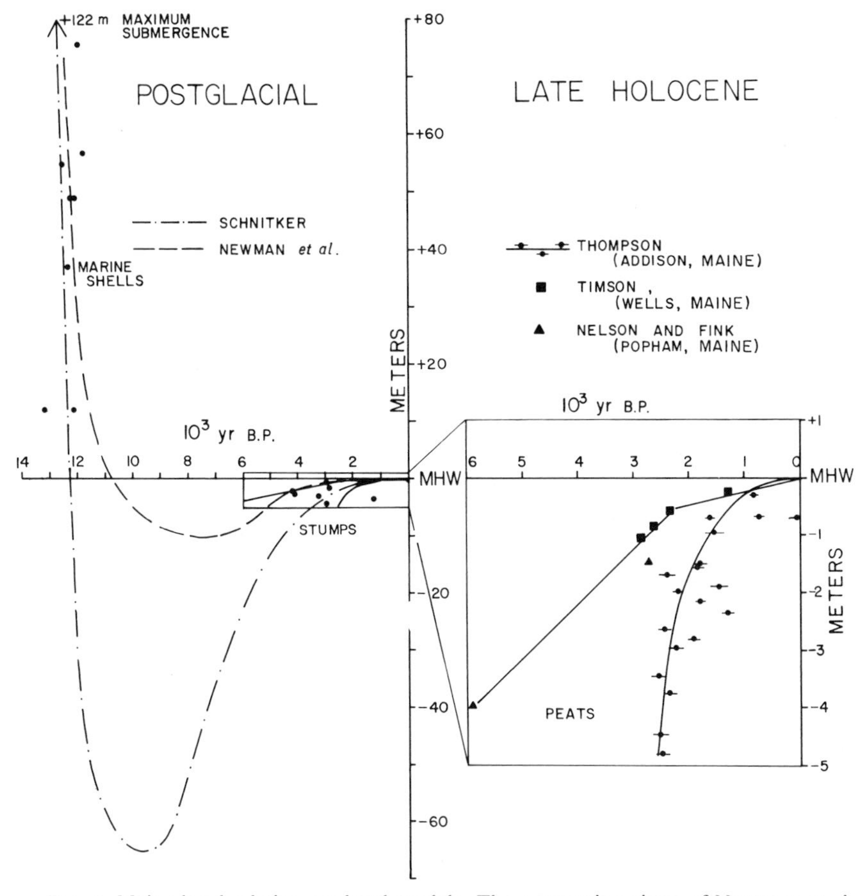

Fig. 2. Maine local relative sea-level models. The contrasting views of Newman *et al.* (1980) and Schnitker (1974a) are based on indirect evidence. Schnitker's (1974a) curve has been reconfigured to a relative sea-level curve. The Holocene data are salt-marsh peats in two contrasting neotectonic zones in Maine: the SW and WC zones (Timson, 1978; Nelson and Fink, 1978) and the eastern EC (Thompson, 1973).

yr B.P. based on depth of weathering within the Presumpscot Formation and by comparison with published "eustatic" curves. Newman and co-workers (1980) reached a similar conclusion from a few dated stumps and comparison with rheological models for isostatic rebound (Cathles, 1980; Clark *et al.*, 1978). Schnitker (1974a), on the other hand, presented geomorphic and seismic stratigraphic evidence for an early Holocene sea-level lowstand 65 m below present in the nearshore Gulf of Maine off Boothbay. This conclusion is supported by Oldale and co-workers (1983), who found evidence for a −47-m lowstand in the drowned Merrimack River paleodelta off Massachusetts; by Birch (1984), who saw similar evidence (> −40 m) off Portsmouth, New Hampshire; by Knebel and Scanlon (1987), who suggest a −40-m lowstand in Penobscot Bay; and by the present study, with topset–foreset contacts on the Kennebec River paleodelta in Maine at −45 to −50 m. The date of this event is unknown, but probably falls between 8000 and 10,000 yr B.P.

Holocene sea-level rise from the lowstand is presently documented only within the past 3000–6000 yr B.P., using deposits less than 6 m below present mean high water for carbon-14 dating. Again, there are several views on the rate of sea-level rise in coastal Maine over this period. Hussey (1959, 1970) used tree stumps as sea-level indicators, showing slow Holocene rise (10 cm/100 yr) in southern Maine. Data by Timson (1978) and Nelson and Fink (1978) support this slow rise. Thompson (1973), however, found a much more rapid sea-level rise of 1.15 m/100 yr at 3000 yr B.P. for Addison, in eastern Maine. He also demonstrated the wide variations and lack of usefulness of the tree-stump data. Recent sea-level curves from around the Bay of Fundy (Scott and Greenberg, 1983; Quinlan and Beaumont, 1981) strongly support Thompson's (1973) curve. Both data sets may be correct. Neotectonic warping of the Maine coast may be causing more rapid relative subsidence of the Eastport area as compared to Rockland or Portland (Anderson *et al.*, 1984). This warping may have widespread implications for rates of coastal erosion and processes of sediment redistribution within Maine's estuaries (Kelley, Chapter 6, this volume).

We speculate that during Holocene sea-level rise the Pleistocene glaciomarine and earlier Holocene sediments were reworked by the transgressing shoreline. Depocenters for sediments moved up estuaries or along retreating bluff and beach shorelines. We are presently evaluating a Holocene estuarine sedimentation model (Kelley, Chapter 6, this volume; Shipp *et al.*, Chapter 8, this volume) in which fine sediments are transferred from seaward to landward temporary storage sites during transgression, with residence times on the order of 1000–5000 yr. In this model, the seaward ends of the estuaries and peninsulas have been partially stripped of sediments, while the heads are loci for deposition in mudflat, marsh, and subtidal settings. The sources of sediment are postulated to be primarily internal: erosion of Pleistocene bluffs and tidal channel margins. Export of sediment to and import from the Gulf of Maine is likely (Schnitker, 1972, 1974b), but the volume and its relative importance is unknown. Effects of coastal erosion on archaeological sites also provide evidence for rates of coastal erosion (Kellogg, 1982). Sediment processes, pathways, and rates of erosion, transport, and accumulation are only beginning to be investigated. Subsidiary questions include the effects of changing hydrographic conditions with changing sea level (e.g., the effect of sills), regional changes in tidal range, effects of human alteration, and variations along the coast in the relative importance of waves, tides, and relative sea-level change.

An ideal stratigraphic section within the coastal zone begins with glacially carved bedrock, capped first with Pleistocene units: lodgement till overlain by deglacial moraines and stratified drift such as glaciomarine deltas, interfingering with and overlain by Presumpscot Formation glaciomarine mud. This sequence is interrupted by an erosional uncomformity, with gullies, slumps, channels, and an oxidized, desiccated zone (particularly noticeable on the Presumpscot Formation). At the mouth of large rivers reworked glacial material has been deposited in deltas, during the latest phases of regression to stillstand. Elsewhere there are no thick accumulations of sediment of this age, and the unconformity is overlain directly by Holocene transgressive deposits. The Holocene transgressive record has variable sand, gravel, and mud deposits of beach and estuarine environments. Marshes can be preserved at the base of the Holocene sequence, as the leading edge of the transgression, or as a cap to the sequence in protected areas today. Other possible units include freshwater bog deposits and other terrestrial environments overridden by the trangression. We also identify gas deposits, produced within thick organic-rich sediments (Kaplan, 1974). Finally, slumping and possible fault offsets occur within the sequence, especially where the Presumpscot Formation overlies bedrock; such slumps are common in coastal Maine today (Thompson, 1979).

III. Methods

One thousand kilometers of high-resolution 3.5-kHz seismic reflection profiling was used in a preliminary survey of shallow marine stratigraphy in Maine's coastal zone. A Raytheon RTT 1000A 3.5/7.0-kHz portable survey system provided resolution of less than 25 cm and penetration greater than 50 m in mud, in water depths from 1 to more than 80 m. The simultaneous use of a 200-kHz fathometer allowed a precise depth trace to compare to the 3.5-kHz subbottom record. Navigation control was by Loran C, radar, and charted navigation points. In the outer estuaries and offshore two 12-m boats were used. Each was equipped with radar and Loran C, which provided accurate navigation. Locations of the profiles are shown in Fig. 3.

Data were analyzed by tracing characteristic acoustic signals. The records were digitized manually and transformed to a uniform orientation and ×50 or ×20 vertical exaggeration. This digitization normalized the records (to account for changes in vessel speed), simplified interpretations, and allowed presentation of a greater number of interpreted lines in a smaller space.

Seismic reflectors were distinguished by acoustic character. Intensity, shape, relative sharpness of initial pulse, and continuity were primary criteria for identification of sediment properties deduced from impedance contrasts. Seismic stratigraphic analysis (see, e.g., Payton, 1977; Sieck and Self, 1977) of sequences, unconformities, and cross-cutting relationships allowed a preliminary geometric reconstruction of the Quaternary stratigraphy. Seismic signatures were compared with known units on shore and observed in submersible and scuba dives, as well as lithologic intervals in vibracores, piston cores, box cores, and bridge borings.

Figure 4 displays the returns from typical examples of the units identified in the study area. Bedrock (Figs. 4a and 4f) gives a strong, sharp 3.5-kHz return either when exposed or when covered by 20 m of mud. It is often spiky in areas of foliated rocks, and shows no internal relectors. Interpretation of bedrock is confirmed from many passes next to surface outcrops and over charted rock bottoms. Acoustic impedance difference between bedrock and till is lower; thus bedrock gives a less distinct return when covered by till. Glacial moraines (composed of poorly sorted stratified drift and till in coastal Maine; Smith, 1982) are identified by a strong sharp return, chaotic, discontinuous internal reflectors, and often a rounded shape (Fig. 4d). Till may also occur as a thin veneer over bedrock, giving a broad strong return with a separate bedrock return beneath (Fig. 4c). The interpretation of till is strengthened from data near shoreline outcrops in Casco Bay, Machias Bay, and Gouldsboro Bay (Kelley *et al.*, 1984, 1986; Shipp *et al.*, 1984, 1985). These exposures occur on moraines that presumably continue into the submarine portions of the bays. Stratified drift is interpreted from moderate to strong returns, with repetitive variations in intensity vertically and laterally (Fig. 4c). Reflectors are often continuous for hundreds of meters, and may be tabular, wedge, or trough shaped. The alternating signal intensity probably represents interbedded

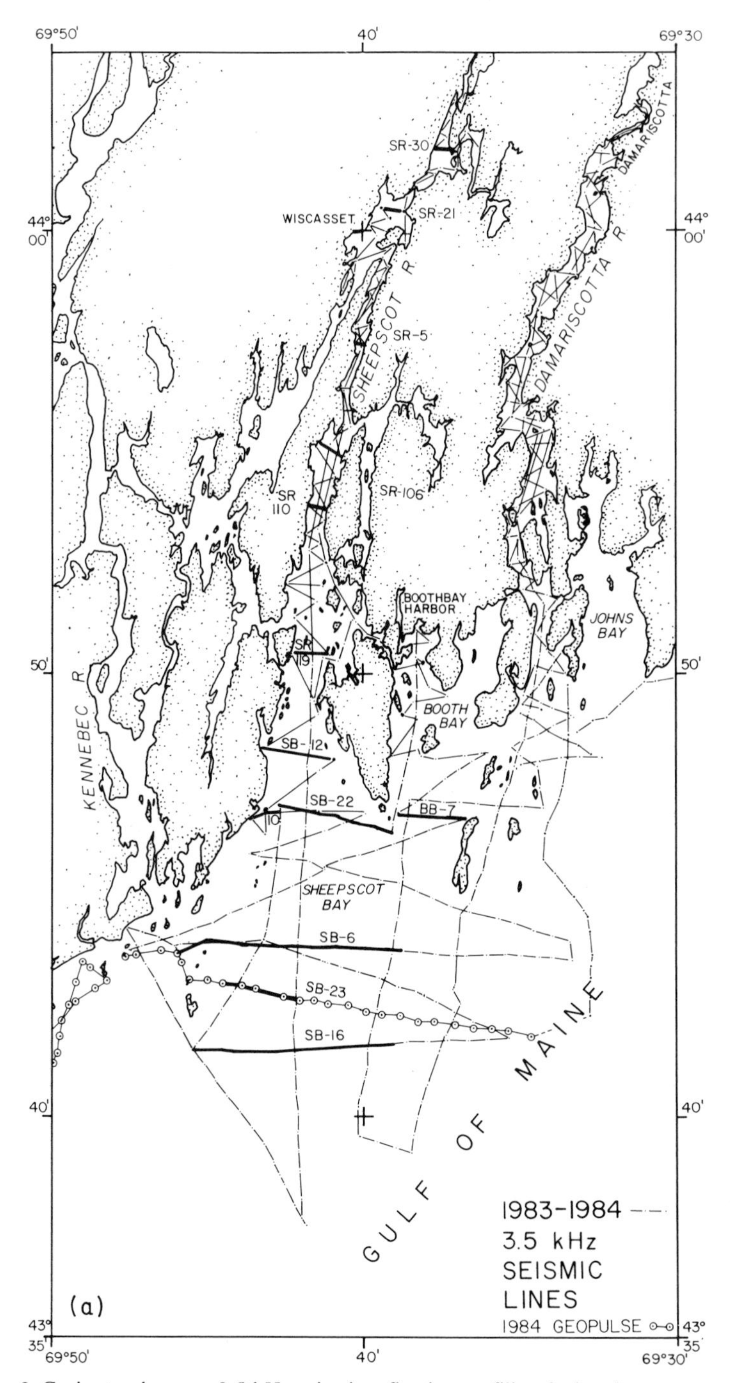

Fig. 3. Cruise track maps, 3.5-kHz seismic reflection profiling during the summer and fall of 1983. Navigation was by Loran C, radar, and navigation markers. In each case seismic lines discussed in the text are heavier and numbered. (a) Cruise tracks for the Sheepscot and Damariscotta estuaries. (*Continued on next page*)

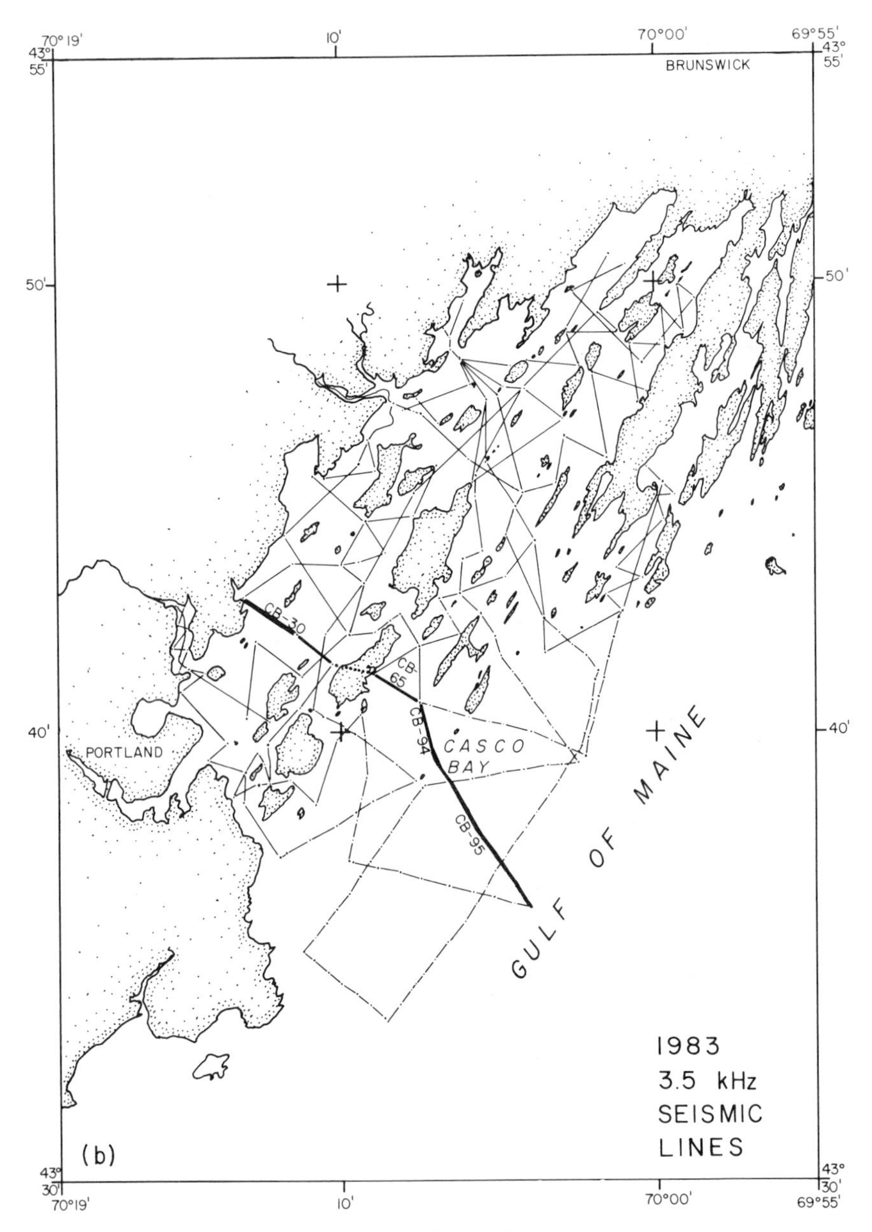

FIG. 3. (b) Cruise tracks for the Casco Bay estuary.

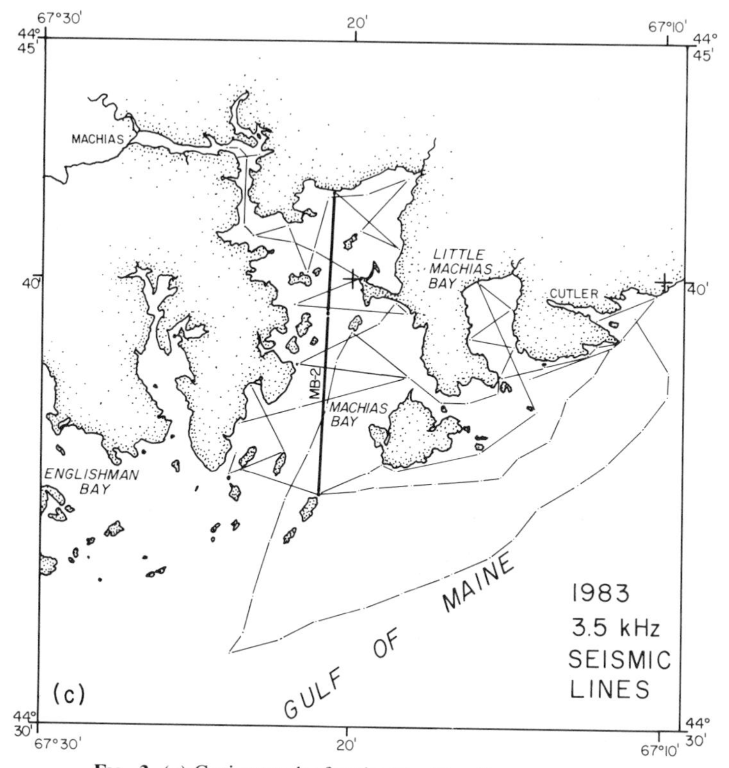

FIG. 3. (c) Cruise tracks for the Machias Bay estuary.

sand and gravel. Stratified drift is exposed on land at Sprague Neck in Machias, less than 6 km away from the seismic line segment of Fig. 4c. Glaciomarine mud of the latest Pleistocene Presumpscot Formation (Fig. 4b) is interpreted from strong, generally flat or gullied surface signals and a weakly bedded to acoustically transparent internal character. At its base, reflectors mimic bedrock topography, suggesting a drape deposit. Presumpscot Formation marine mud has been penetrated by fourteen submarine vibracores on seismic profiles lines, two in the Sheepscot River, six in the Damariscotta River, four in Muscongus Bay, and two in Fox Island Thorofare, North Haven (Belknap *et al.*, 1987; Shipp, 1987).

The top of these pre-Holocene units is a distinct reflector, which commonly demonstrates a sharp contrast between less compacted Holocene and more compacted Pleistocene mud, coarser grained drift, or hard bedrock (Fig. 4b). Over drift and glaciomarine mud an erosional unconformity is clearly indicated by planation and gullying, slumping, or a lag surface

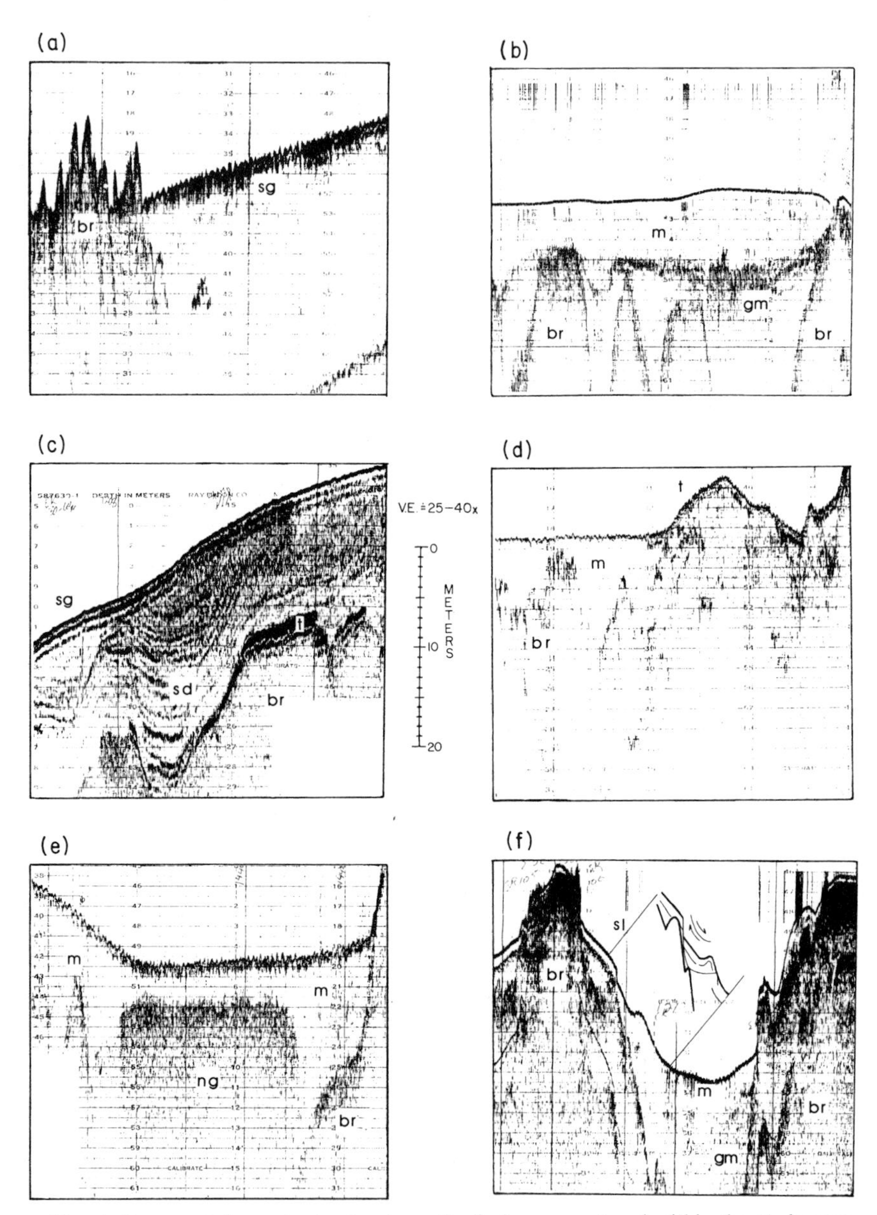

Fig. 4. Representative seismic signatures for facies encountered within the study area. All are 3.5-kHz Raytheon RTT1000A records, ×2 scale, 2 in/min, at speeds near 5–7 knots and power output from −6 to −18 dB. Each also shows enhancement of deeper returns by a bottom-triggered time-variable gain control. A simultaneous 200-kHz record acts as a fathometer, and is printed 0.25 m (chart scale) above the initial 3.5-kHz return. The abbreviations used are br, bedrock; gm, glaciomarine mud; m, Holocene mud; ng, natural gas; sd, stratified drift; sg, sand and gravel; sl, slump; t, till.

produced by winnowing. Each of these unconformity types has been observed on land nearby (Thompson, 1979, 1982).

Slumps are recognized by a headward scarp, coherent rotated blocks, deformed anticlinal toe, and a curved basal fault surface (Fig. 4f). Slumps are exclusively found on steep slopes along estuary margins and involve Holocene sediments and Presumpscot Formation over bedrock. Apparently slumped debris is contributing to the basin fill deposits of the estuaries. Recent slump activity on land has been documented in Maine (Thompson, 1979; Novak *et al.*, 1984) with the characteristics described above, while Birch (1984) suggests similar slumps offshore of New Hampshire.

Holocene muds are identified by a smooth, gently sloping surface on the 200-kHz return and weak to absent 3.5-kHz return (Figs. 4b and 4e). Internally, muds show weak stratification only at the highest gain. They tend to pond in pockets on bedrock ridges, to drape underlying structure, and to accumulate in paleovalleys. Most bedrock ridges are completely barren of mud, apparently swept clear by tidal currents or slumping. Holocene sands and gravels are identified by a strong "ringing" return in the 3.5-kHz signal and by absorption of most of the input energy and consequent masking of most features below more than 2–5 m thickness (Fig. 4a). Slopes on sand and gravel can be steep ($> 20°$ observed), while mud is found on slopes generally less than 1°. Sand and gravel are often found as a thin (< 2 m) cap reworked from stratified drift, either over submerged glaciomarine deltas or near modern beaches (Figs. 4a and 4c). Dives in the submersible *Mermaid II* in 1984 confirmed the surficial sand, gravel, and mud deposits along seismic profile SB-6 (Fig. 5), from Tom Rock to the Sheepscot paleovalley.

Further confirmation of seismic profiles is provided by bridge borings at the Wiscasset River Route 1 bridge (Miller and Baker, 1982; Kelley and Belknap, 1985) and in Casco Bay that penetrate the entire Quaternary section.

One other Holocene feature is a moderate-strength return with a fuzzy surface and convex-up shape (Fig. 4e). It absorbs most of the acoustic energy (i.e., water-bottom multiples abruptly disappear when this signal appears). This return is found only in thick muds within filled channels and valleys. It is interpreted as natural-gas bubbles within the sediment, similar to that found by Schubel (1974), Vilks *et al.*, (1974), and Keen and Piper (1976). This interpretation has been confirmed by piston cores taken by D. Schnitker (personal communication), which penetrated these layers. Caps for these cores were blown off at sea-level pressure by a flammable gas.

Figure 5 is a 3.5-kHz record from a portion of line SB-6, shown as

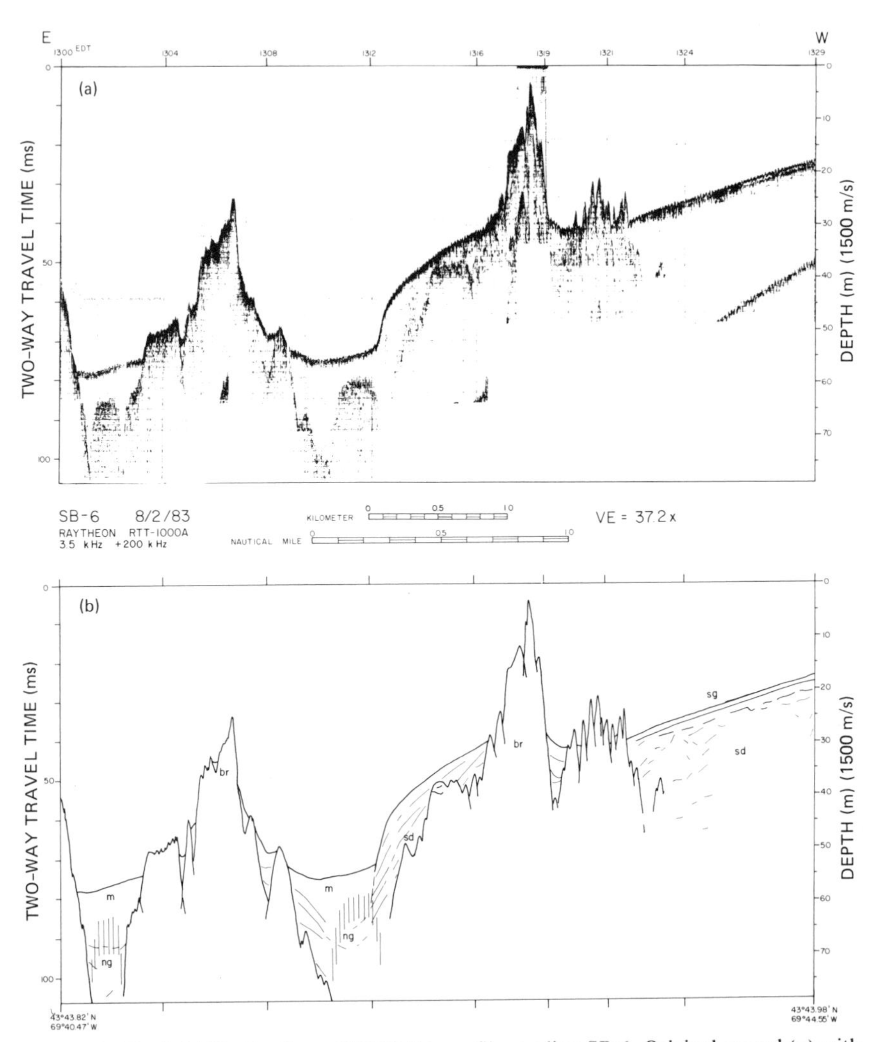

Fig. 5. The 3.5-kHz Raytheon RTT1000A profile, on line SB-6. Original record (a) with the interpreted line drawing (b). Identification of units follows the methods of Fig. 4. Location shown in Fig. 3a.

Fig. 6. ORE-Geopulse boomer profile, on line SB-23. Original record (a) with the interpreted line drawing (b). Identification of units similar to the methods described for Fig. 4, but the lower frequency and deeper penetration emphasize the lower units of major impedence contrasts, while the lower resolution and large bubble-pulse mask finer details near the surface. Location shown in Fig. 3a.

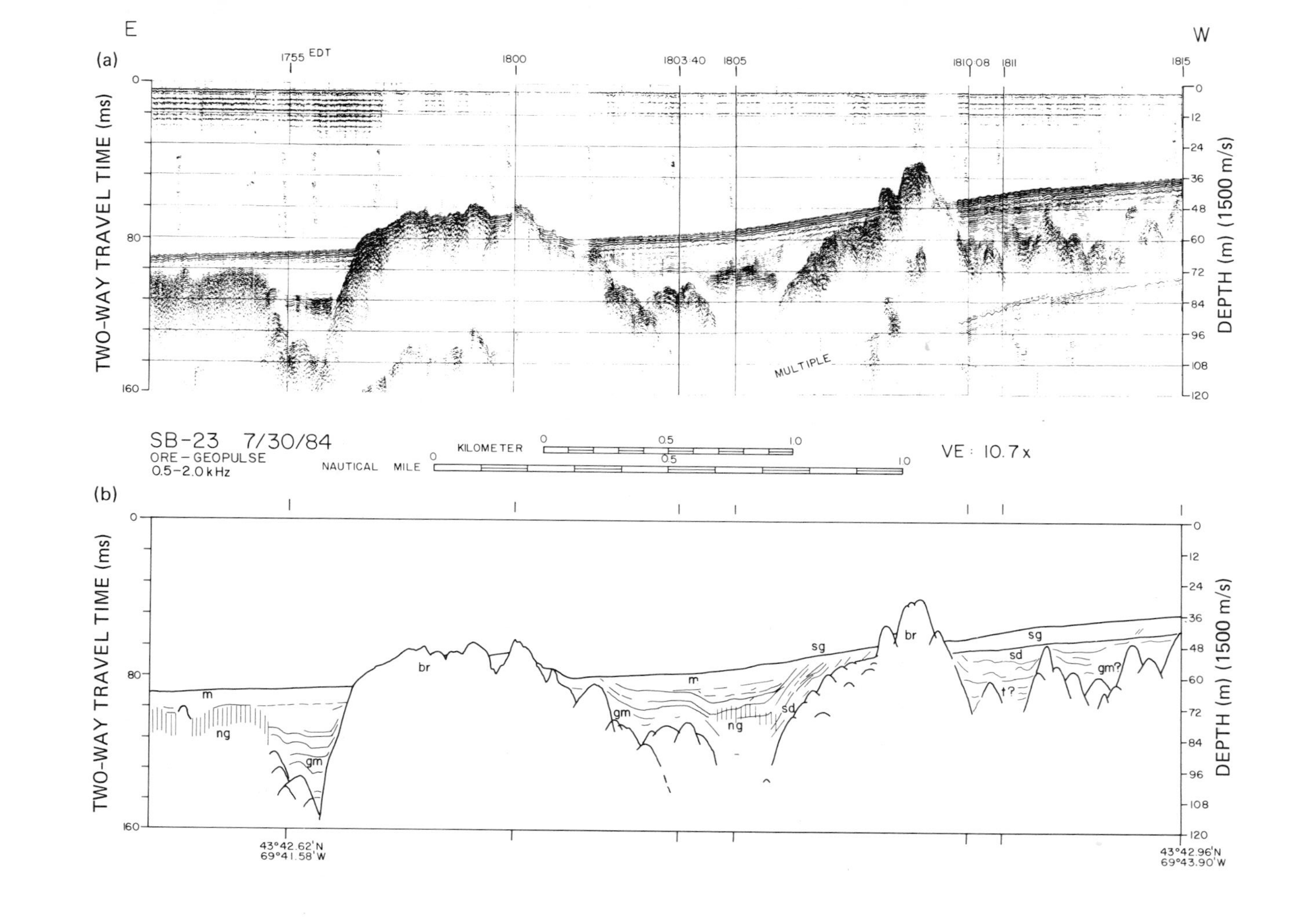
E
W
(a)
1755 EDT
1800
1803:40
1805
1810:08
1811
1815
TWO-WAY TRAVEL TIME (ms)
DEPTH (m) (1500 m/s)
MULTIPLE
SB-23 7/30/84
ORE - GEOPULSE
0.5-2.0 kHz
KILOMETER
NAUTICAL MILE
VE : 10.7x
(b)
m
ng
gm
br
sg
sd
gm?
t?
43°42.62'N
69°41.58'W
43°42.96'N
69°43.90'W

original data and a line-drawing interpretation. Some of the representative interpretive facies from Fig. 4 can be identified here. In comparison, Fig. 6 is an original record and interpretation of an ORE-Geopulse boomer record from line SB-23, taken in 1984. Although little of this data is available at this time, it can be seen that the greater penetration capability of the lower frequency, higher power boomer will allow better mapping of thick sand and gravel units. The 3.5-kHz system provides better resolution of shallow reflectors, however. Figures 5 and 6 differ from the interpreted line drawings of Figs. 7–10 in that they have not been digitized, rectified, compressed, or turned to a common orientation.

After classification and digitization, data were organized into a series of cross sections and longitudinal sections. Ultimately, isopach and structure contour maps will be constructed, but at this time only general seismic facies maps have been completed. For this discussion, representative cross and longitudinal sections from Sheepscot River and Bay, Casco Bay, and Machias Bay have been used to present a preliminary regional overview.

IV. Discussion

A. Sheepscot Estuary

The Sheepscot estuary lies within a glacially sculpted strike valley, where glacial flow was nearly parallel to bedrock strike. It is herein termed a strike-parallel embayment. The estuary is 30 km long, funnel-shaped, and very narrow, with widths ranging from 1 to 3 km. It is 80 m deep in the outer reaches, decreasing to a 10-m thalweg at its head. Freshwater input averages 25 m^3/s (Fefer and Schettig, 1980), and tidal range is 3.3 m at spring range. Waves are unimportant in the Sheepscot River because of the limited fetch, but dominate the shorelines of Sheepscot Bay.

Figures 7 and 8 are a series of ten interpretive line drawings of 3.5-kHz seismic profiles in the Sheepscot River estuary and the contiguous nearshore Gulf of Maine. Locations of these sections are shown in Fig. 3a. These cross sections show that the Sheepscot River portion of the estuary is confined to a single bedrock valley. Near its head the estuary is choked with till and glaciomarine mud to at least half the depth of the bedrock valley. Identification of this valley fill is confirmed by bridge borings (Upson and Spencer, 1964; Miller and Baker, 1982) for the Wiscasset Route 1 bridges and a corresponding 3.5-kHz profile taken 10 m south of the new bridge (Kelley and Belknap, 1987). Below the Eddy, a constriction 1 km south of Wiscasset, the river is narrowly confined by

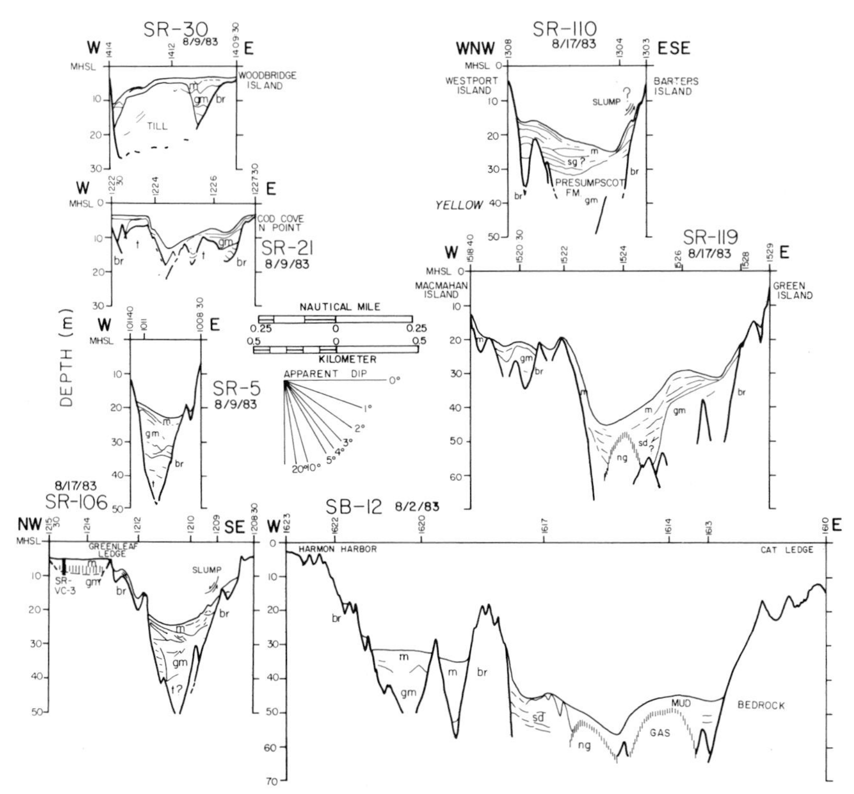

FIG. 7. Line drawings of 3.5-kHz seismic profiles in the Sheepscot River, rectified to a common scale and ×20 vertical exaggeration. Locations of the profiles are shown in Fig. 3a. These are depth sections, with an assumed water velocity of 1.5 km/s and a sediment velocity of 1.6 km/s. Profiles are cross sections, presented with the most headward in the upper left and most seaward in the lower right, aligned vertically over the thalweg in each section. Interpretations are explained in the text and in Fig. 4.

bedrock walls (Fig. 7; SR-S, SR-106, SR-110), and contains till, Presumpscot Formation, and Holocene mud. There are two locations where identified slumps are shown (Fig. 7; SR-106, SR-110). The Holocene fill has an overlapping hummocky appearance, possible due to multiple slumping events that have partially filled the paleovalley.

The lower Sheepscot River opens to include several glacial valleys partially filled with Presumpscot Formation and Holocene sediments, and intervening ridges. Vibracore SR-VC-3 penetrated Presumpscot Formation at 4.7 m depth in the sediment, in Greenleaf Cove on SR-110 (Fig. 7). Other axial and cross-section profiles (not shown) demonstrate the pres-

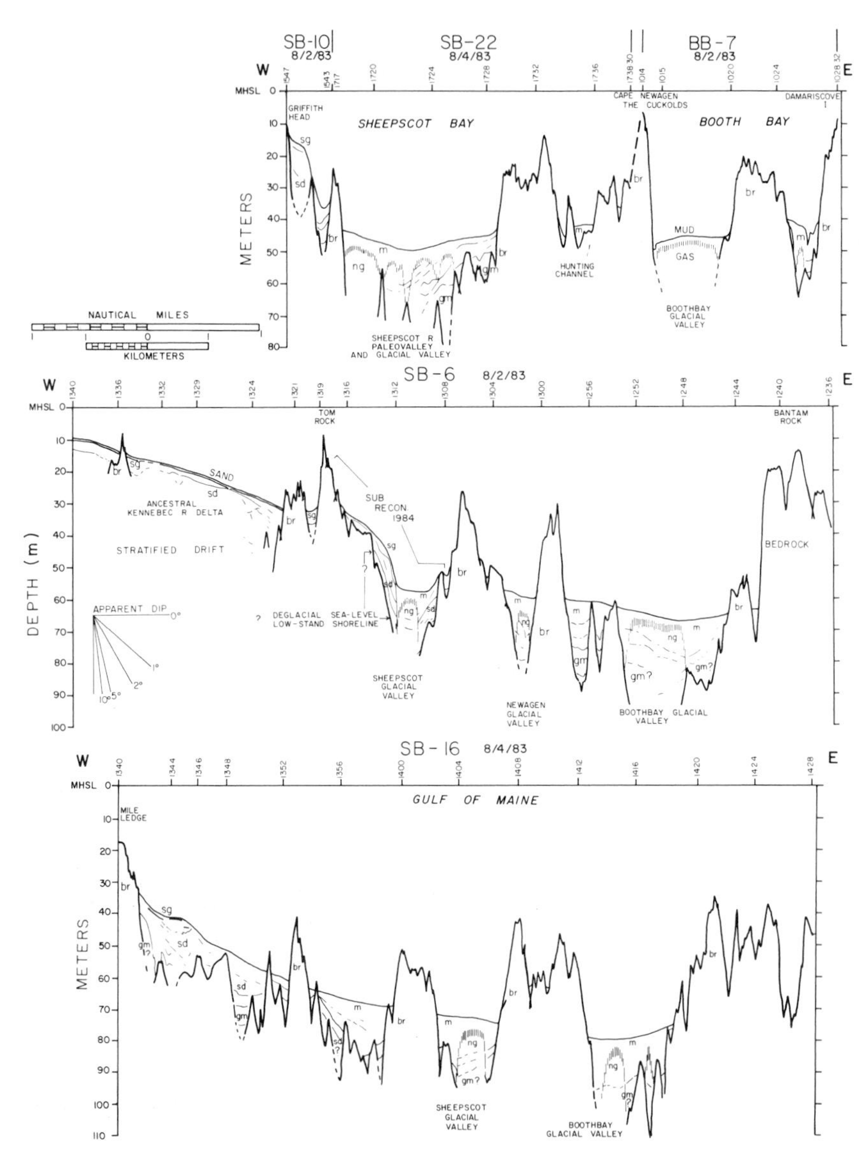

FIG. 8. Line drawings of 3.5-kHz seismic profiles in Sheepscot and Booth Bays and the nearshore Gulf of Maine. Locations are shown in Fig. 3a. Construction of these sections is as in Fig. 5, at ×50 vertical exaggeration.

ence of minor moraines within the valley, correlative with those found on land nearby (Thompson and Borns, 1985). Profiles SR-119 and SB-12 (Fig. 7) show the deeply scoured glacial valley with till, glaciomarine mud, Holocene mud, and natural gas (short vertical dashes on Figs. 7 and 8) confined to the deepest glacial valleys. This location for gas is typical in this estuary and may indicate a unique set of conditions early in deglacial time, with high sedimentation rates of organic-rich material, as shown by studies in marine sediments elsewhere (Claypool and Kaplan, 1974). Alternatively, these may represent low O_2 availability at the time of deposition, an equilibrium between gas production, pore pressure, and permeability where mud of low permeability is required to produce and trap the methane bubbles, or perhaps gas production from buried marshes. Elsewhere in the study area, natural gas is found in shallow mud flats and in filled channels cut in the Presumpscot Formation in inner Casco Bay (Fig. 9).

Sheepscot Bay, on the margin of the open Gulf of Maine, shows traces of the ancestral Sheepscot and Booth Bay estuaries. Figure 8 shows three cross sections normal to bedrock strike and valley trends. The deeper paleovalleys are similar to the present estuaries, but they have thicker mud sections and ridges that are completely barren of sediment. Natural gas is nearly ubiquitous in muds thicker than 15–20 m. A major new feature shown on these profiles is the ancestral Kennebec delta. It is composed of stratified sediments, possibly glaciomarine and glaciofluvial in nature. The upper 2 m is sand and gravel reworked from the stratified units. The delta shows channels, foreset beds, topset beds, and an outer margin apparently graded to a sea-level lowstand at least 50 m below present. Crossen (1983) and Smith (1982, 1985) have described glaciomarine deltas inland of this lithosome at 80 m above present sea level. These glaciomarine deltas at the marine limit within the state are peak relative sea-level deposits; they were never transgressed by the sea (Thompson, 1982). Since it is at least 100 m below the marine limit, the Kennebec feature requires an origin as a subaqueous outwash fan, or as a delta formed during the later stages of regression, lowstand, and early transgression. It is probably both: it may represent rapid subaqueous outwash deposition proximal to the ice sheet—ice shelf grounding line as the ice retreated to the northwest (Smith, 1985; Domack, 1982), followed by deposition in the subaerial deltaic phase during the late to postregression relative sea-level lowstand, when abundant glaciofluvial sediments and large volumes of meltwater were available in the Kennebec and Androscoggin drainage system (Borns and Hagar, 1965). This deltaic phase indicates a sea-level lowering of at least 50 m below present, based on interpreted topset–foreset contacts (Fig. 8, SB-6), and possibly 65 m below present based on the abrupt contact

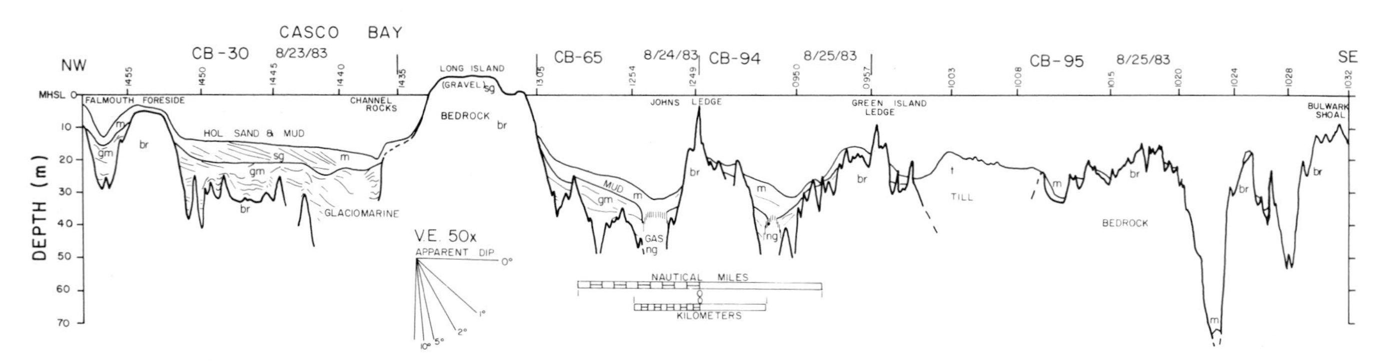

FIG. 9. Line drawing of 3.5-kHz seismic profiles in Casco Bay, combined into a composite axial section normal to bedrock strike, from inner to outer estuary. Location of the profiles is shown in Fig. 3b. Construction of the section is as in Fig. 5, at ×50 vertical exaggeration.

between steep, truncated coarse units and the more horizontal stratified valley fill (Fig. 8, SB-6 and SB-16).

Reworking of coarse sediments from this fan–delta complex during the Holocene transgression probably supplied the sediments for Popham Beach and Reid State Park Beach as well as nearby sandy tidal flats, such as Sagadahoc Bay, described by Bradley (1957). These sandy areas are unique in the central coast peninsulas; elsewhere bare rock promontories and muddy tidal flats are the norm. They may be related to the drainage of the Kennebec, first as an esker-fed (Borns and Hagar, 1965) subaqueous fan, then as a glaciofluvial delta, and later as reworked littoral sediments (Nelson and Fink, 1978).

The overall postglacial stratigraphy of the Sheepscot strike-parallel embayment first indicates removal of Pleistocene sediments (if present) from bedrock ridges, possibly during emergence or during the subsequent Holocene transgression. Early in this phase organic-rich sediments accumulated rapidly in the paleoestuaries. During sea-level rise sediment was temporarily stored at the head of the estuary while the outer margins were stripped and their adjacent valleys were continuing to accumulate sediments. Today, in a climate of continuing sea-level rise, marshes and tidal flats dominate the upper estuary, while the lower estuary is current-scoured and has slumping from its walls. The open Gulf may be continuing to slowly accumulate mud in bathymetric lows (from slumps and turbidity currents) but has tidal current and wave-washed rock ridges. The importance of a residual landward tidal current, noted by Schnitker (1974b), in moving Gulf sediments into the estuary needs further clarification. Waves and tidal currents also keep the ancient delta swept clear of mud. The transgression and evolution of the estuary fit well with our model of evolution of Maine's estuaries, discussed by Kelley (Chapter 6, this volume).

B. Casco Bay

Casco Bay, southwest of the Sheepscot system, is a complex of narrow bays and islands, strongly controlled by bedrock structure. It was sculpted by ice moving at a high angle (70–90°) to the structural grain. It can be divided into an inner, middle, and outer zone, along a profile normal to the bedrock structure. Thus, Casco Bay is herein termed a strike-normal embayment. Freshwater input is low (40 m^3/s), but the Presumpscot and Royal rivers are probably the muddiest in the state, since they drain thick Presumpscot Formation terrains. Waves are considerably damped in the middle and inner estuary due to baffling effects of the numerous islands, but they are the dominant modifying agent on the shorelines of the outer islands.

A sample of the interpreted seismic data for Casco Bay is shown in Fig. 9, a composite section normal to bedrock strike from inner to outer bay. The locations of these profile lines are shown in Fig. 3b. As in the Sheepscot, Casco Bay is bare of sediment over bathymetric highs in the outer portions, but shows no major accumulation of mud in bathymetric lows. The middle bay shows thicker till, Presumpscot Formation, and some Holocene mud. There is also evidence of current scour maintaining the deep channels. These channels are cut to −60 to −70 m, supporting the postulated sea-level lowstand at this level. The inner bay is an accumulation zone, with thick Holocene muds, drowned stratified sands (suggestive of barriers or tombolos), and a Pleistocene section that has been only slightly reworked. Thick till and Presumpscot Formation muds fill the ancestral Fore River and Cousins River valleys, and line CB-1 may cross the trend of the ancestral Androscoggin River.

As shown in the estuarine evolution model (Kelley, Chapter 6, this volume), sediment-accumulation zones in Casco Bay have moved landward with sea-level rise, but unlike Sheepscot Bay there is no thick accumulation of sediment remaining in the outer bay. The lack of a single major river valley and the strike-normal orientation with numerous coast parallel crossing ridges may explain this difference. Kelley *et al.* (1984, 1986) discuss Casco Bay in greater detail.

C. Machias Bay

Machias Bay is the last example we will consider. Figure 3c is the cruise track map for Machias. Machias Bay is in the east–central geomorphic zone and is traversed by the major Pond Ridge recessional moraine (Fig. 1). Machias Bay has a higher tidal range (4.4 m spring), a freshwater input of 25 m^3/s from the Machias River (Fefer and Schettig, 1980), and is within the apparent coastal downwarping zone of Tyler and Ladd (1980). The bay is rectangular, controlled more by glacial scour and deposition than by differential erosion; it is herein termed an equant embayment. Due to its more open exposure, waves and tides are codominant.

The interpreted axial profile of Fig. 10 shows the overwhelming dominance of glacial drift on this estuary. Three distinct moraines separated the basin at different times, with glaciomarine mud and stratified outwash deposited seaward of the ice grounding line of each moraine. Shipp *et al.*, (1984) have tracked the retreat of the ice from outer to inner bay, along a probable calving embayment. Major infill of Machias Bay occured when marine-based ice stranded at the midpoint of the bay, producing the Pond Ridge moraine (now reworked into the Sprague Neck spit), and accumulated Presumpscot Formation glaciomarine muds along its margins. A

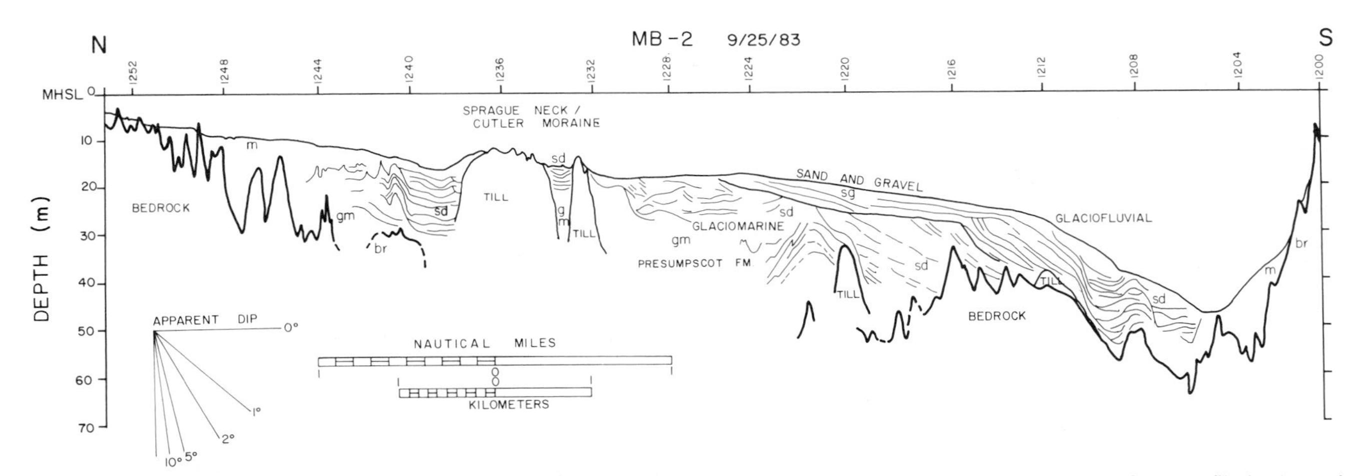

Fig. 10. Line drawing of 3.5-kHz seismic profile in Machias Bay, along the main axis of the estuary. Location of the profile is shown in Fig. 3c. Construction of the section is as in Fig. 5, at $\times 50$ vertical exaggeration.

phase of subaqueous outwash accumulation (similar to that described previously for the Kennebec River system) interfingers with and overlies the glaciomarine mud. As sea level fell to its lowstand position, a prograding glaciofluvial delta covered the earlier subaqueous fan with foreset and topset beds, reaching a position within the present mouth of Machias Bay, at a sea level at least 30 m and possibly as much as 60 m below present. During the subsequent transgression, coastal processes reworked the upper sand and gravel units to a small degree, while bluff erosion around the bay margin supplied mud and coarser sediments to the estuary. The middle and outer parts of the bay are apparently stable, because of the coarse sediments. At the leading edge of the transgression, Holmes Bay and the Machias River are accumulating Holocene mud and marsh deposits, in some locations over 5 m thick, from sediments available from bluff and stream erosion of Presumpscot Formation and till exposures.

V. Conclusions

High-resolution seismic profiling in Maine's estuaries and nearshore shelf have elucidated a Quaternary stratigraphy interpreted as consisting of the following (from latest to earliest):

(1) Holocene estuarine and marine sands, mud, and marsh.

(2) Erosional unconformity, subaerial exposure during sea-level lowstand, and isostatic rebound.

(3) Presumpscot Formation glaciomarine mud.

(4) Stratified outwash (interfingering with and over- and underlain by units 3 and 5).

(5) Till (lodgment, moraine, gravity-flow deposits).

(6) Glacially scoured erosional unconformity on bedrock.

Holocene evolution of the estuaries as determined from seismic stratigraphy follows our model of temporary storage of sediments in estuary heads with erosion of margins and outer bay headlands (Kelley, Chapter 6, this volume), in which the zones move landward with sea-level rise and marine transgression.

Sea level stood 60–70 m below present 8000–10,000 yr B.P. This is in contrast to a rheological model (Cathles, 1980) and speculative sea-level curve (Newman *et al.*, 1980) that suggest only 10 m of lowering at that time.

Three distinct estuarine types were identified: strike-normal embayment, strike-parallel embayment, and equant embayment. A fourth type, arcuate embayments found in southwestern Maine, is presently being investigated. Geologic controls on estuarine evolution in order of importance are

(1) bedrock structure,
(2) direction of glacial scour relative to structure,
(3) glacial deposition,
(4) postglacial erosive processes (subaerial and marine), and
(5) tides, waves, and local relative sea-level change rates.

Natural gas produces an acoustically dense signal over deep mud-filled valleys. It may be related to organic-rich, rapidly accumulated sediments from the headward estuarine zone. Slumps are found on steep slopes on estuary margins, and supply sediments to deeper valleys.

Maine estuaries differ from coastal-plain estuaries in that they are rock-structure–controlled, rather than based in unconsolidated sediments (see, e.g., Weil, 1977; Nelson, 1972), and have been directly affected by glacial erosion and deglacial deposition, rather than an alternation between subaerial and fluvial process with marine transgression (see, e.g., Swift, 1973). Maine's estuaries are similar to those in the Bay of Fundy (Amos, 1978) and Nova Scotia (Piper *et al.*, 1983), having shared a similar recent geologic history, and also resemble the glaciomarine seismic stratigraphy of Scotland (Boulton *et al.*, 1981; Davies *et al.*, 1984), but differ greatly from those of Labrador (Barrie and Piper, 1982), which have experienced continuous postglacial emergence. The model of postglacial evolution developed for the Maine coast has applications to rock-framed estuaries in midlatitudes worldwide. The importance of determining the local relative sea-level curve for understanding the stratigraphic relationships cannot be overemphasized.

Acknowledgments

This research was supported by the Maine Geological Survey under a contract to the U.S. Nuclear Regulatory Commission, and by the National Science Foundation EPSCOR program. Field and data-processing assistance by Michael Dunn, Bradley Bird, Stephanie Staples, and Jane Peplinski is gratefully acknowledged. We would also like to thank the Maine Department of Transportation for access to bridge boring records. This chapter was improved by reviews from Robert N. Oldale of the U.S. Geological Survey, Woods Hole, and other anonymous reviewers.

References

Amos, C. L. (1978). The post glacial evolution of the Minas Basin, N.S.: A sedimentological interpretation. *J. Sediment. Petrol.* **48,** 965–982.

Anderson, W. A., Kelley, J. T., Thompson, W. B., Borns, H. W., Jr., Sanger, D., Smith, D. C., Tyler, D. A., Anderson, R. S., Bridges, A. E., Crossen, K. J., and Ladd, J. W. (1984). Crustal warping in coastal Maine. *Geology* **12,** 677–680.

Barrie, C. Q., and Piper, D. J. W. (1982). Late Quaternary marine geology of Makkovik Bay, Labrador. *Geol. Surv. Can. Pap.* **81-17.**

Belknap, D. F., Shipp, R. C., and Kelley, J. T. (1986). Depositional setting and Quaternary Stratigraphy of the Sheepscot Estuary, Maine. *Geogr. Phys. Quat.* **40,** 55–69.

Belknap, D. F., Kelley, J. T., Borns, H. W., Jr., Shipp, R. C., Jaçobson, G. L., and Stuckenrath, R., Jr. (1987). Sea-level change curves for coastal Maine. *In* "New England Seismotectonic Study Activities During Fiscal Year 1983" (W. B. Thompson and J. T. Kelley, eds.). Maine Geol. Surv., Augusta.

Birch, F. S. (1984). Seismic sedimentary units of the inner continental shelf of New Hampshire. *Geol. Soc. Am. Abstr. Program* **16,** 3.

Bloom, A. L. (1960). "Late Pleistocene Changes of Sealevel in Southwestern Maine." Maine Geol. Surv., Augusta.

Bloom, A. L. (1963). Late Pleistocene fluctuation of sealevel and postglacial crustal rebound in coastal Maine. *Am. J. Sci.* **261,** 862–879.

Borns, H. W., Jr. (1973). Late Wisconsin fluctuations of the Laurentide ice sheet in southern and eastern New England. *In* "The Wisconsinan Stage" (R. F. Black, R. P. Goldthwait, and H. B. Willman, eds.), *Geol. Soc. Am. Mem.* **136,** 37–45.

Borns, H. W., Jr., and Calkin, P. E. (1977). Quaternary glaciation, west central Maine. *Geol. Soc. Am. Bull.* **88,** 1773–1784.

Borns, J. W., Jr., and Hagar, D. J. (1965). Late-glacial stratigraphy of a northern part of the Kennebec River valley, western Maine. *Geol. Soc. Am. Bull.* **76,** 1233–1250.

Boulton, G. S., Chroston, P. N., and Jarvis, J. (1981). A marine seismic study of late Quaternary sedimentation and inferred glacier fluctuations along western Inverness-shire, Scotland. *Boreas* **10,** 39–51.

Bradley, W. H. (1957). Physical and ecologic features of the Sagadahoc Bay tidal flat, Georgetown, Maine. *In* "Treatise on Marine Ecology and Paleoecology" (H. S. Ladd, ed.), *Geol. Soc. Am. Mem.* **67,** 641–682.

Burbank, W. S. (1929). The petrology of the sediment of the Gulf of Maine and the Bay of Fundy. *U.S. Geol. Surv. Open-File Rep.*

Cathles, L. M. (1980). Interpretation of postglacial isostatic adjustment phenomena in terms of mantle rheology. *In* "Earth Rheology, Isostasy and Eustacy" (N.-A. Morner, ed.), pp. 11–43. Wiley, New York.

Clark, J. A., Farrell, W. E., and Peltier, W. R. (1978). Global changes in postglacial sea level: A numerical calculation. *Quat. Res. (N.Y.)* **9,** 265–287.

Claypool, G. E., and Kaplan, I. R. (1974). The origin and distribution of methane in marine sediments. *In* "Natural Gases in Marine Sediments" (I. R. Kaplan, ed.), pp. 99–139. Plenum, New York.

Crossen, K. J. (1983). Glaciomarine delta and moraines, Sebago Lake region, Maine. *Geol. Soc. Maine Guideb. Fieldtrip* **12,** July 31.

Davies, H. C., Dobson, M. R., and Whittington, R. J. (1984). A revised seismic stratigraphy for Quaternary deposits on the inner continental shelf west of Scotland between 55°30′N and 57°30′N. *Boreas* **13,** 49–66.

Domack, E. W. (1982). Sedimentology of glacial and glacial marine deposits on the George V–Adelie continental shelf, East Antarctica. *Boreas* **11,** 79–97.

Emery, K. O., and Uchupi, E. (1972). Western North Atlantic Ocean: Topography, structure, water, life, and sediments. *Am. Assoc. Pet. Geol. Mem.* **17.**

Fefer, S. I. and Schettig, P. A. (1980). An ecological characterization of coastal Maine. *U.S. Fish Wildl. Serv. Rep.* **FWS/OBS-80/29,** 6 vols.

Flint, R. F. (1971). "Glacial and Quaternary Geology." Wiley, New York.

Folger, D. W., O'Hara, C. J., and Robb, J. M. (1975). Maps showing bottom sediments on the continental shelf of the northeastern United States—Cape Ann, Massachusetts to Casco Bay, Maine. *U.S. Geol. Surv. Misc. Invest. Ser.* Map 1-839, 1:125,000.

Goldthwait, L. G. (1949). Clay survey—1948. *In* "Report of the State Geologists, 1947–1948," pp. 63–69. Maine Dev. Comm., Augusta.

Goldthwait, L. G. (1951). The glacial–marine clays of the Portland–Sebago Lake region, Maine. *In* "Report of the State Geologists, 1949–50," pp. 24–34. Maine Dev. Comm., Augusta.

Hitchcock, C. H. (1873). The geology of Portland. *Am. Assoc. Adv. Sci. Proc.* **22,** Part 2, 163–175.

Hughes, T. J. (1981). Numerical reconstruction of paleo-ice sheets. *In* "The Last Great Ice Sheets" (G. H. Denton and T. J. Hughes, eds.), pp. 222–261. Wiley, New York.

Hussey, A. M., II (1959). Age of intertidal tree stumps at Wells Beach and Kennebunk Beach, Maine. *J. Sediment. Petrol.* **29,** 464–465.

Hussey, A. M., II (1970). Observations on the origin and development of the Wells Beach area, Maine. *Maine Geol. Surv. Bull.* **23,** 58–68.

Jackson, C. T. (1837). "First Report on the Geology of the State of Maine." Augusta, Maine.

Kaplan, I. R., ed. (1974). "Natural Gases in Marine Sediments." Plenum, New York.

Keen, M. J., and Piper, D. J. W. (1976). Kelp, methane, and an impenetrable reflector in a temperate bay. *Can. J. Earth Sci.* **13,** 312–318.

Kelley, J. T., and Belknap, D. F. (1987). Seismic profiling search for offshore Quaternary faulting. *In* "New England Seismotectonic Study Activities During Fiscal Year 1983" (W. B. Thompson and J. T. Kelley, eds.). Maine Geol. Surv., Augusta.

Kelley, J. T., and Timson, B. S. (1983). An environmental inventory and statistical evaluation of the Maine coast. *Geol. Soc. Am. Abstr. Program* **15,** 193.

Kelley, J. T., Belknap, D. F., and Shipp, R. C. (1984). The marine geology of Casco Bay, Maine: A seismic reflection investigation. *Geol. Soc. Am. Abstr. Program* **16,** 27.

Kelley, J. T., Kelley, A. R., Belknap, D. F., and Shipp, R. C. (1986). Variability in the evolution of two adjacent bedrock-framed estuarines in Maine. *In* "Estuarine Variability" (D. A. Wolfe, ed.), pp. 21–42. Academic Press, Orlando, Florida.

Kellogg, D. C. (1982). Environmental factors in archaeological site location for the Boothbay, Maine region with an assessment of the impact of coastal erosion on the archaeological record. M. S. Thesis, Inst. Quat. Stud., Univ. of Maine, Orono.

King, L. H. (1969). Submarine end moraine and associated deposits on the Scotian Shelf. *Geol. Soc. Am. Bull.* **80,** 83–96.

Knebel, H. J., and Scanlon, K. M. (1985). Sedimentary framework of Penobscot Bay, Maine. *Mar. Geol.* **65,** 305–324.

Knebel, H. J. (1986). Holocene depositional history of a large glaciated estuarine, Penobscot Bay. *Mar. Geol.* **73,** 215–236.

Knott, S. T., and Hoskins, S. (1968). Evidence of Pleistocene events in the structure of the continental shelf off the northeastern United States. *Mar. Geol.* **6,** 5–26.

Leavitt, H. W., and Perkins, E. W. (1934). A survey of road materials and glacial geology of Maine, V. I, pt. 1—A survey of road materials of Maine, their occurrence and quality. *Maine Technol. Exp. Stn. Bull.* **30.**

Leavitt, H. W., and Perkins, E. H. (1935). A survey of road materials and glacial geology of Maine, V. II, Glacial geology of Maine. *Maine Technol. Exp. Stn. Bull.* **30.**

Mayewski, P. A., Denton, G. H., and Hughes, T. J. (1981). Late Wisconsin ice sheets of North America. *In* "The Last Great Ice Sheets" (G. H. Denton and T. J. Hughes, eds.), pp. 67–178. Wiley, New York.

Miller, D. B., and Baker, G. L. (1982). "The Wiscasset Route 1 Bridge." Maine Dep. Transp., Bangor.

Murray, H. W. (1947). Topography of the Gulf of Maine. *Geol. Soc. Am. Bull.* **58,** 153–196.

Nelson, B. W., ed. (1972). The environmental framework of coastal plain estuaries. *Geol. Soc. Am. Mem.* **133.**

Nelson, B. W., and Fink, L. K., Jr. (1978). "Geological and Botanical Features of Sand Beach Systems in Maine," Planning Rep. 54. Maine Crit. Areas Program, Maine State Planning Off., Augusta.

Newman, W. S., Cinquemani, L. J., Pardi, R. R., and Marcus, L. F. (1980). Holocene deleveling of the United States' east coast. *In* "Earth Rheology, Isostasy and Eustasy" (N. A. Morner, ed.), pp. 449–463. Wiley, New York.

Novak, I., Swanson, M., and Pollock, S. (1984). Morphology and structure of the Gorham, Me. landslide. *Geol. Soc. Am. Abstr. Program* **16,** 53–54.

Oldale, R. N., and Uchupi, E. (1970). The glaciated shelf off northeastern United States. *In* "Geological Survey Research, 1970," *U.S. Geol. Surv. Prof. Pap.* **700-B,** B167–B173.

Oldale, R. N., Uchupi, E., and Prada, K. E. (1973). Sedimentary framework of the western Gulf of Maine and the southeastern Massachusetts offshore area: Sedimentary framework. *U.S. Geol. Surv. Prof. Pap.* **757.**

Oldale, R. N., Wommack, L. E., and Whitney, A. B. (1983). Evidence for postglacial low relative sea-level stand in the drowned delta of the Merrimack River, western Gulf of Maine. *Quat. Res. (N.Y.)* **33,** 325–336.

Osberg, P. H., Hussey, A. M., II, and Boone, G. M. (1984). Bedrock geologic map of Maine. *Maine Geol. Surv. Open-File Rep.* **84-1,** 1:500,000.

Ostericher, C. (1965). Bottom and subbottom investigations of Penobscot Bay, Maine, 1959. *U.S. Nav. Oceanogr. Off. Tech. Rep.* **173.**

Payton, C. E. (1977). Seismic stratigraphy—Applications to hydrocarbon exploration. *Am. Assoc. Pet. Geol. Mem.* **26.**

Piper, D. J. W., Letson, J. R. J., Delure, A. M., and Barrie, C. Q. (1983). Sediment accumulation in low sedimentation, wave dominated, glaciated inlets. *Sediment. Geol.* **36,** 195–215.

Pratt, R. M., and Schlee, J. (1969). Glaciation on the continental margin off New England. *Geol. Soc. Am. Bull.* **80,** 2335–2342.

Quinlan, G., and Beaumont, C. (1981). A comparison of observed and theoretical postglacial relative sea level in Atlantic Canada. *Can. J. Earth Sci.* **18,** 1146–1163.

Schlee, J., and Pratt, R. M. (1970). Atlantic continental shelf and slope of the United States—gravels of the northeastern part. *U.S. Geol. Surv. Prof. Pap.* **529-H.**

Schnitker, D. (1972). History of sedimentation in Montsweag Bay. *Maine Geol. Surv. Bull.* **25.**

Schnitker, D. (1974a). Postglacial emergence of the Gulf of Maine. *Geol. Soc. Am. Bull.* **85,** 491–4494.

Schnitker, D. (1974b). Supply and exchange of sediments in rivers, estuaries, and the Gulf of Maine. *Mem. Inst. Geol. Bassin Acquitane* **7,** 81–86.

Schubel, J. R. (1974). Gas bubbles and the acoustically inpenetrable, or turbid, character of some estuarine sediments. *In* "Natural Gases in Marine Sediments" (I. R. Kaplan, ed.), pp. 275–298. Plenum, New York.

Scott, D. B., and Greenberg, D. A. (1983). Relative sea-level rise and tidal development in the Fundy tidal system. *Can. J. Earth Sci.* **20,** 1554–1564.

Shipp, R. C. (1987). Late Quaternary sea-level fluctuations and geologic evolution of four embayments along the northwestern Gulf of Maine. Ph.D. Thesis, Oceangr. Program, Univ. of Maine, Orono.

Shipp, R. C., Belknap, D. F., and Kelley, J. T. (1984). Inshore seismic stratigraphy along northeastern coastal Maine: Examples from Gouldsboro Bay and Machias Bay. *Geol. Soc. Am. Abstr. Program.* **16,** 63.

Shipp, R. C., Staples, S. A., and Adey, W. A. (1985). Geomorphic trends in a glaciated coastal bay: A model for the Maine coast. *Smithson. Contrib. Mar. Sci.* No. 25.

Sieck, H. C., and Self, G. W. (1977). Analysis of high resolution seismic data. *In* "Seismic Stratigraphy—Applications to Hydrocarbon Exploration," *Am. Assoc. Pet. Geol. Mem.* **26,** pp. 353–385.

Smith, G. W. (1982). End moraines and the pattern of last ice retreat from central and south coastal Maine. *In* "Late Wisconsinan Glaciation of New England" (G. J. Larson and B. D. Stone, eds.), pp. 195–209. Kendall/Hunt, Dubuque, Iowa.

Smith, G. W. (1985). Chronology of late Wisconsinan deglaciation of coastal Maine. *In* "Late Pleistocene History of Northeastern New England and Adjacent Quebec" (H. W. Borns, Jr., P. LaSalle, and W. B. Thompson, eds.). *Geol. Soc. Am. Spec. Pap.* **197,** pp. 29–44.

Stone, G. H. (1899). The glacial gravels of Maine and their associated deposits. *U.S. Geol. Surv. Monogr.* **34.**

Stuiver, M., and Borns, H. W., Jr. (1975). Late Quaternary marine invasion in Maine: its chronology and associated crustal movement. *Geol. Soc. Am. Bull.* **86,** 99–104.

Swift, D. J. P. (1973). Delaware Shelf Valley: Estuary retreat path not drowned river valley. *Geol. Soc. Am. Bull.* **84,** 2743–2748.

Thompson, S. N. (1973). Sea-level rise along the Maine coast during the last 3,000 years. M.S. Thesis, Dep. Geol. Sci., Univ. of Maine, Orono.

Thompson, W. B. (1979). "Surficial Geology Handbook for Coastal Maine." Maine Geol. Surv., Augusta.

Thompson, W. B. (1982). Recession of the late Wisconsinan ice sheet in coastal Maine. *In* "Late Wisconsinan Glaciation of New England" (G. J. Larson and B. D. Stone, eds.), pp. 211–228. Kendall/Hunt, Dubuque, Iowa.

Thompson, W. B., and Borns, H. W., Jr. (1985). Surficial geologic map of Maine. *Maine Geol. Surv.* 1:500,000.

Thompson, W. B., Crossen, K. J., Borns, H. W., Jr., and Andersen, B. G. (1983). Glacial–marine deltas and late Pleistocene–Holocene crustal movements in southern Maine. *In* "New England Seismotectonic Study Activities During Fiscal Year 1982" (W. B. Thompson and J. T. Kelley, eds.), pp. 153–171. Maine Geol. Surv., Augusta.

Timson, B. S. (1978). New carbon-14 dates. *Maine Geol.* **4,** No. 3.

Tucholke, B. E., and Hollister, C. D. (1973). Late Wisconsin Glaciation of the southwestern Gulf of Maine: New evidence from the marine environment. *Geol. Soc. Am. Bull.* **84,** 3279–3296.

Tyler, D. A., and Ladd, J. (1980). Vertical crustal movement in Maine. *In* "New England Seismotectonic Study Activities in Maine, 1980" (W. B. Thompson, ed.), pp. 99–154. Maine Geol. Surv., Augusta.

Upson, J. E., and Spencer, C. W. (1964). Bedrock valleys of the New England coast as related to fluctuations of sea level. *U.S. Geol. Surv. Prof. Pap.* **454-M,** M1–M41.

Vilks, G., Rashid, M. A., and Van der Linden, W. J. M. (1974). Methane in recent sediments of the Labrador shelf. *Can. J. Earth Sci.* **11,** 1427–1434.

Weil, C. F. (1977). Sediments, structural framework and evolution of Delaware Bay, a transgressive estuarine delta. *Del. Sea Grant Tech. Rep.* **DEL-SG-4-77.**

Willis, B. (1903). Ames Knob, North Haven, Maine. *Geol. Soc. Am. Bull.* **14,** 201–206.

CHAPTER 8

Controls and Zonation of Geomorphology along a Glaciated Coast, Gouldsboro Bay, Maine

R. Craig Shipp

Oceanography Program
University of Maine
Orono, Maine

Stephanie A. Staples

Department of Geological Sciences
University of Maine
Orono, Maine

Larry G. Ward

Center for Environmental and Estuarine Studies
Horn Point Environmental Laboratories
University of Maryland
Cambridge, Maryland

The coastal geomorphology along the glaciated shoreline of Gouldsboro Bay, Maine, was defined to assess both the variation and controls of geomorphic form. Seven geomorphic classes are distinguished by unique combinations of sediment source, sediment texture, overall morphology, and/or dominant biota. Pocket beach, fringing beach, and marsh occur in the high intertidal area above mean tide level (MTL), while mixed flat, mud flat, and mussel bar occur in the low intertidal area below MTL. In exposed settings, rock ledge is either confined to the low intertidal area or persists throughout the entire intertidal profile. In more protected sites, rock ledge is restricted to the high intertidal area.

In decreasing spatial scale, the three controls of coastal geomorphic distribution are (1) the framework of Paleozoic bedrock that influences the overall shoreline morphology (regional scale of 10–100 km); (2) the distribution of glacial sediment, which determines abundance and texture of material in one or several embayments (intermediate scale of 1–10 km); and (3) present-day coastal processes, affecting a specific shoreline site or contiguous sites (local scale of < 1 km). Based on the interaction of these controls, Gouldsboro Bay displays three distinct zones of intertidal coastal geomorphology. These zones are (1) an inner (landward) zone composed of mud flats, mussel bars, and marshes; (2) a central zone, characterized by mixed flats and fringing beaches; and (3) an outer (seaward) zone, dominated by rock ledges and pocket beaches. This tripartite zonation of coastal geomorphology is consistent with observations within other embayments along the Maine coast and in the Canadian Maritimes.

I. Introduction

Most of coastal Maine can be characterized as a glacially eroded terrain of Paleozoic bedrock, which is covered with a discontinuous sheet of late Pleistocene glacial sediment. Because of shoreline complexity and remoteness, the coastline of Maine has received comparatively little attention from coastal geologists until recently.

The first attempt to examine the coastal geomorphology of Maine was in D. W. Johnson's classic work, "The New England–Acadian Shoreline" (Johnson, 1925). Only recently has a systematic survey of the geomorphology of the entire coast been completed [Timson, 1977; Maine State Planning Office (MSPO), 1983]. Timson mapped the entire coastal area from the shallow subtidal to the supratidal area, using 53 geomorphic classes to describe all of the coastal environments. Due to limited ground-truth verification and overlap of closely related classes, these maps are only useful on a regional scale. More recently, Kellogg (1982), in a study of archaeological sites along the central Maine coast, identified six shoreline types. These types are defined by three relative energy levels (high, intermediate, and low) and the dominance of either wave or tidal processes. Due to his concern for archaeological site preservation, Kellogg's regional classification scheme emphasizes the supratidal area. In another study, using the maps produced by Timson (1977) and a much simplified classification scheme, Kelley (Chapter 6, this volume) determined the environmental setting for 1670 shore-normal traverses selected along the Maine coast. Applying cluster and Q-mode factor analyses, he showed that three end members (mud flat, marsh, and rock ledge) account for 82.8% of the coastal variation and are the principle shoreline elements of Maine's glaciated coastline.

These recent studies have addressed regional coastal variation. The purpose of this investigation is to identify the intertidal geomorphologic variation within one bay, Gouldsboro Bay. Using a descriptive classification scheme, the entire shoreline of the bay was assigned to one of seven major geomorphic classes. Analysis of the distribution of these classes identified the probable controlling factors as well as a distinct tripartite zonation within the bay.

II. Physical Setting

Gouldsboro Bay is located in eastern coastal Maine along the northwestern margin of the Gulf of Maine (Fig. 1). In addition to the main bay, there are three tributary bays peripheral to Gouldsboro Bay. Joy Bay, West Bay, and Grand Marsh Bay are shallow irregular extensions of the main bay, possessing less than 10% of the total water volume of the system. The mean tidal range at the entrance of bay is 3.2 m, with spring tides exceeding 4.0 m [National Oceanic and Atmospheric Administration (NOAA), 1981]. There is little variation in tidal range within the bay (Shipp, unpublished data). The prevailing winds are consistent over the entire Maine coast, generally blowing out of the northwest in the winter and south to southwest in the summer (Lautzenheiser, 1972). The dominant storm winds are from the northeast (occasionally the southeast) and are usually associated with extratropical cyclones.

The bedrock geology surrounding the bay consists primarily of mid-Paleozoic silicic intrusive rocks, whose relief has been modified by Quaternary glaciations. Most of the bedrock terrain is composed of granodiorites, granites, and quartz monzonites (Chapman, 1962). Scattered mafic intrusive rocks occur as dikes in the main bay and become more abundant and larger in the tributary bays. The overall topography is that of north–south trending bedrock valleys formed by repeated glaciations and stream erosion during the interglacials (Denny, 1982). These oriented valleys are imposed on a weak north–northeast to south–southwest bedrock strike.

Late Wisconsinan glacial sediments, up to several tens of meters thick, cover the area discontinuously (Bloom, 1963; Borns, 1973; Stuiver and Borns, 1975). Late Wisconsinan deglaciation occurred from about 13,800

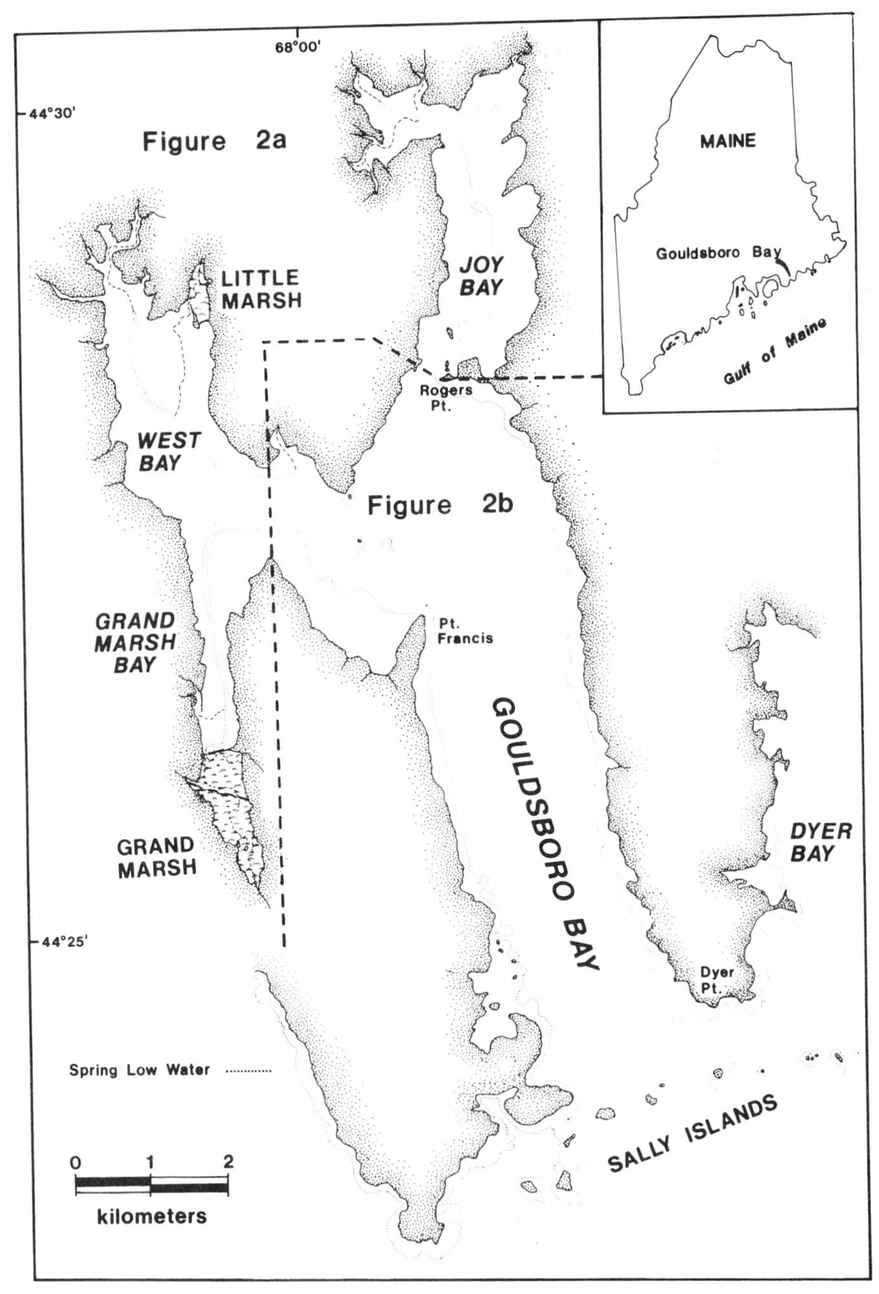

FIG. 1. Location map of Gouldsboro Bay, Maine. Divisions refer to geomorphic maps in Fig. 2.

to 12,500 yr B.P. During deglaciation, a rapid submergence of the present coastal area, which was caused by a rising sea level following the retreating ice, resulted in the deposition of a blanket of marine sediment over glacial deposits. This deposit, composed primarily of a cohesive blue–gray silt, is commonly found throughout coastal Maine and has been named the Presumpscot Formation (Bloom, 1960). As glacial ice retreated northwestward, rapid rebound of the coastal area resulted in reexposure and retreat of the shoreline to a position 10–20 km seaward of the present-day shoreline at an approximate depth of 55–70 m (see, e.g., Schnitker, 1974; Shipp, 1985; Belknap *et al.*, Chapter 7, this volume). Since that time (~9,500 yr B.P.), eustatic sea-level rise and a possible slow depression of the crust (Anderson *et al.*, 1984) have resulted in a general submergence of the coast and a marked "drowned topography."

Based initially on the early work of Jackson (1837), the coast of Maine was divided into four coastal compartments, primarily on the basis of bedrock lithology, bedrock structure, and Quaternary sediment supply. This concept has been further refined and subdivided by the recent works of Fefer and Schettig (1980), MSPO (1983), Shipp *et al.* (1985), Belknap *et al.* (Chapter 7, this volume), and Kelley (Chapter 6, this volume). Due to the small size, equant geometry, and diversity of coastal environments, Gouldsboro Bay is a typical example of the small bay subdivision in the island–bay complex (east–central compartment; see Fig. 1 in Belknap *et al.*, Chapter 7, this volume).

III. Methods

The technique used to characterize the coastal geomorphology in the study area was a systematic survey of geomorphic environments using methods discussed by Hayes *et al.* (1973). The methods applied to Gouldsboro Bay were the study of existing documentation (topographic and surficial sediment maps, nautical charts, and vertical and oblique aerial photographs), aerial reconnaissance, field mapping, and selection of representative profile stations to characterize each geomorphic class. Details of the technique are discussed in Shipp *et al.* (1985).

Using a descriptive classification scheme derived from the field survey of coastal geomorphology, a map was constructed that illustrates the distribution of geomorphic classes (Fig. 2). Field mapping was completed between June and November 1982. Five visits were made to the study area in the winters of 1981–1982 and 1982–1983 to observe winter processes. Grain-size distribution of sediment was determined by visual in-

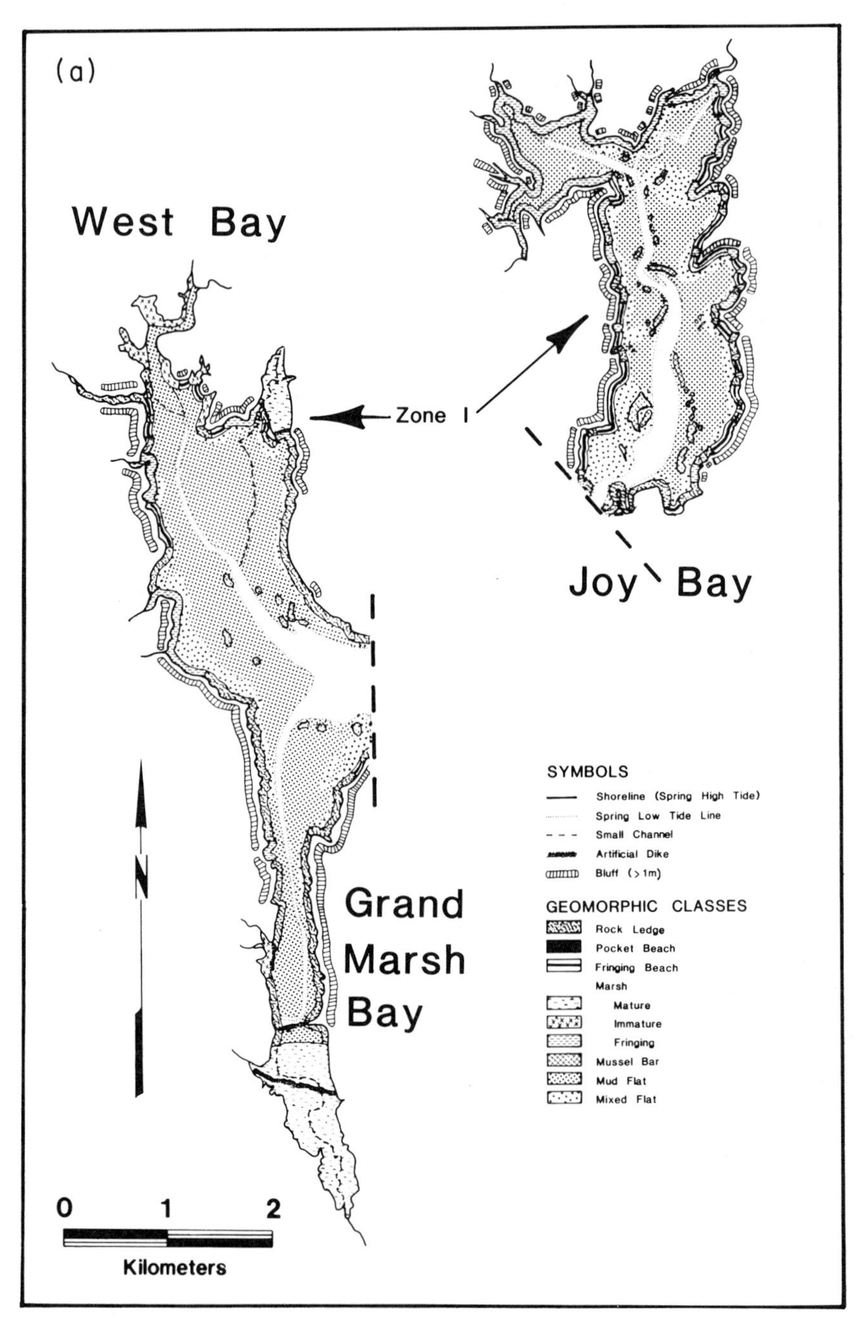

FIG. 2. Distribution of geomorphic classes in (a) zone I and (b) zone II and zone III. Legend in (a) is common to both maps. Location pointer marking coastal lagoon on the western side of zone III in (b) is explained in notes on Table III.

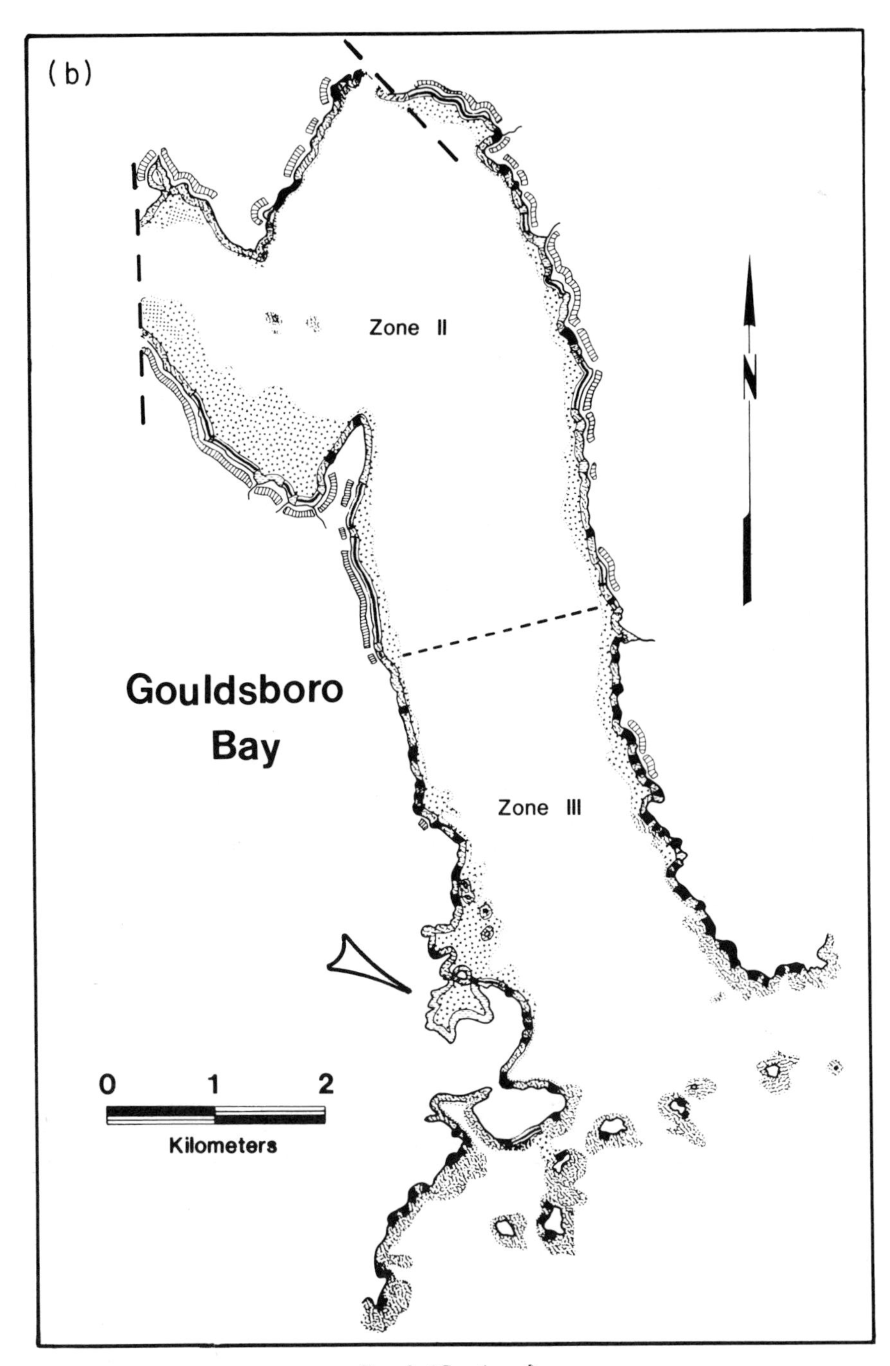

FIG. 2. (*Continued*)

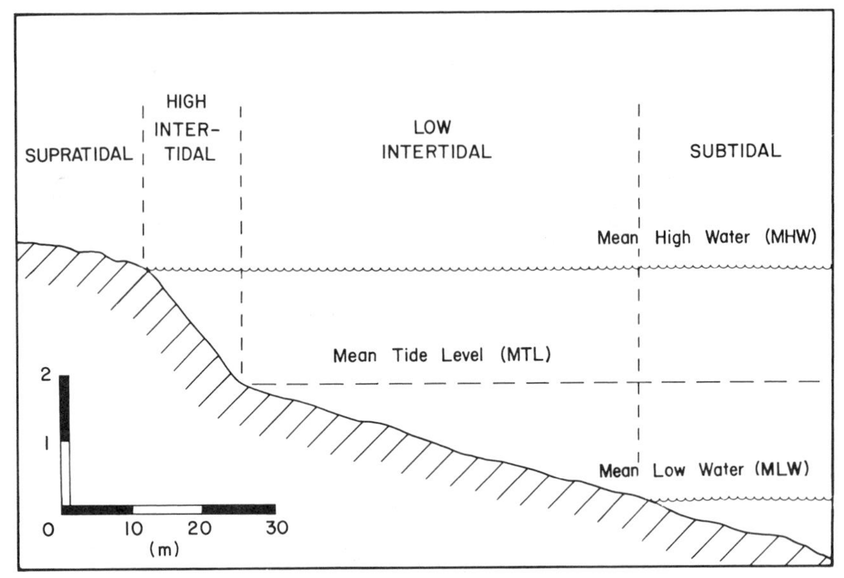

FIG. 3. Ideal shoreline profile, which illustrates the difference between the narrow high intertidal area above MTL and the broad low intertidal area below MTL.

spection of samples. Wave-energy level at specific sites was determined by assessment of fetch, adjacent subtidal depth, and indications along the shoreline (e.g., storm berms, erosional scour, and persistent vegetation). The three relative wave-energy levels used by Kellogg (1982) are retained. Distribution of geomorphic classes was determined by area measurements taken from a field survey map based on aerial photographs.

Within Gouldsboro Bay, intertidal geomorphic classes can be differentiated vertically within the intertidal area. Geomorphic classes common above MTL (high intertidal area) are characterized by steep (slopes $> 2°$) and narrow, shore-parallel bands. Geomorphic classes common below MTL (low intertidal area) are characterized by an extensive area of low relief that rarely exceeds 1.0° slope (Fig. 3).

The majority of shoreline profiles in Gouldsboro Bay have a distinct break in slope at MTL, which causes the geomorphic boundary. Exceptions occur along profiles exposed to extremes in wave energy. A typical profile across a shoreline exposed to continuous high wave energy exhibits a steep, rock ledge with no break in slope across the intertidal area. Conversely, a common profile along a low-wave-energy shoreline, such as a marsh, displays an expansive high intertidal area marked by an abrupt slope change to a steep-walled tidal channel in the low intertidal area.

IV. Geomorphic Classes

The coastal environments of Gouldsboro Bay have been divided into seven major classes, which fit into an overall framework of three geomorphic zones. The major classes are pocket beach, fringing beach, marsh, mixed flat, mud flat, mussel bar, and rock ledge. Pocket beach, fringing beach, and marsh are found only in the high intertidal area, while mixed flats, mud flats, and mussel bar are restricted to the low intertidal area. Rock ledge is either confined to the low intertidal area or persists throughout the entire intertidal profile in exposed settings, while in more protected sites rock ledge is restricted to the high intertidal area. Discrete combinations of classes form the basis for the geomorphic zones, which are exposed to different levels of wave energy. Figure 2 illustrates the distribution of environmental classes within the three geomorphic zones of Gouldsboro Bay. Table I summarizes the salient features of each class.

Pocket beaches found in Gouldsboro Bay are similar to those first described by Bascom (1964) for the U.S. West Coast. The major difference in the Gouldsboro features is their small size, which rarely exceeds 200 m in shore-parallel length. Less than 1% of the entire intertidal surface of the bay is classified as pocket beach (Table II). Most pocket beaches (47) are located in the high-wave-energy setting in the southern half of the main bay (zone III), while only a few (7) are found in the moderate-wave-energy setting in the northern half of the main bay (zone II) (Fig. 2). No pocket beaches are present in the low-wave-energy setting in the tributary bays of zone I.

Fringing beaches found in Gouldsboro Bay are analogous to continuous linear beaches described by Ward *et al.* (1980) for the outer Kenai Peninsula of Alaska. Fringing beaches have a longer shore-parallel extent (up to 1 km) than pocket beaches, and occur seaward of an unconsolidated bluff of glacial sediment. Approximately 1½% of the bay's intertidal surface is covered with fringing beach (Table II). Most are in the more protected zones I (14) or II (32) (Fig. 2). The cooccurrence of beach and bluff suggests that the sandy gravelly sediment in the fringing beach is supplied from an adjacent coastal bluff.

Marshes are assemblages of salt-tolerant vegetation above MTL. Three subclasses of salt marsh are evident in Gouldsboro Bay. Mature marsh covers a wide expanse near mean high water (MHW) and is dominated by *Spartina patens*, with *Juncus gerardi* and *Scirpus* spp. of secondary importance near the terrestrial upland border. Numerous salt pans cover this high marsh surface. Immature marsh is colonized primarily by *Spartina alterniflora*. Only the most landward edge of immature marsh has sufficient

TABLE I. CHARACTERISTICS OF INTERTIDAL GEOMORPHIC CLASSES

Geomorphic class/subclass	Characteristics
Pocket beach	Occurs above MTL
	Arcuate strandline
	Shore-parallel length < 200 m
	Composed of gravel, sand, or a mixture of both
	Sediment rounded to subrounded
	Sediment derived from adjacent headland
	Backed by storm berm or rock ledge in supratidal area
Fringing beach	Occurs above MTL
	Linear strandline
	Shore-parallel length 200–1000 m
	Composed of gravel (mostly boulders) and sand
	Sediment subrounded to subangular
	Sediment derived from adjacent bluffs in supratidal area
Marsh	Colonization of salt-tolerant vegetation above MTL
Mature	Distinct vegetation zonation
	Colonizes large, funnel-shaped embayments
	Narrow, meandering tidal channels with many tributaries
	Numerous salt pans
Immature	Weak vegetation zonation
	Colonizes small, narrow embayments
	Wide, straight tidal channels with few tributaries
	Few salt pans
Fringing	No vegetation zonation (generally single species only)
	Shore-parallel width < 10 m
	Lacks salt pans, tidal channels, or tributaries
Mixed flat	Occupies wide areas below MTL
	Composed of mud mixed with gravel and/or sand
	Abundant bedrock outcrops
Mud flat	Occupies wide areas below MTL
	Composed of predominately mud
	Few bedrock outcrops
Mussel bar	Occupies oval to linear patches near edge of tidal channel
	Long axis < 200 m
	Composed of silty mud and blue mussels *(Mytitus edulis)*
Rock ledge	Bare bedrock commonly covered with loose debris and/or glacial erratics
Exposed	Persists either throughout the entire intertidal area or only in low intertidal area below MTL
Protected	Found only in high intertidal area above MTL

elevation (near MHW) to be vegetated by *S. patens*. In addition, salt pans are rarely present. Fringing marsh is colonized by low-density bands of *S. alterniflora* that are located just above MTL and rarely exceed a width of 10 m. This is the only marsh subclass found in the more seaward zones II and III (excluding the anomalous coastal lagoon indicated on Fig. 2b).

TABLE II. AREAS AND PERCENTAGES OF GEOMORPHIC CLASSES IN GOULDSBORO BAY[a]

Class	Zone I	Zone II	Zone III	Class totals
Pocket beach	0	2.5	11.4	13.9
	(0)	(0.16)	(0.72)	(0.88)
Fringing beach	12.3	11.2	0.9	24.4
	(0.78)	(0.71)	(0.06)	(1.55)
Marsh	121.3	1.9	2.3	125.6
	(7.65)	(0.12)	(0.15)	(7.92)
Fringing	10.6	1.9	2.3	14.8
	(0.67)	(0.12)	(0.15)	(0.94)
Immature	36.1	0	0	36.1
	(2.28)	(0)	(0)	(2.28)
Mature	74.7	0	0	74.7
	(4.71)	(0)	(0)	(4.71)
Mixed flat	246.4	235.2	109.9	591.6
	(15.55)	(14.84)	(6.93)	(37.32)
Mud flat	566.6	11.2	1.7	579.5
	(35.75)	(0.71)	(0.11)	(36.57)
Rock ledge	30.2	15.0	161.7	206.9
	(1.91)	(0.95)	(10.20)	(13.06)
Exposed	0.9	3.4	152.8	157.1
	(0.06)	(0.22)	(9.64)	(9.92)
Protected	29.4	11.6	8.9	49.9
	(1.86)	(0.73)	(0.56)	(3.15)
Mussel bar	42.9	0	0	42.9
	(2.70)	(0)	(0)	(2.70)
Zone totals	1019.8	277.1	287.9	1584.8
	(64.34)	(17.49)	(18.17)	(100.00)

[a]Values on top are intertidal areas in hectares; values in parentheses are areal percentages of the entire intertidal surface within Gouldsboro Bay.

All subclasses of marsh combine to dominate the high intertidal area of zone I, particularly mature and immature marsh, which are indigenous. The two major stands of mature marsh are Grand Marsh in Grand Marsh Bay and Little Marsh in West Bay (Fig. 1). Most of the immature marsh is located in the northern part of either West Bay or Joy Bay (Fig. 2a).

Two classes of flat occupy nearly three-quarters of the intertidal area in Gouldsboro Bay (Fig. 2; Table II). Mixed flats, which are composed of a combination of mud, sand, and gravel, are the dominant geomorphic environment in the low intertidal area of the northern half of zone III and all of zone II. In the tributary bays of zone I, extensive flats composed of mud are dominant in the low intertidal area. These mud flats, particularly in West Bay, are frequently rimmed by mixed flats along the landward edge and support a large commercial harvest of the soft clam *Mya arenaria*.

The most restricted class identified in Gouldsboro Bay is mussel bar. Mussel bars are oval to linear patches of the blue mussel *Mytilus edulis*, which frequently occur near the edges of the subtidal channels. The distribution of mussel bars is confined exclusively to the tributary bays in zone I.

Rock ledge, a widely distributed geomorphic class, spans the entire vertical range in the intertidal area of Gouldsboro Bay (Fig. 2; Table II). Variations in the distribution of ledge are related to degree of exposure in the intertidal area. In the highest energy settings (primarily in the southern half of zone III), rock ledge is either confined to the low intertidal area seaward of pocket beaches or extends throughout the entire intertidal profile (MHW to MLW). In the remainder of the bay, protected rock ledge is present only in the high intertidal area above MTL; below MTL, rock ledge is covered by broad flats.

V. Controls of Coastal Geomorphology

There are three spatial scales of control that determine the distribution of coastal geomorphic environments in Gouldsboro Bay. In decreasing order, they are the framework of Paleozoic bedrock, distribution of glacial sediments, and present-day coastal processes. The bedrock framework influences the overall shoreline geometry within an entire coastal compartment (regional scale of 10–100 km). Glacial sediment distribution determines the abundance and texture of modern deposits in one or several adjacent embayments (intermediate scale of 1–10 km). Modern coastal processes affect one specific shoreline site or contiguous sites within a single embayment (local scale of < 1 km).

A. Bedrock Framework

The regional-scale shoreline geometry of Gouldsboro Bay is controlled by the bay's terrane of granodiorites, granites, and quartz monzonites. These massive rocks weather into convex-upward, hemispherical forms due to the specific dynamics of jointing and sheeting in silicic intrusive rocks (Chapman, 1958). It is this weathering process that has produced the equant shape of Gouldsboro Bay, as well as other embayments within the island–bay complex of the east–central coastal compartment.

Other coastal compartments along the Maine coast display a distinctive regional shoreline geometry, which is controlled by the specific dynamics of denudation for the bedrock lithology and structure within that compartment (Belknap *et al.*, Chapter 7, this volume; Kelley, Chapter 6, this volume). In fact, Denny (1982) maintains that bedrock exercises the primary control on the distribution of landforms over the entire New England landscape, while Quaternary erosion and deposition have only produced minor changes in drainage and topography.

B. Glacial Sediments

Variation in coastal geomorphology on the intermediate scale within Gouldsboro Bay is controlled by the distribution of glacial deposits. The dominant glacial deposits around the perimeter of the study area are recessional moraines (Fig. 4). Two major types of recessional moraines have been described in coastal Maine by Smith (1982a,b). The more numerous type is minor moraines that commonly occur in small clusters resembling washboards. Minor moraines are typically 3–4 m high, 12–18 m wide at the base, and less than 0.5 km long. The less abundant form is large moraines that generally occur individually or in irregularly spaced groups. Large moraines average 10–15 m high, 100–200 m wide at the base, and 0.5–1 km long. When either type of moraine intersects the shoreline, high bluffs (3–15 m) of poorly sorted, weakly stratified material are exposed.

In addition to the abundant bluffs from moraines, lower bluffs (< 3 m) are also present. These bluffs are composed of laminated to massive silt, interpreted as glaciomarine sediment of the Presumpscot Formation, or deposits of diamicton interpreted as glacial till (ground moraine). Regardless of the type of sediment, the various mechanisms of bluff erosion (Kelley and Kelley, 1986, p. 70) introduce bluff-derived sediment into the intertidal area for reworking into coastal geomorphic environments.

The intertidal shoreline in zones I and II is rimmed by near-continuous bluffs. Additionally, minor moraines cutting the coast are numerous in both these zones (at least 25 per zone). In contrast, the terrestrial surficial sediment cover around the perimeter of zone III is thin and has fewer moraines cutting across the coastline than either zone I or zone II (Fig. 4). The zone III shoreline is adjacent to an upland, which is mapped predominately as bedrock covered with undifferentiated, thin glacial material. The reason for the lack of sediment in this outer zone is attributed primarily to wave resuspension and erosion during early Holocene sea-level fluctuations. Only a limited number of minor moraines (11) cut across the coast in this zone and generally form eroding bluffs at the supratidal/high intertidal level. This pattern of glacial sediment distribution is not unique

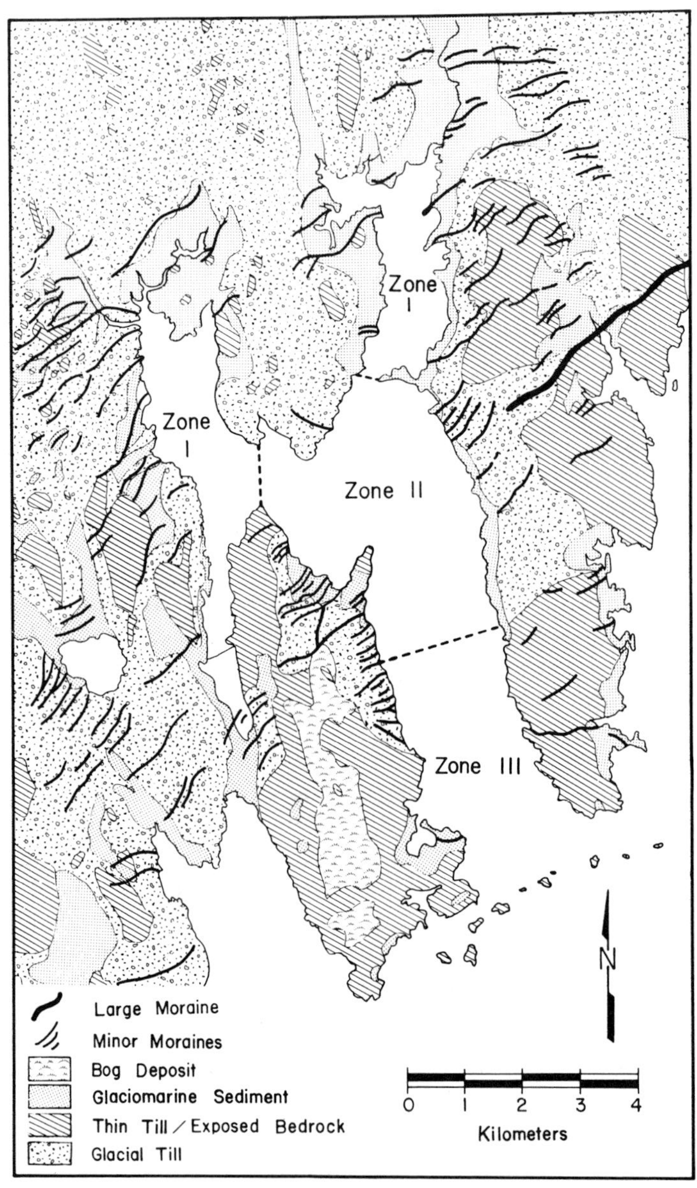

FIG. 4. Surficial geologic map of the Gouldsboro Bay region. Data sources used to construct this map are Borns (1974a,b), Borns and Andersen (1982a,b), Thompson and Borns (1985), aerial photogrammetry, and field survey.

to Gouldsboro Bay, but is observable at other sites along the coast and well illustrated on the surficial geologic map of Maine (Thompson and Borns, 1985).

The effect of the distribution of glacial deposits is to directly control the abundance and texture of sediment available to the intertidal area. In turn, this exerts control on the coastal geomorphology within one or several adjacent embayments. In Gouldsboro Bay, the development of numerous pocket beaches in zone III is enhanced by the paucity of sediment. Any available sediment (mostly gravel) in zone III is reworked by wave energy into pocket beaches along any slight reentrant in the exposed rock ledge. However, similar sites in zones I and II are commonly occupied by fringing beaches due to the large sediment supply provided by numerous bluffs backing the intertidal area. Fringing beaches are longer than pocket beaches, because of the abundant supply of sediment to fill in the irregularities of the underlying rock ledge. Another example of coastal geomorphic control affected by glacial sediment dispersal patterns is the distribution of rock ledge versus flats and marshes. Although wave reworking is a major influence (discussed below), the dominance of ledge and the scarcity of marsh in zone III is partially attributed to insufficient sediment supply, caused by lack of eroding supratidal bluffs.

C. Coastal Processes

Modern coastal processes are important agents in controlling the geomorphology on the local scale of a specific coastal site. Two important coastal processes observed in Gouldsboro Bay were variations in wave energy and the effects of winter ice.

The most important coastal process affecting geomorphology is sediment reworking and redistribution caused by variations in wave energy. The degree of wave exposure for any site along the shoreline is determined by shoreline orientation, fetch, and amount of bedrock outcropping in the low intertidal area. Orientation and fetch determine the intensity of wave energy reaching the intertidal shoreline. In contrast, the height and extent of outcropping ledge determine the degree of protection in the high intertidal area. Examples of wave-energy effects are the extensive rock ledges in the southern half of zone III (mentioned above) and presence of abundant marsh in zones I and II. Even considering the sediment supply differences among the different zones, much of zone III is stripped of sediment due to the greater exposure to wave attack. Only the coarser-sized material remains to be reworked into predominately gravel pocket beaches. With less exposure to wave energy, zones I and II contain less rock ledge (because it is only exposed above MTL) and more marsh than zone III. The distribution of ledge is similar between zones I and II, but the amount of marsh area increases as the wave exposure decreases toward the landward end of zone I (Fig. 2).

Winter ice movement has both erosive and protective processes associated with it. The effects of ice were only documented during two winter seasons, so other processes that have not been observed may be important. The erosive property of ice is most clearly seen in the marshes and flats of zones I and II. At low tide, ice slabs freeze to the intertidal surface. As the tide rises, the slab is lifted from the marsh or flat surface with sediment and/or vegetation frozen to the base of it. This material is then either resuspended in the water column or redeposited elsewhere. This process accounts for the gravel commonly found on the high marsh surface.

The protective effect of ice results from the stacking of ice on the eastern shore of the main bay by prevailing northwesterly winds in the winter. The accumulation of ice on the eastern shore protects many of the intertidal environments from further ice scraping and wave erosion. On the western shore, much less ice is deposited, so the entire intertidal area is exposed to reworking by the infrequent storm winds from the easterly quadrant. This effect is illustrated as a seasonal wind diagram in Fig. 5. An example of the effect of this process is observed in the unequal distribution of fringing marsh between the eastern and western shoreline of the main bay (Fig. 2). Fringing marsh is more common in the eastern margin, where it is protected by a thicker build-up of winter ice. On the western side, fringing marsh is only found in well-protected coves. Protection of marsh by ice has also been observed along the edge of the Minas Basin in the Bay of Fundy (Knight and Dalrymple, 1976).

VI. Geomorphic Zonation

Investigation of the coastal geomphology in Gouldsboro Bay has revealed three distinct geomorphic zones: (1) an inner (landward) zone I in the tributary bays, (2) a central zone II in the northern half of the main bay, and (3) an outer (seaward) zone III in the southern half of the main bay (Fig. 2).

The most common geomorphic classes in the inner zone I are mud flat (55.6%), mixed flat (24.2%), total marsh (11.8%), and fringing beach (11.8%) (Table III). The diagnostic attributes of this zone are the vast coverage of mud flat and presence of three indigenous environments: mussel bar, mature marsh, and immature marsh. Another important characteristic, which also occurs in zone II, is the presence of rock ledge in

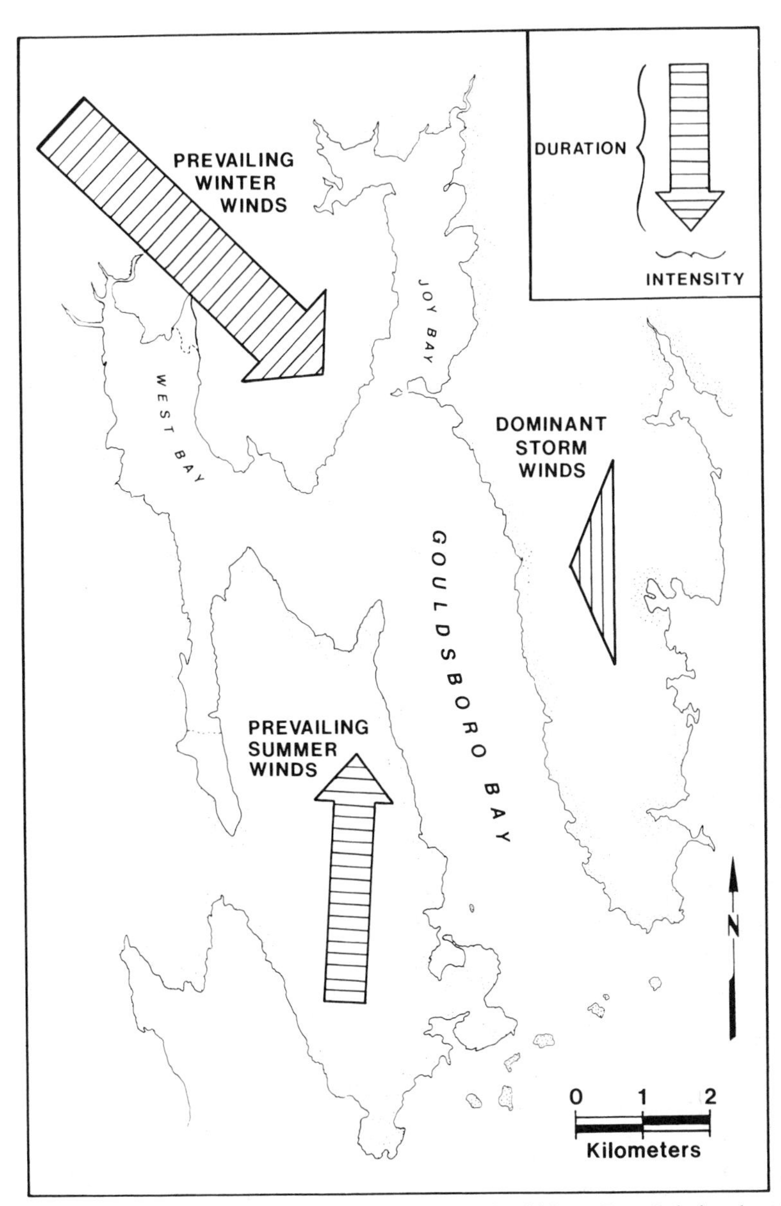

Fig. 5. Summary of annual wind patterns affecting Gouldsboro Bay. Relative durations and intensities are not drawn to actual scale. Based on wind summaries from Lautzenheiser (1972) and Fefer and Schettig (1980).

TABLE III. PERCENTAGES OF GEOMORPHIC CLASSES COMPRISING EACH ZONE

Class	Zone I (%)	Zone II (%)	Zone III (%)
Pocket beach	0	0.9	3.9[a]
Fringing beach	11.8	4.0[a]	0.3
Marsh	11.8	0.7	0.8
Fringing	1.0	0.7[a]	0.8[b]
Immature	3.5[a]	0	0
Mature	7.3[a]	0	0
Mixed flat	24.2	84.4[a]	38.2
Mud flat	55.6[a]	4.0	0.6
Rock ledge	3.0	5.6	56.2
Exposed	0.1	1.2	53.1[a]
Protected	2.9[a]	4.2[a]	3.1
Mussel bar	4.2[a]	0	0
Zone totals	100.0%	100.0%	100.0%

[a]Diagnostic geomorphic environment for that particular zone as summarized on Fig. 6.
[b]High value of fringing marsh due to unusual coastal lagoon on the western side of zone III. Location of lagoon is marked by pointer on Fig. 2b.

the high intertidal area only. Like zone II, most of supratidal shoreline is rimmed with bluffs, except the bluffs in zone I are more extensively vegetated and exhibit much less erosion. While sand and gravel are present as small isolated patches generally above MTL, the dominant sediment size is mud. The effect of wave action is much reduced, because most of this zone is protected from waves by limited fetch and the extensive low intertidal area. This condition acts to dampen wave energy through most of the tidal cycle.

The central zone II is dominated by mixed flat (84.4%), total rock ledge (5.6%), fringing beach (4.0%), and mud flat (4.0%) (Table III). The distinctive characteristics of this zone are the dominance of mixed flat, the presence of numerous fringing beaches, and the presence of only one marsh subclass, fringing marsh. The seaward end of this zone is delimited by the onset of near-continuous bluffs of glacial sediment rimming the supratidal shoreline. A majority of the bluffs in this zone are unvegetated, indicating at least periodic erosion. Gravel, sand, and mud are all present, with gravel and sand more common in the seaward end and mud and sand more abundant in the landward portion. The wave energy in this zone is intermediate in intensity, although somewhat variable because of changes in shoreline orientation.

The intertidal area of the outer zone III is dominated by total rock ledge (65.2%), mixed flat (38.2%), and pocket beach (3.9%) (Table III).

The combination of geomorphic environments that distinguishes zone III is the extensive rock ledge and abundance of pocket beaches. Mixed flats occur only in the more protected parts of zone III. Bluffs of unconsolidated glacial sediment are absent or uncommon along the supratidal shoreline. Gravel is the predominate size class found in this zone of high wave energy. Rarely, deposits of sand and mud are present in sites protected from direct wave exposure. Evidence of high wave energy persists well into the supratidal area in the form of storm berms behind pocket beaches and gravel mixed with drift debris behind rock ledges.

Based on the study of Gouldsboro Bay, a generalized model of geomorphologic variation is proposed (Fig. 6). Zone I is portrayed as an intertidal area of low wave energy, in which accumulation dominates, creating expansive mud flats, mussel bars, and two subclasses of marshes. In contrast, the intermediate wave energy in zone II allows both sediment erosion and accumulation to occur, forming extensive mixed flats, nu-

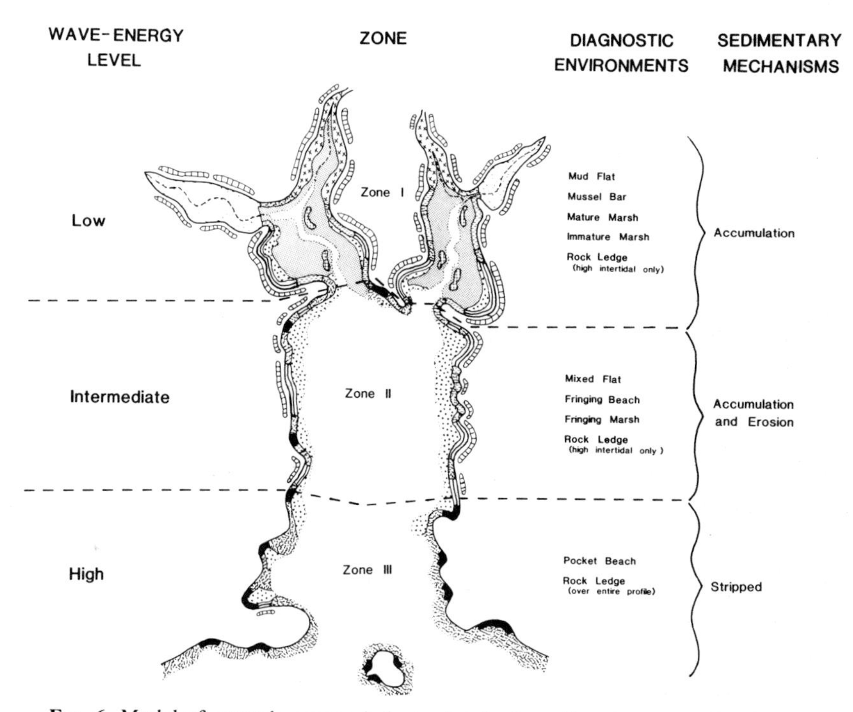

FIG. 6. Model of coastal geomorphologic distribution of environments for the coast of Maine based primarily on Gouldsboro Bay, but corroborated by several other recent studies (see, e.g., Belknap *et al.*, 1986; Belknap and Shipp, 1986; Kelley, Chapter 6, this volume). Legend is the same used in Fig. 2a.

merous fringing beaches, and marsh of the fringing subclass only. Finally, zone III is an intertidal area of high wave energy, which is generally stripped of sediment and characterized by extensive rock ledges and coarse-grained pocket beaches.

Although this proposed zonation is based on the study of only one bay, investigation of three other embayments shows a similar pattern (Belknap and Shipp, 1986; Shipp, unpublished). This concept also agrees with similar zonation schemes proposed for Maine's coastal embayments based on other criteria (Kelley, Chapter 6, this volume; Belknap *et al.*, 1986). Furthermore, a study of the coastal environments of the Makkovik region, Labrador, revealed an analogous zonation based on geomorphic determination of three energy levels (Rosen, 1980). Taken in concert with the present study, these additional studies provide corroborative evidence that suggests a more regional utility of this tripartite zonation scheme.

VII. Conclusions

From the study of intertidal geomorphology in Gouldsboro Bay, Maine, four conclusions are reached concerning the zonation and controls of geomorphic variation.

(1) The intertidal geomorphology of Gouldsboro Bay can be separated into seven major classes. Pocket beach, fringing beach, and marsh are found in the high intertidal area above MTL. Conversely, mixed flat, mud flat, and mussel bar are located in the low intertidal area below MTL. Depending on the degree of exposure, rock ledge can either be confined to the low intertidal area or persist throughout the entire intertidal profile or be restricted to the high intertidal area only.

(2) Three levels of control determine the distribution of geomorphic environments in Gouldsboro Bay. The framework of Paleozoic bedrock dictates the overall shoreline morphology within a coastal compartment (regional scale of 10–100 km). The distribution of late Quaternary glacial sediment controls the abundance and texture of present-day deposits in one or several adjacent embayments (intermediate scale of 1–10 km). Modern coastal processes, particularly wave exposure and winter ice movement, modify deposits at specific shoreline sites within a single embayment (local scale < 1 km).

(3) Gouldsboro Bay can be partitioned into three discrete zones based primarily on intertidal geomorphic variation, but also considering wave energy level, predominant sediment grain size, and supratidal bluff distribution. The inner (landward) zone I is a region of accumulation of mud,

marked by characteristic mud flats, marshes, and mussel bars. The adjacent supratidal area consists of vegetated bluffs. The central zone II is dominated by erosion and accumulation of mud, sand, and gravel and is distinguished by many mixed flats and fringing beaches, which are rimmed by numerous eroding bluffs. The outer (seaward) zone III is stripped of sediment except for occasional deposits of gravelly sediment and is characterized by rock ledge and pocket beaches backed by few bluffs.

(4) In light of the present study and several other recent investigations, this tripartite zonation appears applicable to the entire Maine coast as well as the Canadian Maritimes. This zonation also illustrates the diversity of form caused by variations in bedrock, sediment distribution, and process levels, which is so common along a glaciated coast. With particular reference to glaciation, the Maine coast has been affected by erosion of bedrock, deposition of glacial material (primarily morainal deposits and glaciomarine sediment of the Presumpscot Formation), and subsequent sediment availability controlled by bluff erosion at the shoreline.

Acknowledgments

Support of this study was provided by grant NA81AA-D-CZ076 from the Office of Coastal Zone Management of the National Oceanic and Atmospheric Administration to the Marine Systems Laboratory of the Smithsonian Institution. The authors wish to thank W. H. Adey and P. Gaston for assistance in the field. N. Fisher, D. M. FitzGerald, P. McLaren, and P. S. Rosen are acknowledged for their critical reviews. W. H. Adey, D. F. Belknap, and J. T. Kelley reviewed several drafts of the manuscript and were particularly helpful in contributing their ideas to the geomorphic classification of the Maine coast.

References

Anderson, W. A., Kelley, J. T., Thompson, W. B., Borns, H. W., Jr., Sanger, D., Tyler, D. A., Anderson, R. S., Bridges, A. E., Crossen, K. J., Ladd, J. W., Andersen, B. G., and Lee, F. T. (1984). Crustal warping in coastal Maine. *Geology* **12,** 677–680.

Bascom, W. (1964). "Waves and Beaches." Anchor Doubleday Garden City, New York.

Belknap, D. F., and Shipp, R. C. (1986). Quaternary geology of the Damariscotta River. *In* "Guidebook of the 78th Meeting of the New England Intercollegiate Geologic Conference" (D. W. Newberg, ed.), Field Trip B-9. Bates Coll., Lewiston, Maine.

Belknap, D. F., Shipp, R. C., and Kelley, J. T. (1986). Depositional setting and Quaternary stratigraphy of the Sheepscot estuary: A preliminary report. *Geogr. Phys. Quat.* **40,** 55–69.

Bloom, A. L. (1960). "Late Pleistocene Changes of Sea Level in Southwestern Maine." Maine Geol. Surv., Augusta.

Bloom, A. L. (1963). Late Pleistocene fluctuations of sea level and postglacial crustal rebound in coastal Maine. *Am. J. Sci.* **261,** 862–879.

Borns, H. W., Jr. (1973). Late Wisconsin fluctuations of the Laurentide ice sheet in southern and eastern New England. *Geol. Soc. Am. Mem.* **136,** 37–45.

Borns, H. W., Jr. (1974a). Reconnaissance surficial geology of the Bar Harbor quadrangle, Maine. *Maine Geol. Surv. Open-File Rep.* **74-1,** 1:62,500.

Borns, H. W., Jr. (1974b). Reconnaissance surficial geology of the Petit Manan quadrangle, Maine. *Maine Geol. Surv. Open-File Rep.* **74-8,** 1:62,500.

Borns, H. W., Jr., and Andersen, B. G. (1982a). Reconnaissance surficial geology of the Cherryfield quadrangle, Maine. *Maine Geol. Surv. Open-File Rep.* **82-2,** 1:62,500.

Borns, H. W., Jr., and Andersen, B. G. (1982b). Reconnaissance surficial geology of the Tunk Lake quadrangle, Maine. *Maine Geol. Surv. Open-File Rep.* **82-5,** 1:62,500.

Chapman, C. A. (1958). Controls of jointing by topography. *J. Geol.* **66,** 552–558.

Chapman, C. A. (1962). Bays-of-Maine igneous complex. *Geol. Soc. Am. Bull.* **73,** 883–887.

Denny, C. S. (1982). Geomorphology of New England. *U.S. Geol. Surv. Prof. Pap.* **1208,** 1–18.

Fefer, S. I., and Schettig, P. A. (1980). "An Ecological Characterization of Coastal Maine," Vol. 1. Northeast Reg. U.S. Fish Wildl. Serv., Boston, Massachusetts.

Hayes, M. O., Owens, E. H., Hubbard, D. K., and Abele, R. W. (1973). The investigation of form and processes in the coastal zone. *In* "Coastal Geomorphology: Proceedings of the Third Annual Geomorphology Symposia Series" (D. R. Coates, ed.), pp. 11–41. Dowden, Hutchinson, and Ross, Stroudsburg, Pennsylvania.

Jackson, C. T. (1837). "First Report on the Geology of the State of Maine." Smith & Robinson Publ. Co., Augusta, Maine.

Johnson, D. W. (1925). "The New England–Acadian Shoreline." Hafner, New York.

Kelley, J. T., and Kelley, A. R., eds. (1986). "Coastal Processes and Quaternary Stratigraphy, Northern and Central Maine," Guide to the Annual Field Trip of the Eastern Section of the Society of Economic Paleontologists and Mineralologists, Orono, Maine.

Kellogg, D. C. (1982). Environmental factors in archaeological site location for the Boothbay, Maine region with an assessment of the impact of coastal erosion on the archaeological record. M.S. Thesis, Univ. of Maine, Orono.

Knight, R. J., and Dalrymple, R. W. (1976). Winter conditions in a macrotidal environment, Cobequid Bay, Nova Scotia. *Rev. Geogr. Montreal* **30,** 68–85.

Lautzenheiser, R. E. (1972). Climates of Maine. *In* "Climates of the States" (National Oceanic and Atmospheric Administration, ed.), Vol. 1, pp. 136–156. Water Inf. Cent., Port Washington, New York.

Maine State Planning Office (MSPO) (1983). "The Geology of Maine's Coastline." Maine State Exec. Dep., Augusta.

National Oceanic and Atmospheric Administration (NOAA) (1981). "Tide Table 1982—High and Low Predictions: East Coast of North and South America." U.S. Dep. Commer., Washington, D.C.

Rosen, P. S. (1980). Coastal environments of the Makkovik region, Labrador. *In* "The Coastline of Canada" (S. B. McCann, ed.), *Geol. Surv. Can. Pap.* **80-10,** 267–280.

Schnitker, D. (1974). Postglacial emergence of the Gulf of Maine. *Geol. Soc. Am. Bull.* **85,** 491–494.

Shipp, R. C. (1985). Late Quaternary evolution of the Wells embayment. *Geol. Soc. Am. Abstr. Program* **17,** 63.

Shipp, R. C. (1987). Late Quaternary sea-level fluctuations and geologic evolution of four embayments along the northwestern Gulf of Maine. Ph.D. Thesis, Univ. of Maine, Orono.

Shipp, R. C., Staples, S. A., and Adey, W. H. (1985). Geomorphic trends in a glaciated coastal bay: A model for the Maine coast. *Smithson. Contrib. Mar. Sci.* **25.**

Smith, G. W. (1982a). End moraines and the pattern of the last ice retreat from central and southern coastal Maine. *In* "Late Wisconsinan Glaciation of New England" (G. J. Larson and B. S. Stone, eds.), pp. 195–209, Kendall/Hunt, Dubuque, Iowa.

Smith, G. W. (1982b). DeGeer Moraines of the Maine coastal zone. *Northeast. Southeast. Sect. Meet. Geol. Soc. Am. Abstr. Program* **14,** 57.

Stuiver, M., and Borns, H. W., Jr. (1975). Late Quaternary marine invasion in Maine: Its chronology and associated crustal movement. *Geol. Soc. Am. Bull.* **86,** 99–104.

Thompson, W. B., and Borns, H. W., Jr., eds. (1985). "Surficial Geologic Map of Maine." Maine Geol. Surv., Augusta. 1:500,000.

Timson, B. S. (1977). "A Handbook of Coastal Marine Geologic Environments of the Maine Coast," Open-File Rep. (111 maps). Maine Geol. Surv., Augusta.

Ward, L. G., Moslow, T. F., Hayes, M. O., and Finkelstein, K. (1980). "Oil Spill Vulnerability, Coastal Morphology, and Sedimentation of the Outer Kenai Peninsula and Montague Island," Task D-4 of the Outer Continental Shelf Program. Nat. Oceanic Atmos. Adm., Washington, D.C.

CHAPTER 9

Sediment Accumulation Forms, Thompson Island, Boston Harbor, Massachusetts

PETER S. ROSEN
AND
KENNETH LEACH

Department of Geology
Northeastern University
Boston, Massachusetts

Submergence of drumlinoid topography in Boston Harbor, Massachusetts, has lead to a reworking of Pleistocene sediments by longshore processes and the deposition of accumulative shore forms. The spit processes on Thompson Island characterize accumulative processes on the Boston Harbor Islands, and are an example of low-wave-energy gravel beach processes. Most deposition is at longshore drift convergences caused by refraction of waves around the island. Resulting cuspate spits represent a reorientation of the shoreline into locally dominant wave approach directions.

On gravel spits, storm ridges form the emergent platforms. In areas of high longshore sediment input, multiple symmetrical accumulative ridges form. In areas of lower sediment input, a single asymmetrical overtopping ridge migrates landward during storm events.

An overwash system existed for about 6 years. The overwash processes on this gravel beach differs from those on sandy shorelines. Overwash is not primarily a storm event. The overwash throat is similar to an inlet in that it is permanent, and migrates in the downdrift direction. This migration has left a overwash fan/platform, which is comparable to a flood tidal delta complex behind a sandy barrier. The gravel overwash deposits are composed of a characteristic gravel-over-sand sequence.

I. Introduction and Setting

Thompson Island is one of nearly 30 islands in Boston Harbor created by submergence of Pleistocene drumlinoid topography (Kaye, 1976). The Boston Harbor region is the only location in the United States where a drumlin field intersects the coast, but other examples are found along the coast of Nova Scotia and western Ireland (Shepard and Wanless, 1970). The unconsolidated tills provide a source for beach sediments. The diminishing rate of sea-level rise in the past 3000 yr (Kaye and Barghoorn, 1964) has led to reworking of the eroding glacial sediments by longshore transfers and the deposition of shore accumulation forms. This accumulation of coastal sediments has caused the central Massachusetts coastline to begin to take on secondary characteristics, including a general straightening of the shore (Shepard, 1937). The development of accumulation forms gave rise to the harbor itself, as the embayment is separated from Massachusetts Bay by two flanking tombolo-and-headland complexes [Winthrop–Deer Island to the north and Nantasket–Hull to the south (Fig. 1)]. The development of the Nantasket complex has been methodically reconstructed by Johnson (1910). Modern wave processes within this low-wave-energy setting have reworked the eroding, typically unsorted gravel to create a variety of accumulative landforms, including longshore spits, cuspate spits, and tombolos.

The islands within this protected harbor, including Thompson, are subjected to coastal processes characterized by extremely low wave energy resulting from fetch limitations. Boston Harbor mouth opens into Massachusetts Bay with a direct northeast exposure. However, Thompson Island is sheltered from the harbor mouth, so locally generated seas can play a role in affecting net drift directions. Along the eastern shore of the island, net drift is toward both the north and south directions, representing influence from both the prevalent southwest and dominant northeast winds. Longshore drift on this island, as well as others, is affected by wave refraction patterns around the islands. Oncoming waves refract around the island, so a drift convergence can form at the leeward side resulting in the formation of cuspate spits.

There is no onshore–offshore cycling of sediments on the beaches. The nearshore zone is composed of sandy silts derived from the fine fraction of eroding Pleistocene bluffs. On accumulative beaches, small-scale gravel ridges migrate up the beachface with rising tides.

The accumulative shoreforms on Thompson Island are models for several processes typical of several other harbor islands. The major accumulative forms on Thompson Island are two cuspate spits and one longshore spit.

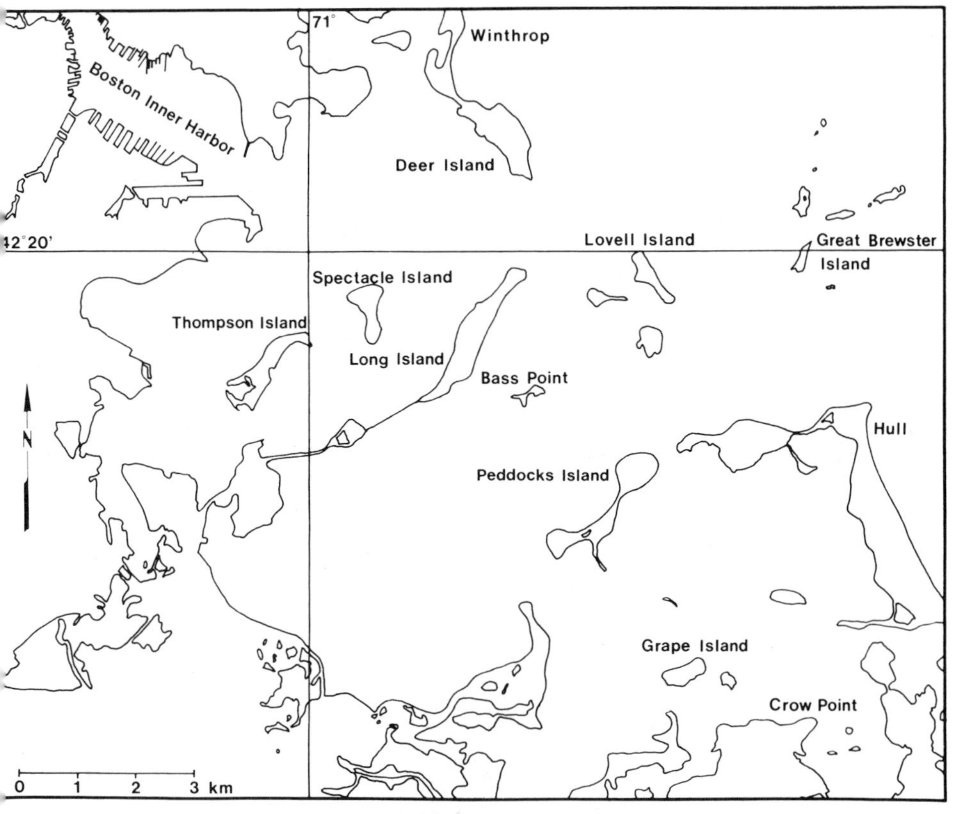

FIG. 1. Map of Boston Harbor, showing Thompson Island and the Harbor Islands.

II. South Cuspate Spit: Drift Convergence and the Initiation of a Tombolo

The South Cuspate Spit is one of the few accumulative forms in Boston Harbor composed of a high percentage of sand. The sources of this sand are the possible glaciomarine deltaic deposits that comprise part of the south end of the island (Caldwell, 1984). The emergence and preservation of this landform results from progressive vegetated dune ridge growth capping a gravel beach. An aerial mosaic of the island (Fig. 2) shows wave refraction patterns resulting from a northerly wind. The crests are encircle both sides of the island and converge at the South Cuspate Spit. This photograph also demonstrates that the cuspate form represents a re-orientation of the shore into two opposing wave approach directions (Ro-

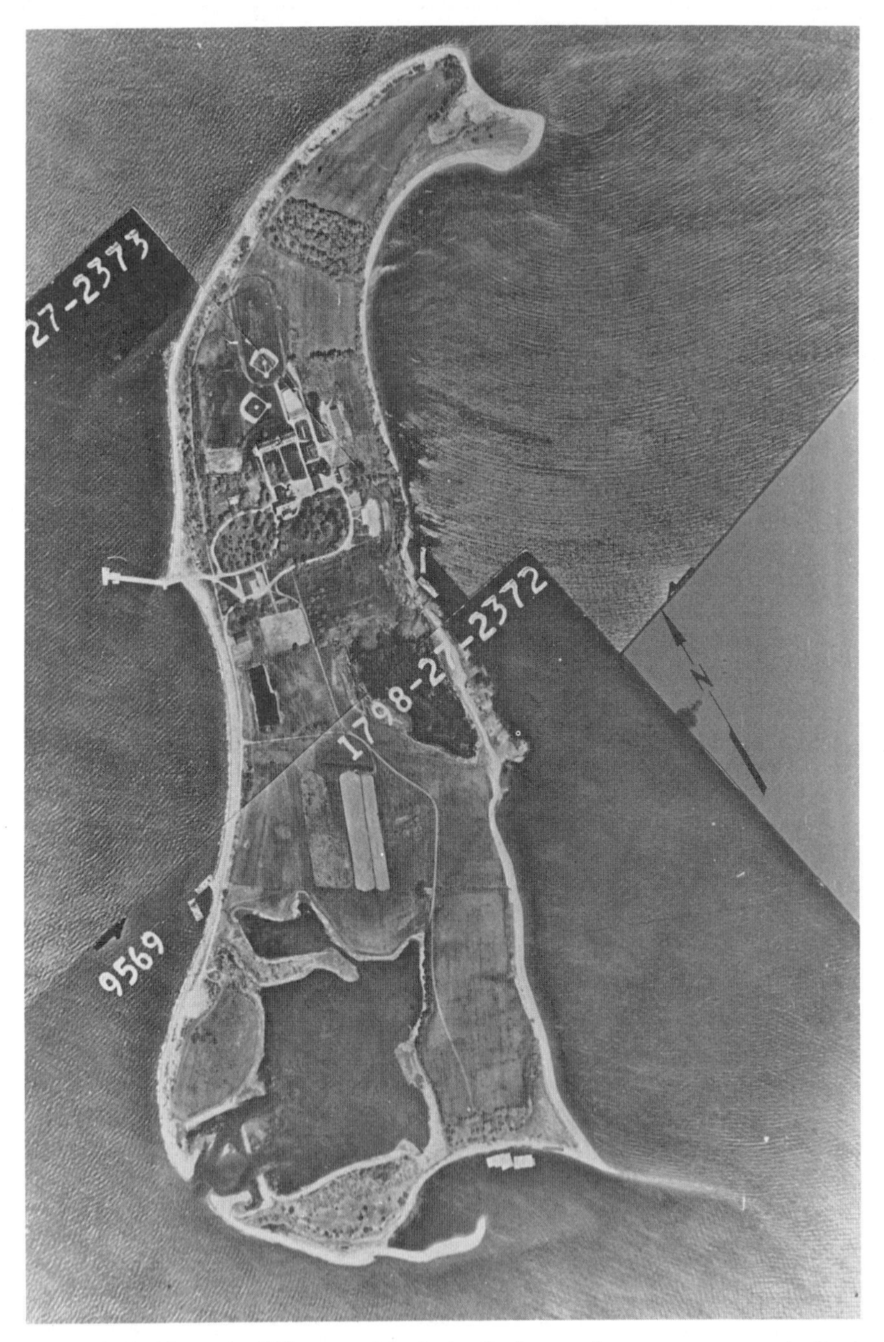

FIG. 2. Aerial mosaic of Thompson Island, showing intersecting wave patterns at the South Cuspate Spit.

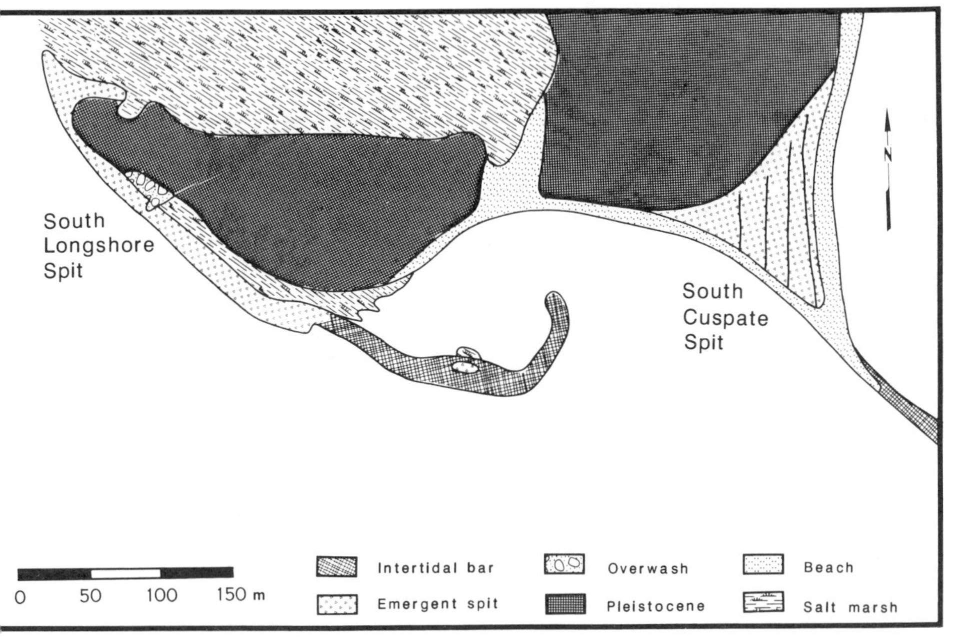

FIG. 3. Geomorphology of south Thompson Island, including the South Cuspate Spit and South Longshore Spit.

sen, 1975). Similar cuspate forms occur at the south ends of Peddocks, Grape (Fisher and Jones, 1982), and Spectacle islands.

Regardless of the relative energy in these opposing drift systems, there is a dearth of longshore sediment from the southwest, due to an updrift tidal inlet and spit, and there is an abundance of sediments from the north, due to an eroding sandy bluff. This has resulted in the gradual accumulation of the spit in the northward direction. This trend is recorded in a succession of preserved dune ridges parallel to the northern shore of the spit, and truncation of these ridges along the southern shore (Fig. 3). Dune ridge elevations gradually increase going from southwest to northeast (older to younger), which may reflect rising water levels during formation.

While the preserved dune ridge form suggests a migrating system, historic analysis of shoreline positions has revealed that there has been no discernable change in spit shoreline position since at least 1847. As the bluff to the north (the source area) has retreated at rates up to 30 cm/yr over the same period, the accumulation rate on this spit appears to be in equilibrium with sea-level rise.

The spit ends as an intertidal bar that extends across a channel to Squantum Head on the mainland. The longshore convergence on this spit, resulting in the formation of this bar, appears to represent an initial stage of tombolo formation, which is a more common accumulation form on other Boston Harbor islands. Many other islands (i.e., Spectacle, Peddocks, Lovell, Long, and Great Brewster) are nearby drumlinoids tied by tombolos in this manner. It has been suggested that this "island tying" is not solely a function of wave processes. Underlying glacial topography may serve as a shallow platform that promotes the formation and preservation of these deposits (A. C. Redfield, 1971, personal communication). While this intertidal bar has been in about the same position since the 1700s, it has not emerged to form a tombolo as in nearby areas. This is due to scouring by tidal currents, as this channel plays a role in the flushing of Dorchester Bay. Since the bar has not shifted since at least 1847, there appears to be a balance between longshore input and tidal scour.

III. North Cuspate Spit: Migrating Spits and Gravel Ridge Forms

The northern half of Thompson Island has been classified as the remains of a drumlin composed of silty gravel (Caldwell, 1984). The longshore sediments derived from this source have a low sand content and there is no dune formation, which is more typical of Boston Harbor accumulation forms.

The gravel comprising the North Cuspate Spit has a high percentage of Cambridge argillite (Billings, 1976) pebbles, which may underly the glacial sediments in the area. This lithology results in a dominance of flat (disk or blade) shapes. As is typical of gravel beaches, flat shapes are preferentially transported landward, while round shapes (spheroids or rods) are transported seaward to the low water line (Bluck, 1967). Storm wave activity results in net onshore transport and the formation of symmetrical accumulative ridges composed of flat pebbles above the high water line. These storm ridges form the emergent portion of the spit (Fig. 4). The ridges have been deposited successively on the southern face of the spit, at least partially due to the abundance of source material from eroding bluffs in that direction. Tracings of the ridge crests from aerial photographs have shown that the spit was initially more parallel with the original shoreline of the submerged glacial materials (Fig. 5). Successive storm-ridge accumulation has reoriented the southern shoreline of this spit to a position nearly parallel to oncoming southwesterly waves.

The north-facing shore of this spit truncates the gravel ridges, indicating long-term retreat. This shore is exposed to higher wave energy and lacks

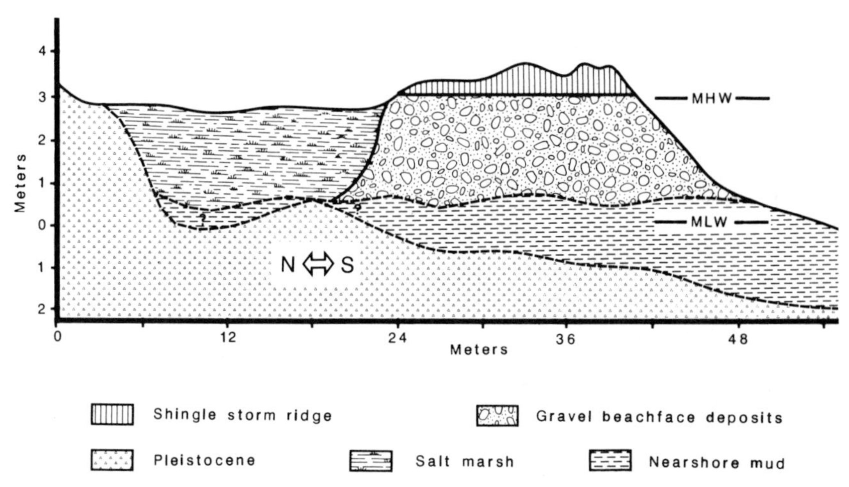

FIG. 4. Cross section of North Spit based on profiles and marsh soundings. Storm ridges form emergent portions of spit, which was deposited over nearshore muds.

a significant longshore sediment source. The backbeach in this area forms an overtopping ridge, or accumulation of gravel that is periodically overtopped by storm waves causing landward migration (Orford and Carter, 1982). The overtopping does not form distinct overwash channels, so a landward-dipping slipface exists along the entire length of the ridge. Since the ridge is composed of flat pebbles, and currents during overtopping are primarily unidirectional, the internal geometry of the ridge is dominated by seaward-dipping imbricated pebbles.

The occurrence of symmetrical accumulative ridges or asymmetrical overtopping ridges in the backshore corresponds to long-term shoreline changes. Since 1847, the northern shoreline of this spit has retreated an average of 15 cm/yr, while the southern shoreline has accreted at an average rate of 10 cm/yr. Therefore, this landform is mobile and is migrating in a southerly direction. This mobility has been observed on other cupate spits in the harbor, such as Bass Point on Long Island.

The North Cuspate Spit encloses a small salt marsh. The peat has been dated at 2250 yr B.P. (Kaye and Barghoorn, 1964), which represents a minimum age for the spit. This date correlates closely with the diminution of the rate of sea-level rise (about 3000 yr B.P.) identified by Kaye and Barghoorn for the Boston area. The beginning of accumulation forms in Boston Harbor from the reworking of Pleistocene sediments may have initiated at that time.

The connection of the marsh with salt water is through an overwash

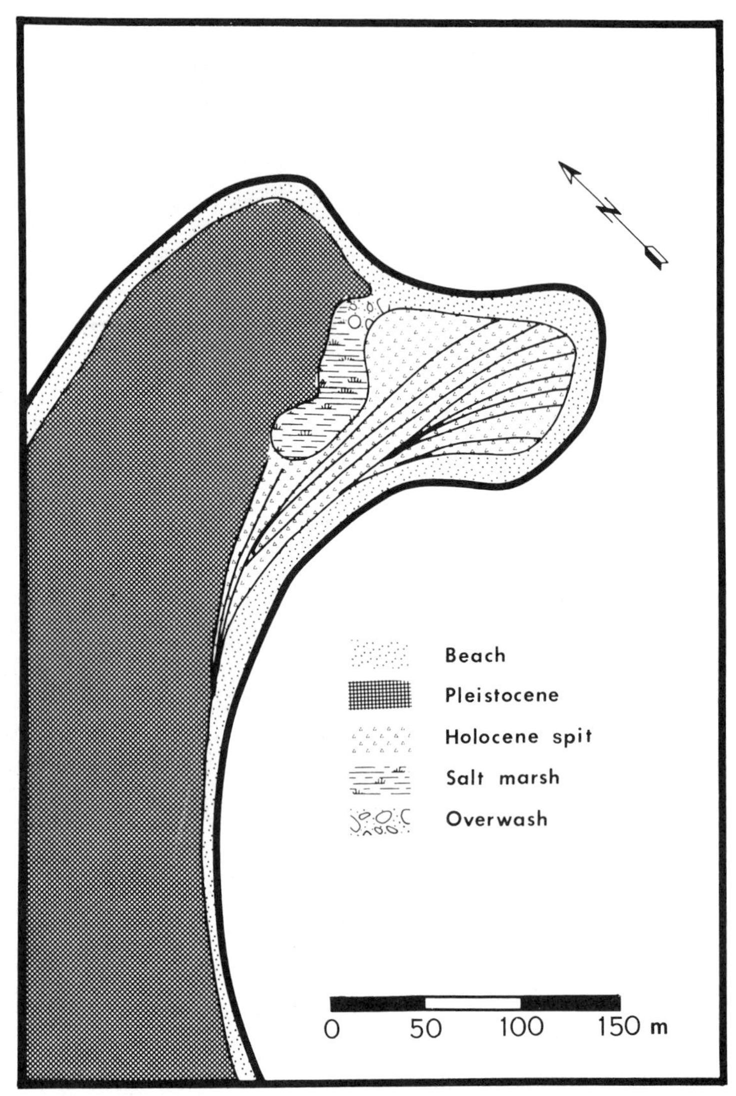

FIG. 5. Geomorphology of north Thompson Island, including the North Cuspate Spit. Lineations on the spit surface traced from air photos represent storm-ridge crests. These outline a reorientation of the shoreline as accumulation progressed.

channel. Since ridge growth is the response of intertidal gravel to most wave activity, the channel has only opened three to four times per year for the past 4 years. The brackish conditions presently in the salt marsh have led to an active *Phragmites* sp. population overlying *Spartina* peat. Soundings in the marsh show a maximum depth of about 3 m, including

FIG. 6. Shore ice at the North Cuspate Spit. Note sediment layers entrained in ice.

a basal mud deposit that served as a platform for the accumulation of this spit. A steep slope exists at the boundary between the marsh and the first ridge of the enclosing spit. This steep slope is comparable to the present form of the downdrift end of the South Longshore Spit. The South Longshore Spit is probably an analog of the initial form of this spit.

Throughout the salt marsh are several random cobble deposits resulting from winter sediment transport by ice. Freeze-up around the island is not uncommon due to the low wave energy. The high tide range (mean = 2.8 m) causes continual floating and grounding of shore ice while freezing is taking place. This is effective for the entrainment of beach materials into the ice (Rosen, 1979). Since freeze-up is not usually complete throughout the harbor, waves continually redistribute the shore ice. This has resulted in ice ridges up to 1.7 m in height around the North Cuspate Spit (Fig. 6). However, gravel ice-push ridges, typical during freeze-ups on more exposed coasts (Rosen, 1978), have not been observed. Although shore ice entrains numerous layers of gravel, beach profiles before and after freeze-up in 1981 on the south shore of the spit showed no measureable change in the beach. The major impact of ice on the Thompson Island shoreline is the redistribution of salt-marsh peat, which is not uncommon in New England (Redfield, 1972). The ice-transported peat blocks have established a fringe marsh along much of the eastern shore of the island.

IV. South Longshore Spit: Gravel Overwash and Bar Emergence

The South Longshore Spit differs from other spits on the island and most spits in the harbor in that it does not represent a regional longshore drift convergence point. This is due to the sheltering from the east by South Cuspate Spit. The longshore sediment source has been cut off for the past few decades by a tidal inlet updrift of the spit (Figs. 2 and 3).

The spit is composed of two geomorphically distinct regions. The updrift end of the spit has existed prior to 1770. This portion of the spit is fully emergent. The supratidal areas are composed of flat pebbles forming a continuous overtopping ridge. This updrift region borders on a narrow salt marsh, and is close to welding onto the adjacent shoreline. The landward migration rates between 1982 and 1984 have averaged 4.6 m/yr. This rate may be influenced by the lack of a longshore sediment source.

A gravel overwash throat had existed on this portion of the spit since at least 1978, and may have opened as a result of the February 9, 1978, nor'easter (Blizzard of 1978). It closed in summer 1984. Other gravel barriers that have overwash throats are the northeast tombolo shoreline on Peddocks Island and the north tombolo shoreline on Lovells Island. The processes associated with this overwash system differ greatly from overwash on sandy barriers. The overwash channel is a semipermanent feature, and migrates in the downdrift direction similar to a tidal inlet. Migration rates increased in winter months, and averaged 7.5 m/yr between 1981 and 1984. The cross-sectional area of the throat also increased during winter months, presumably due to larger volumes of overwash flow (Rosen, 1984). The migration of the throat has led to the deposition of a 40-m-

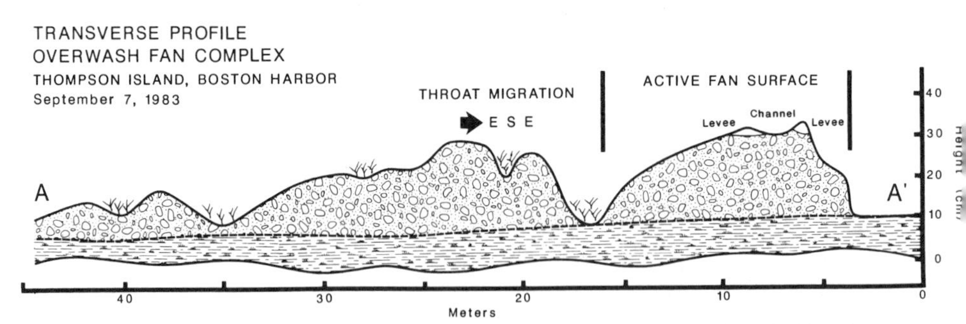

FIG. 7. Shore-parallel profile of the gravel overwash fan/platform landward of South Longshore Spit showing successive depositional areas separated by vegetated clefts, and an active fan surface with levees enclosing an active overwash channel. Profile location shown as A–A′ on Fig. 8.

long fan/platform overlying the adjacent salt marsh. Preserved on this platform were convex mounds separated by vegetated (*Sudea maritima*) clefts (profile shown in Fig. 7, location shown in Fig. 8). These mounds appear to represent overwash deposition associated with successive positions of the throat.

Overwash is not primarily a storm event. The maximum throat elevation was typically near mean high water. Most tides above that level, or neap tides during higher wave energy events, do overwash. Accumulative gravel ridges regularly migrated up the beachface and block the throat, but were readily breached by seepage (Carter and Orford, 1984) and runup.

Sediment transport during overwash was driven by wave bores entering the throat and downslope flow of water. Wave bores were not observed to be effective at gravel initiation without accompanying flow. Since there is a lag of about ½ hr in the filling of the marsh (by way of the lagoon) during rising tides, water levels are higher on the seaward side of the spit. The first stage of overwash is seepage through the gravel barrier. The elevation of the backbarrier was about 10–15 cm higher updrift of the throat as compared to downdrift areas due to prior deposition of the fan/platform. Seepage was consistantly first observed adjacent to, and downdrift of, the throat position. While the volume of water that entered the marsh as seepage was not significant in filling the marsh basin, it attained sufficient velocities to winnow and transport sand from the gravel ridge and deposit seepage lobes landward of the ridge slipface (Carter and Orford, 1984).

The flow through the overwash throat took place from about ½ hr before high tide, and ended at high tide. By high tide, the marsh basin had filled so water no longer flowed through the throat (Fig. 9). Overwash flow and

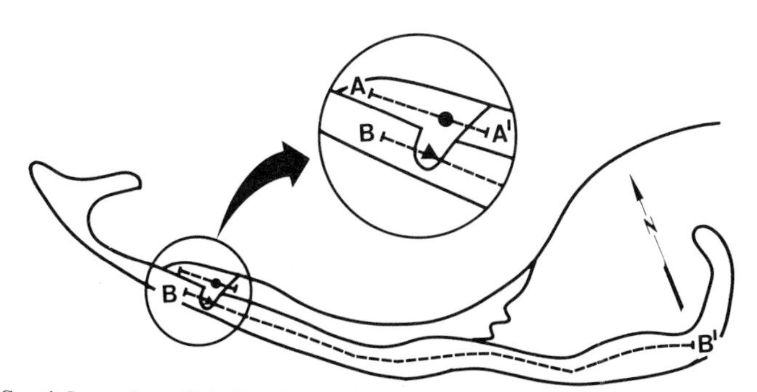

FIG. 8. South Longshore Spit data locations. A–A′, transverse profile, overwash fan (Fig. 7); B–B′, spit crest profile (Fig. 12); solid circle, overwash fan (Fig. 9); triangle, overwash throat (Fig. 9).

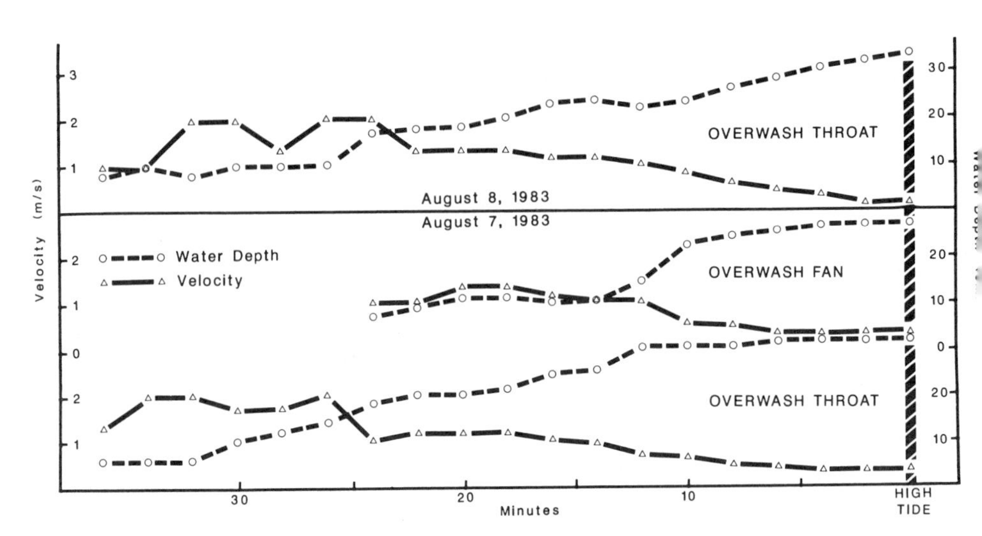

FIG. 9. Velocity and depth relationships in Thompson Island overwash flow on August 7–8, 1983. Note that overwash flow ends by high tide, when water levels landward and seaward of the spit are equalized. Overwash throat and overwash fan stations shown in Fig. 8.

accompanying wave bores transported all sizes of material through the throat. Transport across the fan platform was dominated by down-gradient flow. While coarse material tended to move landward, sand-sized material accumulated as levees on the channel margins. The levees were best developed when wave action was low (which was most typical). As water levels increased on the fan/platform, the levees consistently were first overtopped and breached at the same position: the downdrift side of the throat, which is adjacent to the salt-marsh surface. This area had the steepest gradient, as the existing fan/platform raised basin elevations both landward and updrift of the throat position. Once a crevasse formed in the levee, some flow was diverted from the existing landward-oriented channel. A microdelta composed entirely of sand encroached laterally on the marsh, while all transport of coarser material was landward. The wave bore moving landward through the throat apparently played a significant role in the initiation and transportation of coarser material. The area of sand deposition through the levee crevasse is the same general area where sand seepage lobes were deposited earlier in the event. In future overwashes, the throat will migrate downdrift and the cobble layer will be deposited over the sand (Fig. 10).

The characteristic depositional unit resulting from this gravel overwash

FIG. 10. Nature of gravel barrier overwash, August 7, 1983. (a) Initial encroachment of water through throat during rising tide. Flow is contained by sandy levees flanking channel. Note seepage through barrier at left center of photo. (b) Breaching of levees on downdrift side of overwash channel (bottom center of photo). Sand accumulates near breach; pebbles are transported down main channel. (*Continued on next page*)

Fig. 10. (c) Overwash event ends by high tide when barrier lagoon water levels equalizes with harbor.

is a reverse-graded, gravel-over-sand sequence (Fig. 11). The gravel layer may show faint coarse–fine layering. This probably represents wave setup, which causes high-velocity surges through the throat observed with periods of 2–5 min. This layering also has been observed on overwash on sandy, ocean-facing shorelines (Leatherman and Williams, 1983). The sandy levees were not found preserved in the sequence.

The downdrift end of the South Longshore Spit may have accreted since 1893, as it is not shown on a survey of that date. It is covered during normal high tides (therefore, technically it is a bar; see Schwartz, 1982), although its crest elevation is typically near mean high water (Fig. 12). This portion of the spit has retreated at an average rate of 0.48 m/yr between 1982 and 1984. Since this gravel spit/bar does not support dune grasses, nor is there any aeolian transport, a mechanism for emergence of this feature above mean high water is not obvious. Most of the length of this downdrift segment of the spit does not border on salt marsh, as does the updrift emergent segment. The downdrift segment encloses a small lagoon comprised of subtidal muds.

One position on the downdrift end of the spit is regularly emergent above mean high water (Fig. 12). The emergent gravel "mound" is adjacent to an isolated marsh clump in the lagoon. The emergent mound was up

Fig. 11. Resin peel of reverse-graded overwash sequence from active overwash fan complex shown in Fig. 10. Scale units = 2 cm.

to 15 cm higher in the spring and summer when *Spartina* grasses behind it were tall, and lower or nonexistent in the winter when grasses were not present. Since the emergent updrift segment of the spit ends abruptly where the adjacent salt marsh ends, and the mound location corresponds to the presence of lagoonal salt marsh, the salt marsh appears to be one controlling factor in the emergence of this gravel spit.

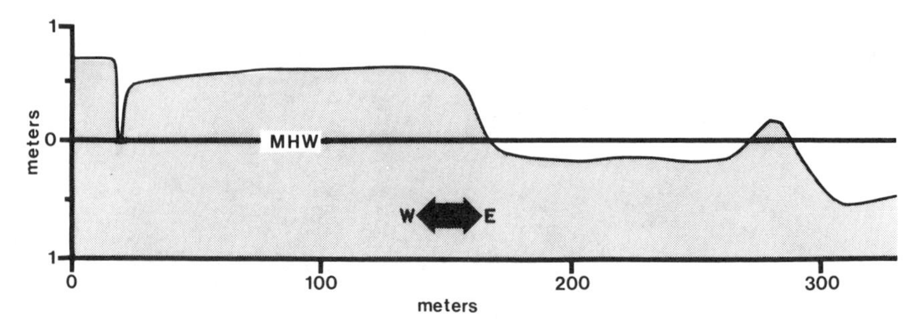

Fig. 12. Profile along crest of South Longshore Spit; showing emergent mound that corresponds to salt marsh in lagoon at west end, and overwash throat near the mean high water line at east end. Profile location shown as B–B′ in Fig. 8.

V. Conclusions

Sediment accumulation forms on Thompson Island are representative of many of the processes on the submerging drumlins inside Boston Harbor. Most coastal accumulation is associated with longshore drift convergences. Most of these drift convergences form tombolos that tie headlands together, but some remain as cuspate spits.

The emergence of a gravel accumulation to form a storm ridge appears to be related to the characteristics landward of the active beachface. The presence of salt marsh or other ridges landward is associated with emerging ridges, while intertidal lagoonal muds landward are associated with an intertidal gravel bar that approaches high water, but does not pass through it.

Storm ridges form the emergent portions of gravel spits. These ridges can be symmetrical accumulative forms that are fixed, or asymmetrical overtopping ridges that migrate landward. The form of storm ridge is associated with longshore sediment supply.

While overtopping is the most common form of storm inundation, gravel overwash channels occur and have a distinct set of processes and deposits that sets them apart from their counterpart on sandy barriers. Overwash is not primarily a storm event. An overwash throat can be similar to a tidal inlet in that it is a permanent feature, migrates in the downdrift direction, and results in a continuous delta–fan deposit landward of the spit. These gravel overwash deposits are composed of a characteristic gravel-over-sand sequence.

ACKNOWLEDGMENTS

The logistical support provided by the Thompson Island Education Center is gratefully acknowledged. Ernest Waterman, Marcia Berman, and Laura Quinn ably assisted in the field. William Perkins assisted with plant identifications.

REFERENCES

Billings, M. P. (1976). Bedrock geology of the Boston Basin. *In* "Geology of Southeastern New England," New England Intercollegiate Geological Conference Guidebook, (B. Cameron, ed.), pp. 28–45.

Bluck, B. J. (1967). Sedimentation of beach gravels: Examples from South Wales. *J. Sediment. Petrol.* **37,** 128–156.

Caldwell, D. W. (1984). Surficial geology and archaeology on Thompson Island, Boston Harbor, Massachusetts. *In* "Geology of the Coastal Lowlands, Boston, MA to Kennebunk, ME," New England Intercollegiate Geological Conference (L. Hanson, ed.), pp. 402–416.

Carter, R. W. G., and Orford, J. D. (1984). Coarse clastic barrier beaches: A discussion of the distinctive dynamic and morphosedimentary characteristics. *Mar. Geol.* **60,** 377–389.

Fisher, J. J., and Jones, J. R. (1982). Origin and development of a cuspate foreland at Grape Island, Boston Harbor, Massachusetts. *In* "Economic Geology in Massachusetts" (O. Farquar, ed.), pp. 511–518. Univ. of Massachusetts Press, Amherst.

Johnson, D. W. (1910). The form of Nantasket Beach, Massachusetts. *J. Geol.* **18,** 162–189.

Kaye, C. A. (1976). Outline of the Pleistocene geology of the Boston Basin. *In* "Geology of Southeastern New England," New England Intercollegiate Geological Conference Guidebook (B. Cameron, ed.), pp. 46–63.

Kaye, C. A., and Barghoorn, E. S. (1964). Late Quaternary sea-level change and crustal rise at Boston, Massachusetts, with notes on the autocompaction of peat. *Bull. Geol. Soc. Am.* **75,** 63–80.

Leatherman, S. P., and Williams, A. T. (1983). Vertical sedimentation units in a barrier island washover fan. *Earth Surf. Processes Landforms* **8,** 141–150.

Orford, J. D., and Carter, R. W. G. (1982). Crestal overtop and washover sedimentation on a fringing sandy gravel barrier coast, Carnsore Point, Southeast Ireland. *J. Sediment. Petrol.* **52,** 265–278.

Redfield, A. C. (1972). Development of a New England salt marsh. *Ecol. Monogr.* **42,** 143–172.

Rosen, P. S. (1975). Origin and processes of cuspate spit shorelines. *In* "Estuarine Research" (L. E. Cronin, ed.), Vol. 2, pp. 72–92. Academic Press, New York.

Rosen, P. S. (1978). Degradation of ice-formed beach deposits. *Marit. Sediments* **14,** 63–68.

Rosen, P. S. (1979). Boulder barricades in central Labrador. *J. Sediment. Petrol.* **49,** 1113–1124.

Rosen, P. S. (1984). Gravel spit processes, Thompson Island, Boston Harbor, Massachusetts. *In* "Geology of the Coastal Lowlands, Boston, MA to Kennebunk, ME," New England Intercollegiate Geological Conference (L. Hanson, ed.), pp. 25–38.

Schwartz, M. L., ed. (1982). "The Encyclopedia of Beaches and Coastal Environments." Dowden, Hutchinson & Ross, Stroudsburg, Pennsylvania.

Shepard, F. P. (1937). Revised classification of marine shorelines. *J. Geol.* **45,** 602–624.

Shepard, F. P., and Wanless, H. R. (1970). "Our Changing Coastlines." McGraw-Hill, New York.

CHAPTER 10

Source of Pebbles at Mann Hill Beach, Scituate, Massachusetts

BENNO M. BRENNINKMEYER
AND
ADAM F. NWANKWO

Coastal Research Institute
Department of Geology and Geophysics
Boston College
Chestnut Hill, Massachusetts

Mann Hill Beach, Scituate, Massachusetts, is a baymouth bar composed of shingle with an average size of 43 mm. The largest boulder on the berm is 330 mm in size. The largest pebbles are found on the top of the beach. More than 75% of the gravel on the beach is flat in shape. The shingle consists of 13 lithologies of rocks exposed no less than 7.5 km and no more than 15 km to the north and west. The gravel was deposited probably as ground moraines in patches both onshore and offshore. One gravelly patch extending 650 m offshore to a depth of 12 m forms the immediate source for the shingle on the beach. The coarsest surface sediment offshore from Mann Hill is 16 mm.

Threshold criteria for Airy waves indicate that the most common waves (6-s period, 0.9 m in height) at Scituate can move the observed offshore sediment from a depth of 9.4 m. The same wave can move the coarsest size found on the beach from a depth of 9 m. Under severe storm conditions these depths increase to 25 and 19 m, respectively.

The larger flat pebbles in the offshore deposit are selectively transported shoreward and deposited on the highest part of the beach. More spherical pebbles, if transported shoreward, roll more readily and are carried seaward by the backwash.

I. Introduction

Roughly half of the world's beaches are covered with shingle material at least partially rounded and sorted by marine processes and falling in the diameter range of 4–256 mm. Compared to sandy beaches, shingle beaches have received little attention. The only real exceptions are Checil Beach on the southern coast of England (see, among others, Cornish, 1898; Carr, 1969, 1971, 1975; Gleason and Hardcastle, 1973), in Wales (Orford, 1975, 1977) (Bluck, 1967), and of Ireland (Orford and Carter, 1982a,b). Two of the most enduring questions concerning shingle beaches are, where does the material come from, and are shingle beaches relict features? Are they due to occasional storms, or are they active in that normal processes affect shingle as they do sand?

With the exception of Neate (1967), very few people have looked at the water depth from which shingle has been transported. This chapter looks at the ultimate origin of the pebbles at Mann Hill Beach in Scituate, Massachusetts, as well as the water depths from which they are transported, and then shows that this beach is not a relict feature but that even small waves can and do move the pebbles both from offshore to onshore as well as vice versa, and also alongshore. Mann Hill Beach is one of the pure shingle beaches along the coast in Massachusetts. It is a 2-km-long bay-mouth bar located in Scituate, 48 km south of Boston (Fig. 1).

A. Glacial History of the Region

The Pleistocene deposits of the area surrounding Boston are surprisingly poor in till and rich in outwash deposits. The area seems to have been located at the margin of two major ice lobes. The direction of ice flow varies from southwest to east through an arc of 135 degrees. However, the flow directions can be divided into five separate groups (Kaye, 1976). These are, in approximate order of occurrence, (a) lobate, spreading to south and east (Back Bay Readvance), (b) south (Beacon Hill Advance), (c) southwest, (d) south and southwesterly (mainly 32, 22, and 16 degrees), and (e) easterly and southeasterly (variable, 80 to 38 degrees). The oldest of the glacial indicators, the deep wide grooves, are easterly.

Many of the drumlin axes in the Boston area show the same range in orientation. In several places the long axis of the drumlin diverges from the striations exposed in the bedrock beneath them.

Mann Hill Beach fronts an area composed of sediments of Pleistocene and Holocene age (Chute, 1965). Glacial overburden, both ground moraine and kame terraces, is abundant. These glacial features are thought to be

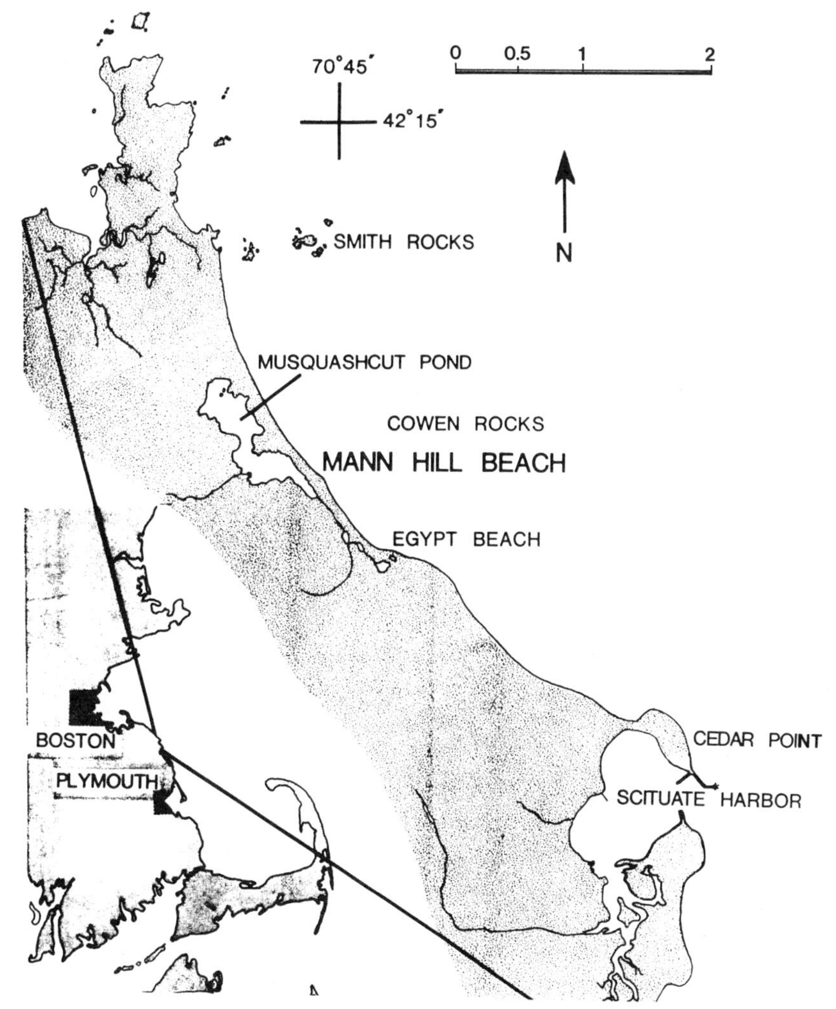

FIG. 1. Location map of Mann Hill Beach in Scituate, Massachusetts.

the result of ice stagnation and wastage, rather than remnants resulting from glacial advance or a steady recession of the ice front through that area (Flint, 1971).

B. BEDROCK OF THE REGION

The only exposure of bedrock in the immediate vicinity of Mann Hill Beach is the Precambrian Dedham Granodiorite at Cowen Rocks, 0.7 km offshore, at Smith Rocks offshore to the north, and at some submerged exposures roughly halfway between these two rocks.

The Dedham Granodiorite underlies the area south of Boston Basin separated from the basin by the Ponkapoag Fault (Fig. 2). The Dedham is more a cartographic unit than a lithologic one. Recent rare-earth analysis (Brenninkmeyer and Dillon, 1984) shows that the Dedham can definitively be divided along the lines suggested by Chute (1966) and Zartman and

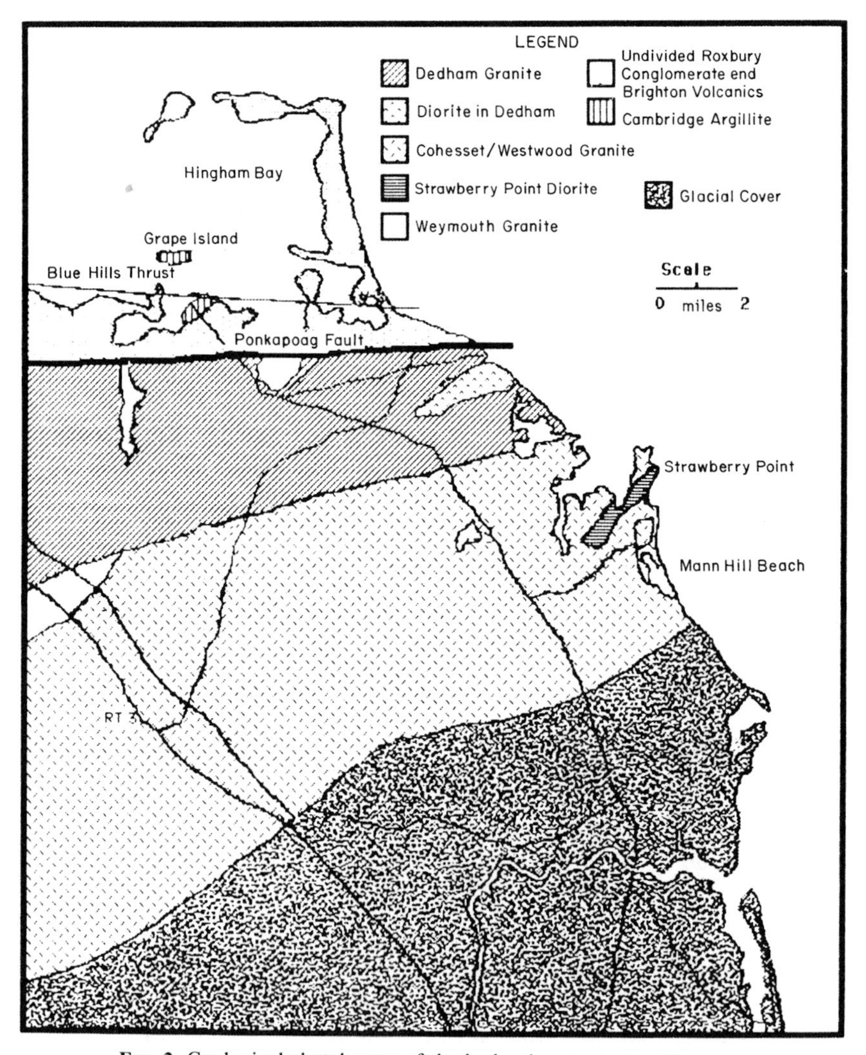

Fig. 2. Geological sketch map of the bedrock area south of Boston.

Naylor (1984). The Dedham has classically been defined as coarse biotite granodiorite (Crosby, 1893; Emerson, 1917). The actual Dedham only crops out for 4 miles south of the Ponkapoag Fault. In the Dedham there are many xenoliths of fine-grained amphibolite. South of that is a finer granite, called the Cohasset–Westwood. The only other granite that is exposed locally is the Weymouth Granite, which crops out in Hingham near Route 3. The Weymouth Granite is a foliated biotite granite.

Two diorites are associated with the granites. There is a dark-green diorite associated with the Dedham and a finer grained diorite associated with the Cohasset at Strawberry Point.

Only north of the Ponkapoag Fault, within the Boston Basin, is found the Boston Bay Group, which is divided into the Cambridge Argillite and the Roxbury Conglomerate. The argillite consists almost exclusively of gray argillite, in which the beds range in thickness from 0.1 to 7.6 cm. Within the argillite is a 120- to 150-m-thick quartzite bed called the "Milton" by Billings (1929). The Roxbury Conglomerate is divided into three members. They are, in ascending order, the Brookline, Dorchester, and Squantum members. The Brookline is a white to pink matrix-supported conglomerate. The pebbles range in size from 1 to 15 cm and are composed of quartzite, quartz monzonite, granite, and felsite. The Brookline member averages 60% conglomerate, 20% sandstone, and 20% argillite. The Dorchester is similar, except it is 15% conglomerate, 25% sandstone, and 60% argillite (Billings, 1976, 1982).

Scattered throughout the Boston Basin are the Mattapan–Brighton volcanics. These are dense, hard, white–pink and red rhyolites as well as "melaphyres." Melaphyres are altered basalts and andesites. These are dark to light green in color.

II. Mann Hill Beach

Mann Hill Beach is believed to have been formed by the devastating storm of November 1898 (Spayne, 1977). Before then, Musquashcut Pond was probably a small embayment. At low tide a sand bar stretches across the entire length of the beach. The shingle reaches 10 m above low water and shows well-developed imbrication, several berm lines, and cusps. The pebbles and cobbles of the shingle are both size and shape sorted.

The sediments at Mann Hill range from fine sand to boulders up to 330 mm in size. On the lower foreshore fine sand is dominant. On the middle foreshore, halfway between low and high tide, pebbles start to make their appearance. On the upper foreshore only shingle is found. The surface shingle is free from any sand matrix. Only at depth and close to the middle foreshore is there a sand matrix. The depth to the sand is time dependent. Right after a storm it is close to the surface. With time, it is lowered by backwash flushing.

The average of all the shingle sizes, based on the 30,000 measurements of the largest diameter, is 43 mm. The average intermediate diameter is 31 mm and the smallest diameter is 16 mm. There is a gradual increase in size from the upper foreshore to the berm. The largest sizes are found at the berm. Generally, a decrease in size is found landward of the berm. If more than one berm is present, as is often the case, there is a small decrease in size just landward of the berm in the swale.

Along the beach there is a gradual decrease of shingle size from north to south. In the north, the average size is 53 mm, while in the south it is 31 mm. The only places where there are local variations of this are within the cusps. Inside the cusp bays the coarsest sediment is found in the landward end and on the north side.

A. Shingle Lithology

There are 13 common lithologies found in the pebbles on Mann Hill Beach. As can be seen in Table I, granites are quite common. Three types of these are found. The most common is the foliated, light pinkish grey medium-grained Dedham Granodiorite. Less commonly found is the Cohasset–Westwood Granite. Even rarer is the Weymouth Granite.

The siltstone is a greenish to brownish grey, fine-grained, thin-bedded argillite that contains quartz, sericite, and opaque minerals (probably graphite) in a finer matrix too small to be identifiable. The siltstone probably belongs to the Boston Bay Group, specifically the Cambridge Argillite or possibly the Braintree Argillite.

The sandstone is obviously coarser, but has the same appearance as the argillite. It is not half as common as the argillite. The sandstone owes its origin to the Boston Bay Group in the Roxbury Conglomerate and Cambridge Argillite. Both of these are found only north of the Ponkapoag Fault.

Both the rhyolite and andesite are part of the Mattapan–Brighton volcanic complex. The rhyolite is the most common rock found on Mann Hill. The rhyolite is predominantly red and purple, but may be brown. Most of the rhyolite is so fine grained it is probably a devitrified tuff. The andesite is the second most common lithology found on Mann Hill Beach. It is bluish to greenish grey with small phenocrysts of sericitized plagioclase, quartz, and chlorite in groundmass of very fine plagioclase laths. Secondary epidote is common, giving the rock a greenish tinge.

Although less than 1% of all the rocks on Mann Hill are recognizable

TABLE I. SHAPE DIFFERENTIATION BETWEEN THE 13 COMMON PEBBLE AND COBBLE LITHOLOGIES AT MANN HILL BEACH (IN PERCENT)

Lithology	Blade	Ellip plate	Ellipsoid	Needle	Plate	Short rod	Sphere	Thick blade	Thick plate	Total
Dedham Granodiorite	0	1	27	1	1	18	15	1	36	10
Cohasset–Westwood Granite	0	1	25	1	1	17	16	2	36	5
Weymouth Granite	0	0	31	2	3	15	10	1	38	3
Diorite in Dedham	0	2	35	1	6	10	7	2	37	7
Strawberry Point Diorite	0	2	37	2	9	9	3	2	36	4
Mattapan Andesite	1	3	36	2	5	11	5	2	36	16
Amphibolite in Dedham	0	2	26	4	9	17	6	4	33	1
Basalt	1	6	34	2	8	10	3	3	33	6
Roxbury Conglomerate	1	5	17	1	9	11	6	6	42	1
Milton Quartzite	0	1	31	2	3	17	10	1	36	11
Mattapan Rhyolite	1	1	32	2	3	14	7	1	39	19
Cambridge Siltstone	1	10	36	1	13	5	2	7	25	14
Cambridge Sandstone	1	8	30	2	9	11	4	3	33	4
Total	0	3	33	2	5	12	7	2	35	

as Roxbury Conglomerate, there should be many more. Many of the Dedham, andesite, rhyolite, quartzite, and amphibolite pebbles are probably second generation, having first been present as pebbles in the Roxbury.

The basalt is a medium to dark greyish green, fine-grained, massive rock. It frequently has subophitic texture. The basalt probably came from the diabase dikes so prevalent throughout the neighboring quadrangles to the northwest.

There are two diorites found on Mann Hill. The first type of diorite is a medium- to coarse-grained massive dark rock, greenish grey in color, consisting predominantly of plagioclase, which is almost completely altered to epidote. The second diorite is very similar except that it is much finer grained and even darker in color. Based on rare-earth patterns, the first diorite is the diorite within the Dedham. The second is the Strawberry Point Diorite.

The quartzite is light grey to white, massive quartzite with minor quantities of biotite. It probably owes its origin to the "Milton Quartzite" of Billings (1929) or the quartzite pebbles in the Roxbury Conglomerate.

Least common of the rocks is the greyish green amphibolite, in which the layers have fine alterations of felsic and mafic minerals producing a striped appearance. The amphibolite is common only as xenoliths in the Dedham Granodiorite.

On Mann Hill Beach, all of the lithologies are not admixed equally. There are more rhyolite and quartzite on the lower fringes of the shingle and more siltstone and andesite on top of the berm. What is remarkable is the size distribution of the different lithologies. Each lithology has virtually the same distribution of intermediate axes (Table II). The only one that is different is the finer grained Strawberry Point Diorite.

B. Ultimate Origin of the Pebbles on Mann Hill Beach

In the over 30,000 measured and identified pebbles, neither the distinctive grey to bluish grey Quincy Granite nor the fine-grained riebeckite granite (Chute, 1969) has been encountered. The only exposure of both of these granites is 15 km to the northwest of Mann Hill Beach in between the Blue Hills Thrust and Ponkapoag Faults. The closest that the Mattapan volcanics are exposed to Mann Hill Beach is 7.5 km to the north. If the Ponkapoag Faults were extended out into the ocean, the closest the Mattapan could be found is 6 km from Mann Hill. Of all the pebbles on Mann Hill, those of the Weymouth Granite are the farthest from their source. These pebbles came from 15 km due west. The majority of pebbles on

TABLE II. PERCENTAGE DISTRIBUTION OF THE GRAIN SIZE (IN CENTIMETERS) OF THE 13 COMMON DIFFERENT LITHOLOGIES IN THE PEBBLES AND COBBLES AT MANN HILL BEACH (IN PERCENT)[a]

Lithology	1	1–2	2–3	3–4	4–5	5–6	6–7	7–8	8–9	9–10	10
Dedham Granodiorite	1	10	35	30	13	6	3	2	1	1	1
Cohasset–Westwood Granite	1	17	39	23	11	5	2	1	1	1	0
Weymouth Granite	1	13	28	28	14	7	3	4	2	1	0
Diorite in Dedham	1	11	36	28	15	5	2	2	1	1	1
Strawberry Point Diorite	0	1	46	28	14	8	3	1	1	0	0
Mattapan Anderite	1	18	36	25	12	5	2	1	1	1	1
Amphibole in Dedham	2	19	35	20	13	6	4	2	0	0	0
Basalt	1	17	39	26	11	5	1	1	1	0	1
Roxbury Conglomerate	1	11	30	34	14	5	1	3	1	0	0
Milton Quartzite	1	18	40	23	10	5	3	1	0	1	1
Mattapan Rhyolite	1	20	38	24	11	4	1	1	1	1	0
Cambridge Siltstone	1	19	39	23	11	5	1	1	1	1	1
Cambridge Sandstone	0	13	33	28	15	6	3	1	1	1	1

[a] The grain sizes are based on measurements of the long axes.

Mann Hill Beach owe their origin to the Back Bay readvance, coming primarily from the north and to a lesser degree from the west. However, the pebbles did not travel more than 15 km. Oldale and O'Hara (1975) as well as Raytheon (1972) report a patchy distribution of gravel offshore from Scituate. These patches form the immediate origin of the pebbles on the beach.

C. SHINGLE SHAPE

Anyone who has visited a shingled beach has seen the differences in the shape of the pebbles between those at midtide and those at the top of the berm. A description of shape or geometric form involves several different but related concepts. On one side is the particle with respect to standard Cartesian coordinates. On the other hand is the sediment's angularity or roundness.

In 1935 Zingg showed that if the ratio of intermediate to the maximum lengths (B/A) of a pebble is plotted against the ratio of the shortest to the intermediate lengths (C/B), the particle may be classified according to its shape. Zingg utilized four shape classes. Flemming (1964) has modified this to nine shapes (Fig. 3).

The flat shapes ($C/B < 0.64$) comprise more than 75% of all the pebbles on Mann Hill Beach. This number is surprising in that 55% of all the

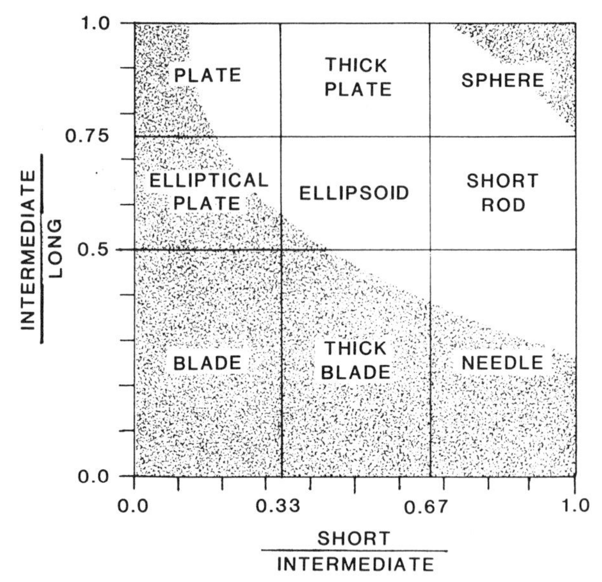

Fig. 3. Modified Zingg diagram of pebble shape classification, after Flemming (1964). The shaded areas are the shapes not commonly found on Mann Hill Beach.

pebbles on the beach are from nonlayered rocks composed of equidimensional grains (granite, diorite, rhyolite, and quartzite). There are two reasons for this: faulting history of the surrounding area and ease of transport.

The Boston Basin and the surrounding areas to the south are characterized by thrust faults, tear faults, and normal faults. Especially in the southern part of the basin, the structure consists of numerous thrusted anticlines, which constitute an imbricate block structure with minor thrusts and rock slices (Billings, 1929). Locally numerous shears are present, spaced at intervals between 1 and 100 cm apart. Moreover, two types of cleavages are prevalent. These ruptures are uniformly perpendicular to the bedding and to the imbricate blocks. Also nearly ubiquitous is a set of remarkably parallel joints spaced at intervals of 1–15 m. When two or three sets of joints are closely spaced, the rocks break into parallelopipeds 5–50 cm on a side (Billings, 1976). These, after being eroded by glacier ice and deposited in drumlins or outwash, are rounded by wave-induced transport to form the flat shapes. The Wadell (1933) operational sphericity follows the frequency of occurrence pattern. The white area of Fig. 3 is bordered on the bottom by a sphericity of 0.5 on the bottom and by 0.9 on the top.

The distribution of the shapes on the beach is lithology dependent. The Cambridge Argillite has the most plates and elliptical plates, and the granites and quartzites have the most spheres (Table I). There is, however, very little differentiation in size among the various lithologies (Table II). The majority of the pebbles could not have been transported very far either by the ice or by waves. The remarkable similarity of the different lithologies within each size shows there is relatively little abrasion and that the sizes are inherited from the faulting and jointing.

The second reason there are so many flat shapes on the beach is ease in transport. Wang (1986), in his flume studies of pebbles over a rough bed, has shown that nonspherical particles, especially blades and plates, are easier to transport than spheres at low fluid velocities. At higher fluid velocities there is no differentiation in shape. Each shape moves as rapidly as the other. Therefore, at depth when waves first have the capacity to transport pebbles, the flat shapes should move shoreward. Closer to the breaker zone, where the velocities are higher, all shapes should move shoreward. At the highest point of the swash, the flat particles are again selectively transported. This is indeed the case. At the top of the berm, over 85% of the pebbles are flat (Brenninkmeyer, 1982). Moreover, if spherical particles are deposited on the beach, they are selectively transported back offshore by the backwash. Spheres and short rocks roll more readily than flat pebbles. They have a lower pivotability (Winkelmolen, 1978). Therefore the spherical pebbles are selectively carried offshore by the combination of low pivotability and backwash.

III. Offshore Sediment

A sediment sampling program of the onshore and nearshore areas at Mann Hill was carried out on August 19, 1981. Offshore, samples were collected from 56 locations with a Shipek sampler. Figure 4 shows the sample locations.

Of the offshore samples, 19 were found to consist of gravel, 14 of a mixture of sand and gravel, and the rest were entirely composed of sand.

Figure 5 is a map of the mean size of the surficial sediments both on the beach and offshore. In order to determine the size, these samples were seived. The diameters were not measured. Pebbles with an average size of 24 mm in diameter are preponderant on the upper part of the foreshore. Offshore in the northwestern and central portions of the study area, mean grain sizes of up to 14 mm are common, while in the entire southern portion, mean grain sizes of less than 0.5 mm are dominant.

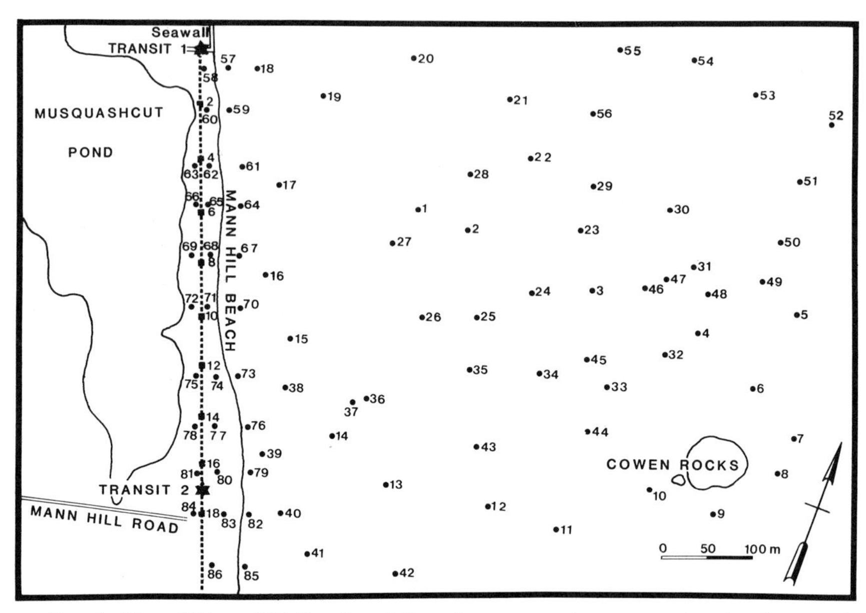

Fig. 4. Map of Mann Hill Beach and the adjacent nearshore area, showing the sample locations.

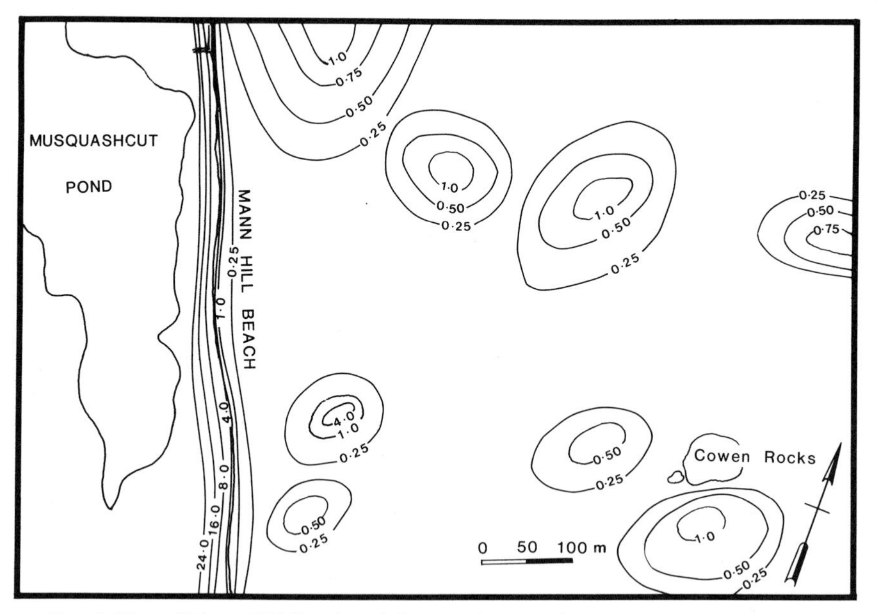

Fig. 5. Map of Mann Hill Beach and the nearshore environments, showing the trend of sediment distribution based on the observed mean sediment sizes. The values of the contours are in millimeters.

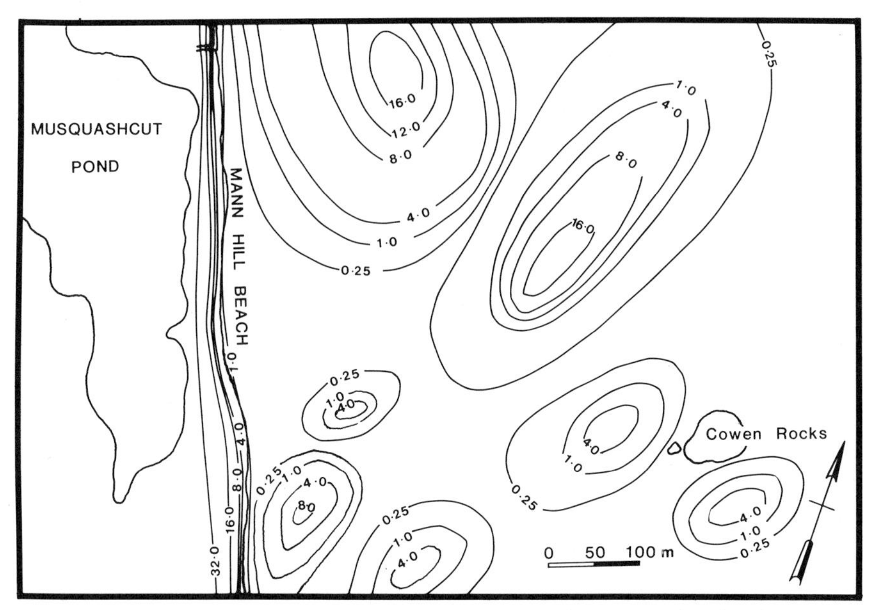

FIG. 6. Map of Mann Hill Beach and the nearshore environments, showing the coarsest sediment sizes observed. The values of the contours are in millimeters.

Figure 6 depicts the coarsest sediment size in each sample. Here, the overall trend is identical to that of the mean sediment sizes. Pebbles of up to 32 mm in diameter are found along the entire length of the upper foreshore. Pebble sizes of up to 16 mm are common in the northwestern and central portions of the study area. In the southwestern portion, the maximum gravel size drops to 4 mm. The coarsest sizes in the nearshore area occur in the zone that stretches from the northwestern corner southeastward, toward Cowen Rocks.

Figure 7 shows the values of Folk's (1968) inclusive graphic standard deviation. In general, the beach is well and moderately well sorted; offshore the sediments are very well to well sorted. Only in the northwestern area and in patches leading out from there to Cowan's Rocks do very poorly sorted sediments prevail.

Sediment skewness was found to be generally positive except in the northwestern and central portions of the study area, where the sediment samples have a negative skewness. The negative skewness values in these areas are probably the result of a tail of coarse material that was swept off the submerged gravelly sand deposit by approaching waves. Fox *et al.* (1966), in a profile of the moment measures of sediment samples taken perpendicular to a shoreline in South Haven, Michigan, came to a similar

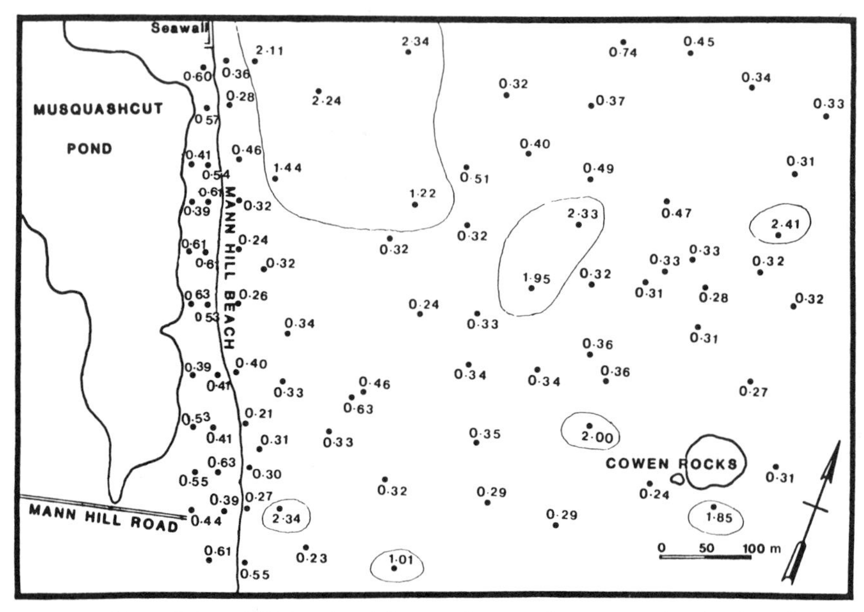

FIG. 7. Map of Mann Hill Beach and the nearshore environments, showing the values (in units) of the inclusive graphic standard deviation obtained by the size analyses of the samples from the various sample locations. The contour values are based on the categories of Folk (1968), with 0.5, 0.7, 1, and 2 delineating areas of well, moderately well, moderately, and poorly sorted sediments.

conclusion. They pointed out that the negative skewness values associated with the offshore bar sediment samples represent long tails of coarse material, which usually contain less than 5% of the sample.

A. SEDIMENT SHAPE

The large pebbles found in the offshore area are predominantly ellipsoids and spheres composed of granite and rhyolite. This contrasts with the flat pebbles that dominate the upper foreshore at Mann Hill Beach. Spheres are not as readily transported, and if they are transported landwards, spheres and rollers roll more readily than flat pebbles. Therefore, they are carried further seaward by the backwash. It is possible, therefore, that the pebbles obtained from nearshore locations 40, 44, and 50 (located in a zone where the samples are dominantly sand) may have come from the upper foreshore. This is shown at stations 23 and 44, a seaward distance of 420 and 480 m, respectively, where two gold-painted pebbles were found. These were 16 and 11.2 mm in size. Gold paint had been sprayed on the pebbles in a square foot area at all the sampling areas of the backshore and upper foreshore in the summer of 1980 (Brenninkmeyer, 1982).

IV. Waves

The most severe storms in the New England area are the relatively frequent northeasters and the comparatively infrequent tropical storms. Figure 8 shows the frequency of the wave periods and wave heights ob-

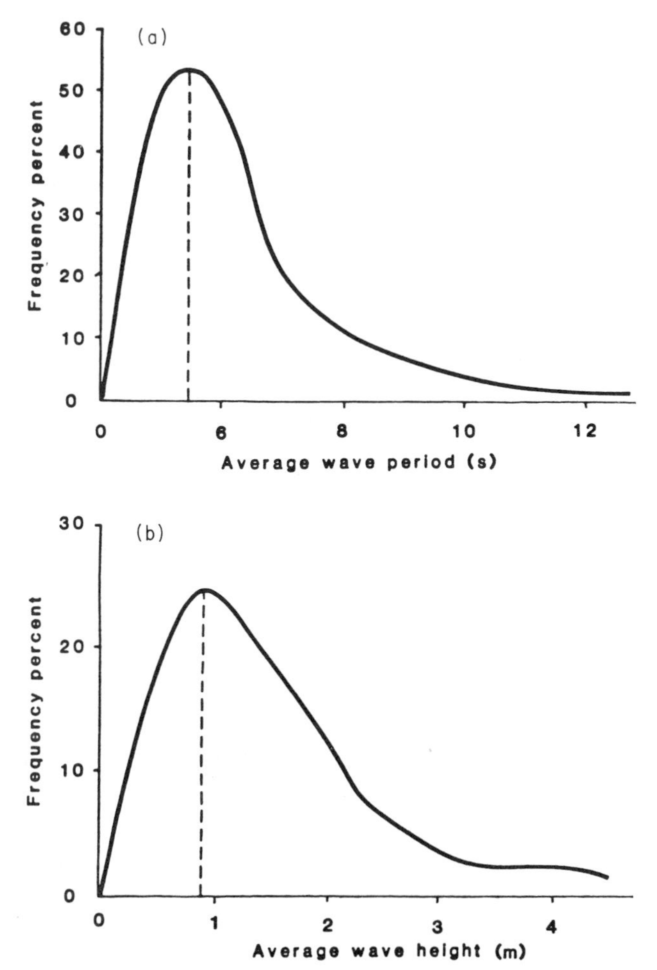

FIG. 8. (a) Wave periods and (b) wave heights for Massachusetts. [From U.S. Naval Weather Service Command (1970).]

served along the coast of Massachusetts. As can be seen, wave periods of 6 s and wave heights of 0.91 m are the most common. These waves reach Mann Hill only from ESE to NNE; all other directions are fetch-bound.

For this study, the wave refraction computer program by Rabe (1975) was utilized. The water depths used in the program were provided by averaging the National Ocean Survey Hydrographic Sounding Sheets at 100-m intervals. Figures 9 and 10 show the refraction pattern of a northeast wave approaching Mann Hill Beach with a wave period of 6 s and a deep-water wave height of 0.91 m at low and high tide, respectively. These figures show that the wave orthogonals continue along a straight course from deep water until they reach the breaker zone, where some show a slight southerly divergence. Thus, there should be a southerly longshore transport during the fair weather waves; Brenninkmeyer (1982), in a study of shingle movement at Mann Hill Beach, observed a dominant southerly transport during normal conditions. These two figures also show that there is no significant difference between the wave orthogonals at high and low tides.

The wave refraction pattern for a wave climate similar to that of the blizzard of February 1978, a 75-yr storm, was also investigated to provide an insight into the effects of such a storm on sediment transport. Figure 11 shows the refraction pattern for such a northeast storm with a deep-water wave height of 4.57 m and a wave period time of 8.5 s. These are the probable height and period for waves based on hindcasting for that February blizzard. During a nor'easter, the orthogonals show a slight northerly trend around the breaker zone. Therefore, there should be a northerly longshore transport during severe storms.

The actual net longshore transport will depend on the storm frequency. Historically there is relatively little net littoral transport in this area (Chute, 1949). From Strawberry Point to Smith Rocks (see Fig. 1), the shoreline is composed of rock and is deeply indented. There is no or at best very little longshore transport beyond each pocket beach. To the south, Egypt Beach is sandy.

The overall straightness of the wave orthogonals from deep water to shore confirms the visual observation that the beach is straight and suggests a simple bottom topography with nearly parallel, evenly spaced bottom contours. Wave energy can therefore be said to be evenly dissipated at the breaker zone along the entire length of the beach, and there is neither a zone of intensive erosion nor a zone of deposition by the most common directions of wave approach.

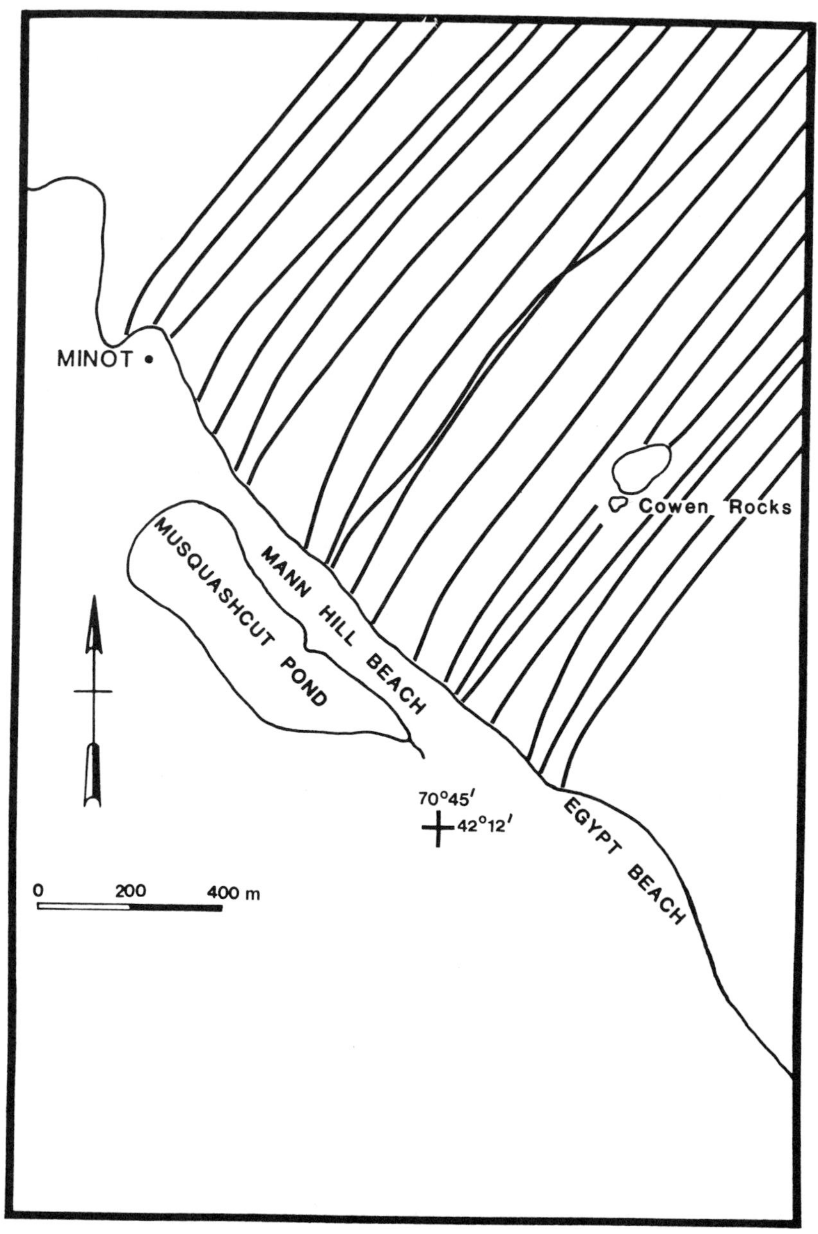

FIG. 9. Wave refraction pattern at low tide of waves approaching Mann Hill Beach from the northeast with a wave period of 6 s and a deep-water wave height of 0.91 m.

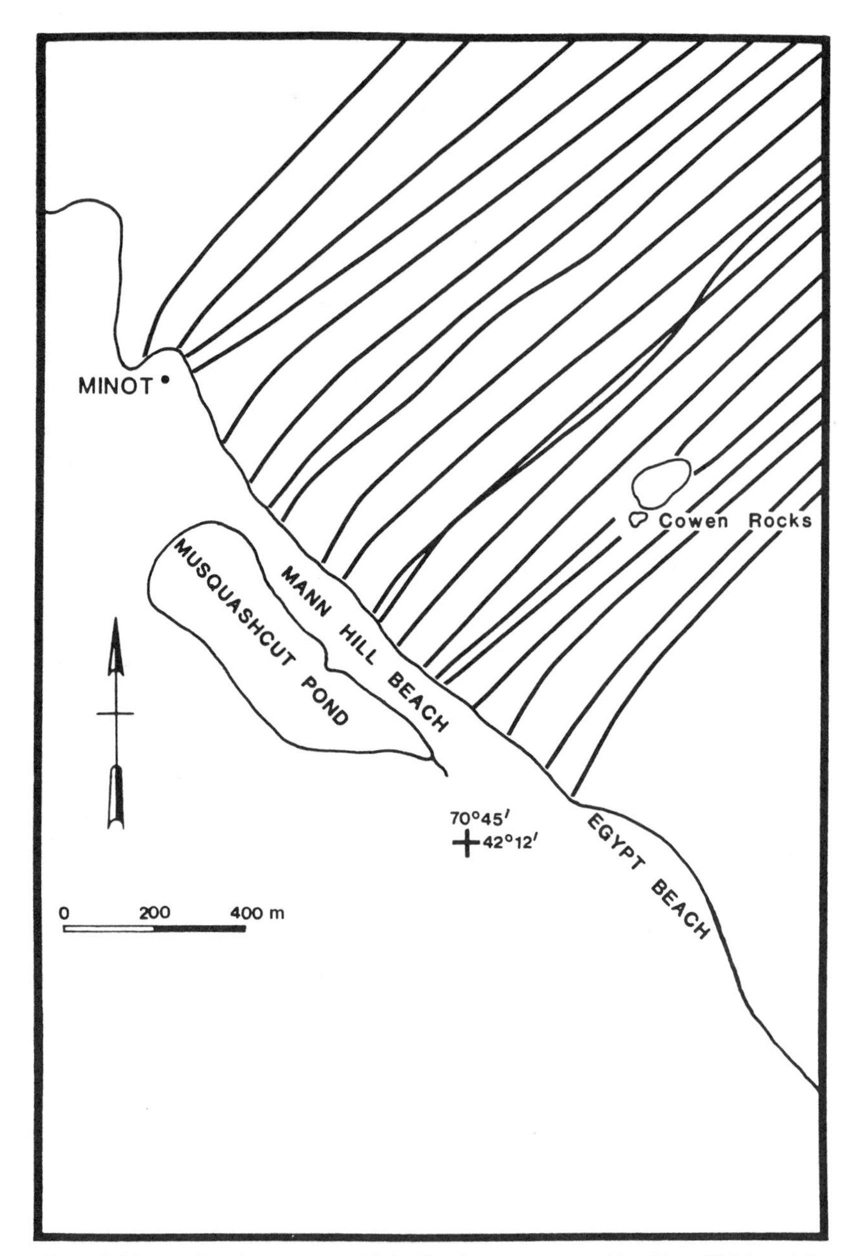

FIG. 10. Wave refraction pattern at high tide of waves approaching Mann Hill Beach from the northeast with a wave period of 6 s and a deep-water wave height of 0.91 m.

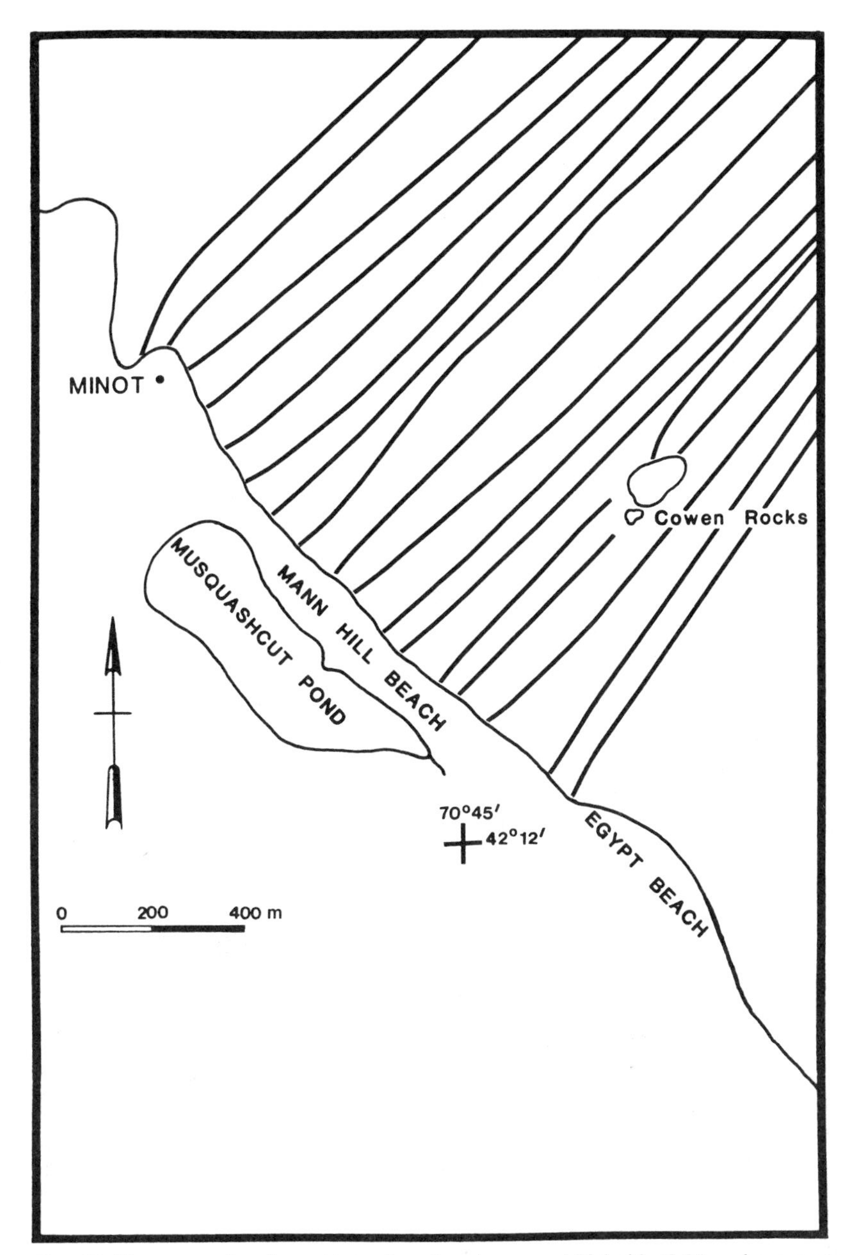

Fig. 11. The wave refraction pattern of northeast waves at high tide (3.96 m above mean low water) for a period of 8.5 s and a deep-water wave height of 4.57 m. These wave statistics are representative of those that accompanied the blizzard of 1978.

V. Wave-Induced Sediment Transport

A computer program for calculating the velocity of water motion in Airy waves required to entrain sediment was provided by Komar and Miller (1973). This program was used to study the movement of the coarsest sediment size (i.e., 16 mm) observed in the offshore area, with the aim of determining whether the approaching waves could actually move such pebble sizes at the depths at which they were located. The threshold computer program was also run for the coarsest pebble size (32 mm in diameter) observed on the beach during the sediment sampling for this project and again for a boulder of 330 mm diameter, which is the maximum gravel size ever found at Mann Hill Beach. The objective of the last two runs was to determine the water depth at which such sediment sizes could be set in motion. Beginning with a wave period of 3 s that increased at an interval of 0.5 s, the program calculates the corresponding water depth and wave height at which the given grain diameter is set in motion.

Table III is a sample output of the threshold of motion for a pebble with diameter 16 mm and for a wave period of 3 s. In this example, values of wave height greater than 2.20 m were not printed because, beyond this wave-height value, the wave steepness is too great for the wave to be stable. Airy wave theory requires such a wave to break. As a result, the program automatically increases the wave period to 3.5 s and repeats the computations.

At Mann Hill Beach, where the prevailing seas have a wave period of 6 s and a wave height of 0.91 m (see Fig. 8), a sediment size of 16 mm will be entrained at a water depth of 9.43 m. Since coarse particles of 16 mm diameter were found in nearshore sampling locations 20, 23, 24, 40, and 50, with corresponding water depths of 5.2, 8.6, 8.4, 2.4, and 13.8 m, it follows that the coarsest sediment particles found in all offshore locations except sampling location 50 can be entrained by the type of waves prevalent along Mann Hill Beach.

A very destructive storm, like the blizzard of 1978 with an estimated wave period of 8.5 s and a wave height of 4.57 m, will entrain the same sediment size at an offshore depth of 18.4 m. It follows, therefore, that the coarsest sediment particle found in sampling location 50, where the water depth is 13.8 m, could be entrained at that depth, given such storm conditions.

A 6-s wave with a wave height of 0.91 m will entrain the sizes found on the upper beach (32 mm) at a water depth of 9.22 m. A wave with a wave period of 8.5 s and a wave height of 4.57 m will do the same at a water depth of 22.7 m. On the other hand, a 6-s wave with a wave height of 0.91 m will set a boulder of 330 mm diameter in motion at a water depth

TABLE III. EXAMPLE OF THE OUTPUT ON COMPUTATIONS OF THRESHOLD FOR VARIOUS GRAVEL SIZES UNDER VARYING WAVE CONDITIONS.[a]

Wave height (cm)	Water depth (cm)	Wave steepness	Ratio of wave height to water depth
10.0	0.224 E +03	0.001	0.04
20.0	0.227 E +03	0.003	0.09
30.0	0.230 E +03	0.006	0.13
40.0	0.235 E +03	0.011	0.17
50.0	0.241 E +03	0.016	0.21
60.0	0.248 E +03	0.022	0.24
70.0	0.256 E +03	0.029	0.27
80.0	0.263 E +03	0.036	0.30
90.0	0.272 E +03	0.043	0.33
100.0	0.281 E +03	0.051	0.36
110.0	0.289 E +03	0.059	0.38
120.0	0.298 E +03	0.066	0.40
130.0	0.307 E +03	0.074	0.42
140.0	0.315 E +03	0.082	0.44
150.0	0.324 E +03	0.089	0.46
160.0	0.332 E +03	0.097	0.48
170.0	0.340 E +03	0.105	0.50
180.0	0.348 E +03	0.112	0.52
190.0	0.356 E +03	0.120	0.53
200.0	0.364 E +03	0.128	0.55
210.0	0.372 E +03	0.135	0.57
220.0	0.379 E +03	0.143	0.58

[a]Grain diameter = 1.600 cm; wave period = 3.00 s; orbital velocity, U_m = 102.66 cm/s.

of 8.98 m, while an 8.5-s wave with a wave height of 4.57 m will entrain the same material at a depth of 18.8 m. Table IV shows the water depths and distances offshore at which waves prevalent at Mann Hill Beach and the type of waves encountered on the same beach during a major storm similar to that of 1978 can entrain the three sediment sizes in question. The distances offshore at which these materials are entrained are based on the seaward distances at which the threshold water depths correspond to the water depths on the hydrographic maps.

Based on the maximum wave steepness after which a given wave becomes unstable, the theoretical maximum water depth and hence the maximum offshore distances at which the given wave can entrain a gravel of known diameter can also be determined. Such data for gravel diameters 16, 32, and 330 mm based on the prevailing waves at Mann Hill Beach under wave conditions similar to the blizzard of 1978 are shown in Table

TABLE IV. DEPTH AND DISTANCE REQUIRED TO ENTRAIN A RANGE OF GRAVEL SIZES UNDER THE WAVE CONDITIONS AT MANN HILL BEACH FOR 6 AND 8.5 WAVES, WHICH ARE 0.91 AND 4.57 M HIGH, RESPECTIVELY.

Sediment diameter (mm)	Wave period (s)	Wave height (m)	Wave steepness (H/L)	Water depth (m)	Approximate distance from shore (m)
Prevailing wave condition					
16	6.0	0.91	0.006	9.43	560
32	6.0	0.91	0.005	9.22	550
330	6.0	0.91	0.002	8.98	540
Severe storm condition					
16	8.5	4.57	0.032	25.10	1400
32	8.5	4.57	0.029	22.70	1200
330	8.5	4.57	0.014	18.80	1060
Maximum depths and distances at which a range of gravel sizes can be set in motion when the 6 and 8.5 waves are at their steepest					
Prevailing wave condition					
16	6.0	8.4	0.145	19.1	1070
32	6.0	9.0	0.149	17.2	980
330	6.0	9.1	0.122	11.7	660
Severe storm condition					
16	8.5	16.6	0.144	42.8	2800
32	8.5	17.0	0.146	38.4	2300
330	8.5	17.5	0.162	27.6	1680

IV. Thus, for the coarsest gravel (diameter 16 mm) observed in the offshore area, a 6-s wave with a wave height of 8.4 m will entrain the gravel at a depth of 19.1 m and a corresponding offshore distance of approximately 1070 m, while an 8.5-s wave with a wave height of 16.6 m will entrain the same gravel at a depth of 42.8 m and a corresponding offshore distance of about 2800 m. However, the possibility of such conditions is remote because only 10–15% of ocean waves exceed 6 m in height and virtually none will be that steep.

Figure 12 is obtained by plotting the water depths at which the three gravel sizes under investigation can be entrained at specified wave periods and wave heights and, finally, drawing lines through the water-depth values to represent the 5-m, 10-m, 15-m, 20-m, 25-m, 30-m, and 35-m water-depth curves. From this figure, one can thus determine the approximate water depth at which any of the three gravel sizes can be set in motion when the wave period and wave height are specified. For Mann Hill Beach,

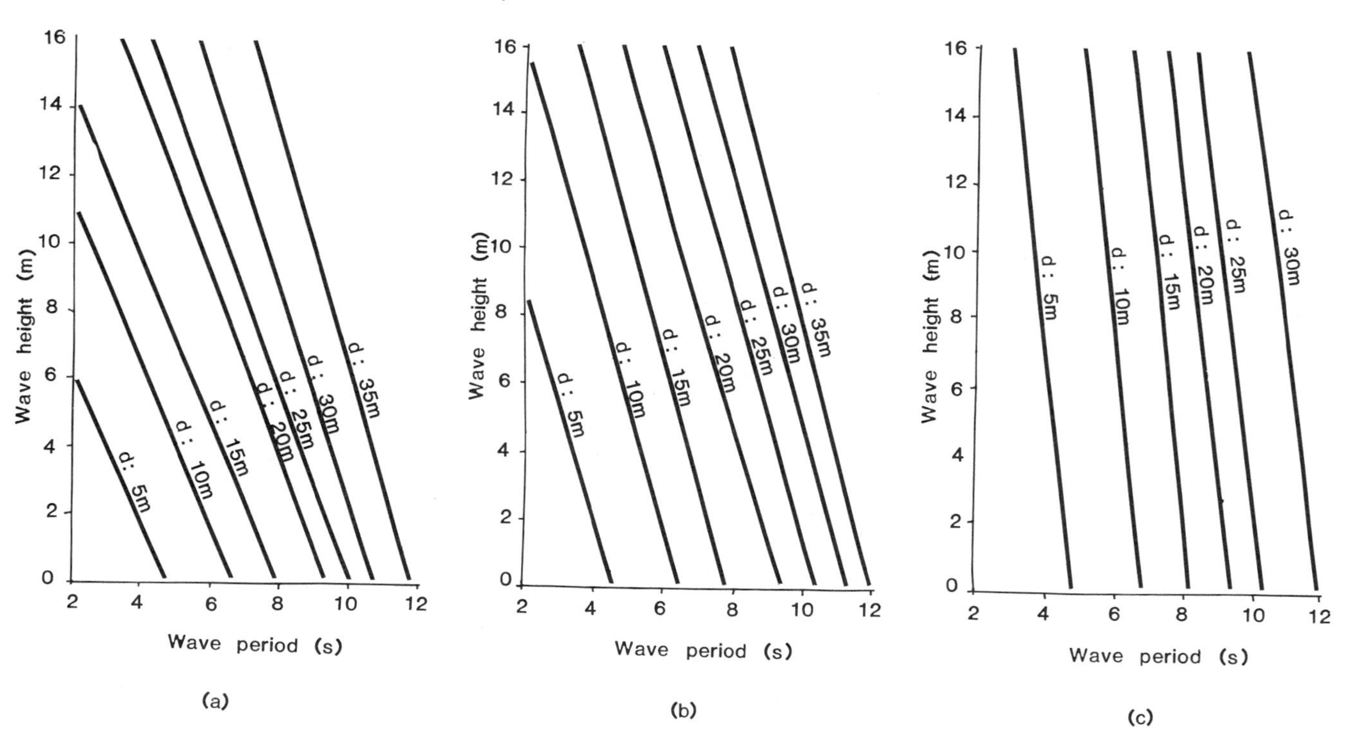

FIG. 12. Depths (5, 10, 15, 20, 25, 30, 35 m) at which threshold velocities are reached for sediment diameters of (a) 16 mm, (b) 32 mm, and (c) 330 mm at any wave period and wave height.

where a wave period of 6 s and a wave height of approximately 1 m are common, Fig. 12 shows that a gravel of 16 mm in diameter will be entrained at a water depth of 9 m, while a gravel of 32 mm in diameter will be entrained at 8.8 m. But the advantage of Fig. 12 is even more evident at higher wave heights. A wave with a height of 6 m and a period of 6 s will entrain a gravel with a diameter of 16 mm at a depth of ~17 m, while a pebble of 32 mm will be entrained at ~14.5 m. This practical means of evaluating the water depth at which the threshold of sediment motion is attained is based on a progressive increase in threshold water depth engendered by increasing wave period and wave height. The higher the wave height, the greater the water depth at which a given sediment size can be entrained. Also, the longer the wave period, the greater the water depth at which a sediment particle is placed in motion. Thus, for a wave period of 6 s, the maximum water depth at which a gravel of 16 mm in diameter is placed in motion is 19.1 m, while that of an 8.5-s wave acting on the same gravel size is 43 m. The pebbles could be moved; however, the waves to move them are too steep and, consequently, unstable.

VI. Source of the Shingle on Mann Hill Beach

The exposed foreshore sediment has more pebbles, is more poorly sorted, and is more consolidated at the northern end of the beach. The shingle size is also coarser on the northern part of Mann Hill Beach. The grain-size distribution of the offshore samples together with diver observation indicates the presence of an extensive zone of gravelly sand offshore from the northern end of the beach. This gravelly zone extends seaward toward Cowen Rocks a distance of about 650 m. It is this zone that forms the primary source of the Mann Hill Beach shingle.

As can be seen in Fig. 13, the water depths of this gravelly sand extend from mean low water to 12 m. As calculated, the waves most common at Mann Hill can move the coarsest sizes from most of this gravel patch. Only from depths deeper than 9 m can the common waves not move the larger clasts. These larger clasts at the seaward end of the gravel patch are moved shoreward during storms.

Still, it is possible that some pockets of submerged gravelly sand exist further offshore in the Mann Hill Beach area (Oldale and O'Hara, 1975), and gravels from such submerged gravelly sand can be moved shoreward by severe storms. The threshold measurements indicate that the coarsest pebbles (330 mm) found on the beach could have been transported from an offshore distance of up to 1680 m for the steepest waves. However, a distance of only 1060 m is more likely. This is the maximum distance

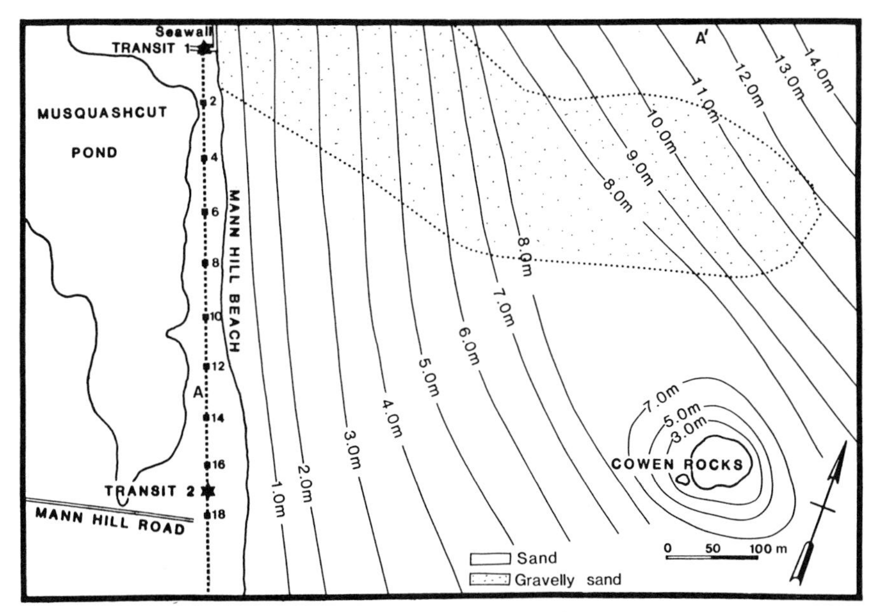

FIG. 13. Nearshore bathymetry and the submerged poorly sorted gravelly sand around the northern end of Mann Hill Beach, Scituate. The bottom contours are in meters.

storm waves similar to the blizzard of 1978 could have moved the coarsest beach clasts.

During storms from deeper depths and from shallower depths in fair weather, whatever sediment is encountered is transported landward. Over a rough bottom, the larger flat pebbles are transported farther than the smaller ones. The smaller pebbles are left in the intersticies. The large flat pebbles are deposited at the swash line. It is the swash-line location that determines where the larger pebbles are deposited. During a storm it is close to the top or over the beach. During fair weather it is lower down the beach. Round shingles, if they are transported, are less likely to be deposited on the beach, but rather, because they roll more readily, are brought back down the beach by the backwash. Any sand that is deposited with the shingle is slowly washed through the gravel by the backwash. This leaves a beach deposit of flat shingle without an admixture of sand on the upper beach and a sand layer at the lower beach.

REFERENCES

Billings, M. P. (1929). Structural geology of the eastern part of the Boston Basin. *Am. J. Sci.* **18,** 97–137.

Billings, M. P. (1976). Geology of the Boston Basin. *Geol. Soc. Am. Mem.* **146,** 5–30.
Billings, M. P. (1982). Ordovician cauldron subsidence of the Blue Hills Complex, eastern Massachusetts. *Geol. Soc. Am. Bull.* **93,** 909–920.
Bluck, B. J. (1967). Sedimentation of beach gravels: Examples from South Wales. *J. Sediment. Petrol.* **37,** 128–156.
Brenninkmeyer, B. M. (1982). Erosion and shingle movement at Mann Hill Beach, Scituate, Massachusetts during the blizzard of February 1978. *In* "Geotechnology in Massachusetts" (O. C. Farquhar, ed.), pp. 479–499. Univ. of Massachusetts Press, Amherst.
Brenninkmeyer, B. M., and Dillon, P. M. (1984). Rock lithology and glacial transport southeast of Boston. *In* "Geology of the Coastal Lowlands, Boston MA to Kennebunk, ME," New England Intercollegiate Geological Conference (L. Hanson, ed.), pp. 417–435.
Carr, A. P. (1969). Size grading along a pebble beach: Checil Beach, England. *J. Sediment. Petrol.* **39,** 297–311.
Carr, A. P. (1971). Experiments on longshore transport and sorting of pebbles: Checil Beach, England. *J. Sediment. Petrol.* **41,** 1084–1104.
Carr, A. P. (1975). Differential movement of coarse sediment particles. *Proc. Int. Coastal Eng. Conf., 14th* pp. 851–870.
Chute, N. E. (1949). Preliminary study of beach erosion of the shoreline between Nantasket Beach and Duxbury Beach, Massachusetts. *U.S. Geol. Surv. Open-File Rep.*
Chute, N. E. (1965). Geologic map of the Scituate quadrangle, Plymouth County, Massachusetts. *U.S. Geol. Surv. Map* GQ-467.
Chute, N. E. (1966). Geology of the Norwood quadrangle, Norfolk and Suffolk Counties, Massachusetts. *U.S. Geol. Surv. Bull.* **1163B.**
Chute, N. E. (1969). Bedrock geologic map of the Blue Hills quadrangle, Norfolk, Suffolk and Plymouth Counties, Massachusetts. *U.S. Geol. Surv. Map* GQ-796.
Cornish, V. (1898). On sea beaches and sand banks. *Geogr. J.* **7,** 528–543, 628–657.
Crosby, W. O. (1893). Geology of the Boston Basin, Nantasket and Cohasset. *Boston Soc. Nat. Hist. Occas. Pap. 4* **1,** Part 1, 1–77.
Emerson, B. K. (1917). Geology of Massachusetts and Rhode Island. *U.S. Geol. Surv. Bull.* **597.**
Flemming, K. C. (1964). Tank experiments on the sorting of beach material during cusp formation. *J. Sediment. Petrol.* **34,** 112–122.
Flint, R. F. (1971). "Glacial and Quaternary Geology." Wiley, New York.
Folk, R. L. (1968). "Petrology of Sedimentary Rocks." Hemphill's, Austin, Texas.
Fox, W. T., Ladd, J. W., and Martin, M. K. (1966). A profile of the four moment measures perpendicular to a shoreline, South Haven, Michigan. *J. Sediment. Petrol.* **36,** 1126–1130.
Glèason, R., and Hardcastle, P. J. (1973). The significance of wave parameters in the sorting of beach pebbles. *Estuarine Coastal Mar. Sci.* **1,** 11–18.
Kaye, C. A. (1976). Outline of the Pleistocene geology of the Boston Basin. *In* "Geology of Southeastern New England," New England Intercollegiate Geological Conference Guidebook (B. Cameron, ed.), pp. 46–63.
Komar, P. D., and Miller, M. C. (1973). The threshold of sediment movement under oscillatory water waves. *J. Sediment. Petrol.* **43,** 1101–1110.
Neate, D. J. (1967). Underwater pebble grading of Checil Beach. *Proc. Geol. Assoc.* **78,** Part 3, 419–426.
Oldale, R. N., and O'Hara, C. J. (1975). Preliminary report on the geology and sand and gravel resources of Cape Cod Bay, Massachusetts. *U.S. Geol. Surv. Open-File Rep.*
Orford, J. D. (1975). Discrimination of particle zonation on a pebble beach. *Sedimentology* **22,** 441–463.

Orford, J. D. (1977). A proposed mechanism for storm beach sedimentation. *Earth Surf. Processes* **2,** 381–400.
Orford, J. D., and Carter, R. W. (1982a). Crestal overlap and washover sedimentation on a fringing sandy gravel barrier coast, Carnsore Point, south-east Ireland. *J. Sediment. Petrol.* **52,** 265–278.
Orford, J. D., and Carter, R. W. (1982b). Geomorphological changes on the barrier coasts of south Wexford. *Geogr. Soc. Irel.* **15,** 70–84.
Rabe, K. (1975). "The Delaware–Dobson Wave Refraction Model," Environmental Prediction Research Facility Computer Programming Note 21. U.S. Dep. Navy, Washington, D.C.
Raytheon Co. (1972). "Final Report of Massachusetts Coastal Mineral Inventory Survey."
Spayne, R. W. (1977). "Shingle Beach, North Scituate, Massachusetts: A Geomorphological Study. Boston State Coll., Boston, Massachusetts.
U.S. Naval Weather Service Command (1970). "Summary of Synoptic Meteorological Observations: North America Coastal Marine Areas," Vol. 2. Washington, D.C.
Waddell, H. A. (1933). Shericity and roundness of rock particles. *J. Geol.* **41,** 310–331.
Wang, S. C. (1986). Shape effects on the transport of particles of equal mass and volume. Thesis, Dep. Geol. Geophys., Boston Coll., Boston, Massachusetts.
Winkelmolen, A. M. (1978). Size, shape and density sorting of beach material along the Holderness coast Yorkshire. *Proc. Yorkshire Geol. Soc.* **42,** 109–141.
Zartman, R. E., and Naylor, R. S. (1984). Structural implications of some radiometric ages of igneous rocks in southeastern New England. *Geol. Soc. Am. Bull.* **95,** 522–539.

CHAPTER 11

Shoreline Development of the Glacial Cape Cod Coastline

JOHN J. FISHER*

Department of Geology
University of Rhode Island
Kingston, Rhode Island

The Cape Cod eastern coastline developed initially from an extensive glacial outwash plain. It is the longest unbroken unconsolidated coastline of glacial origin in New England. Without a single jetty, groin, seawall, or revetment or influence of bedrock topography, it is an excellent example of a glacial coastline developed completely by marine processes. This subsequent marine action has developed spits, barrier spits, and barrier islands growing both north and south from a nodal zone along the eroding glacial sea cliff headlands.

A synoptic shoreline sediment sampling study indicated the boundaries of these various shoreline segments as well as the nodal zone. Grain size analysis indicates longshore transport is most effective along the spit shoreline segment, slightly less effective along the sea cliff headlands, and least effective along the barrier-spit and barrier-island shorelines. Interestingly, grain size increased northward in the direction of longshore drift, due perhaps to winnowing action by winter storms. Beach sediment patterns divided the above shoreline segments into a pattern similar to other mid-Atlantic Bight barrier coastlines. This pattern consists of a regular shoreline segment sequence of (1) a spit or cape, (2) an eroding mainland or headland with a nodal zone, (3) a barrier spit or a continuous barrier island, and finally (4) discontinuous barrier islands separated by inlets.

*The author wishes to thank J. Barrett for drafting and R. Kranes for sediment sampling and analysis.

I. Introduction

Pleistocene glacial deposits are important in determining the character of the coastal geology of southern New England, and for Cape Cod they are the major control. Embayed, barrier, baymouth bar, and spit coastlines are common in this region, as are the unconsolidated cliffed shorelines, and are the result of postglacial submergence and wave action on the glacial deposits. The dramatic unconsolidated cliffed shoreline along the Atlantic coast section of the Cape Cod National Seashore is the result of wave erosion, during a rising sea level, on the thick glacial outwash plain deposits. The rising sea level resulted from worldwide melting of the Pleistocene glacial ice sheet. Longshore transport along this coastline has developed extensive spit shorelines, with the Provincetown Spit to the north and the Nauset and Monomoy Island spit complex to the south (Fig. 1).

The regional coastal geomorphology of the northern Cape Cod outer shoreline has been well known for years, as a result of the pioneering coastal studies of Davis (1896). Davis developed his model of spit development in the Cape Cod Provincetown Spit area. This model included a migrating "fulcrum" point between the erosional marine seacliff and the depositional spit and spit ridges. The model was also applied by Davis to the Sandy Hook Spit of New Jersey and in greater detail somewhat later by Johnson (1919). There are, however, some interesting problems remaining concerning details of this geomorphology as it relates to the remainder of the Cape Cod shoreline, the dynamics of shoreline erosion and regional sediment distribution. Certain specific questions are:

(1) Does the past extensive erosion at Coast Guard Beach represent a migration southward of a possible southern fulcrum point?

(2) Where is the location of the Cape Cod shoreline nodal point, where longshore current changes direction, and is it also migrating southward with the fulcrum point?

(3) Can the location of this nodal point and these fulcrum points be determined from coastal morphology, wave dynamics, or sediment patterns?

Regional coastlines, such as the outer Cape Cod coastline, when analyzed are usually found to be composed of segments of discrete shorelines. The geomorphologic pattern of the Cape Cod coastline is composed of elements that have developed under the wind–wave dynamics of the New England–Middle Atlantic region. The complex regional Cape Cod coastline can be compared to other regional coastlines along the Atlantic coast to determine geomorphological similarities and differences. Morphological

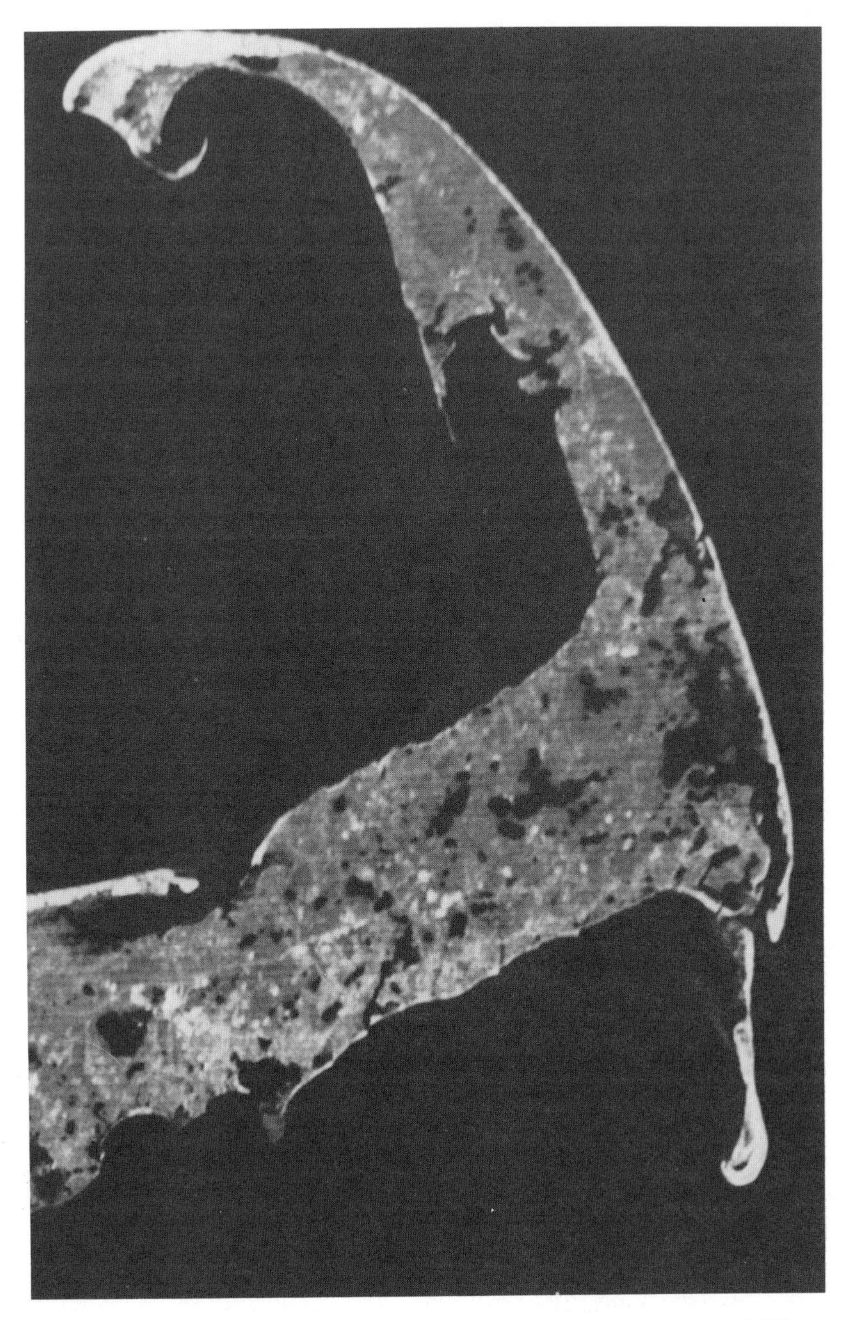

Fig. 1. Landsat satellite image of outer Cape Cod, indicating spectral tonal difference between glacial deposits (gray tone) and recent beach and dune deposits (white tone). Marine oceanic waters, as well as estuaries and kettle-hole ponds, shown in black. Provincelands Spit, as well as Nauset Beach barrier spit and Monomoy Island, visible as white tonal patterns. (NASA photo.)

similarities can then indicate similar histories of development under similar dynamic marine conditions.

II. Glacial Geology of Cape Cod

Cape Cod is the easternmost extent of the major prominent late Wisconsinan end moraine system of the eastern United States. Along this southern New England coast, the Wisconsinan terminal moraine forms the major portion of Long Island and the offshore islands of Block Island, Martha's Vineyard, and Nantucket Island. In contrast, the Wisconsinan recessional moraine extends in part along the northern coast of western Long Island, the southern coast of Rhode Island, continuing as the Elizabeth Islands, and then north along Buzzards Bay to the Cape Cod canal. It continues as a prominent ridge eastward along the Cape Cod Bay shoreline as the Sandwich moraine to the outer Cape shoreline. Both the recessional and terminal moraines then continue offshore under water to the east as the productive Georges Banks fishing grounds (Fig. 2).

The Sandwich moraine itself is composed primarily of stratified drift, with lesser amounts of till, while boulders are common within the moraine. Eastward on the Cape, the Sandwich moraine decreases in height and is subdued in character. The slightly rolling terrain around Orleans, together with boulders in the adjacent fields, indicates glacial ice deposition, but most of the eastern Sandwich moraine is an area of stratified drift deposited on the margin of the ice sheet and let down by collapse. To the north along Cape Cod Bay is a lowland region considered in the past as a ground moraine, but that recent mapping has found to be proglacial deltas. These deltas formed in the glacial lake that occupied the Cape Cod Bay depression as the ice lobe in the bay melted. South of the Sandwich moraine is the sandy Harwich outwash plain. Bordering Nantucket Sound on Cape Cod mainland is the Hyannis coastline, a series of small estuaries with sandy baymouth bars. The bays are primarily drowned preglacial stream valleys that carried south the meltwater drainage from the ice sheet that formed the Sandwich moraine. Postglacial rise of sea level has submerged this coast, and the result is the embayed and baymouth barrier along most of the south shore of Cape Cod.

Western Cape Cod is composed of two intersecting recessional moraines formed by deposition from a Buzzards Bay ice lobe to the west and the Cape Cod Bay lobe to the north. In contrast, the eastern outer Cape is a system of outwash plains, formed in an interlobate area between two ice lobes, the Cape Cod Bay lobe on the west and the South Channel lobe

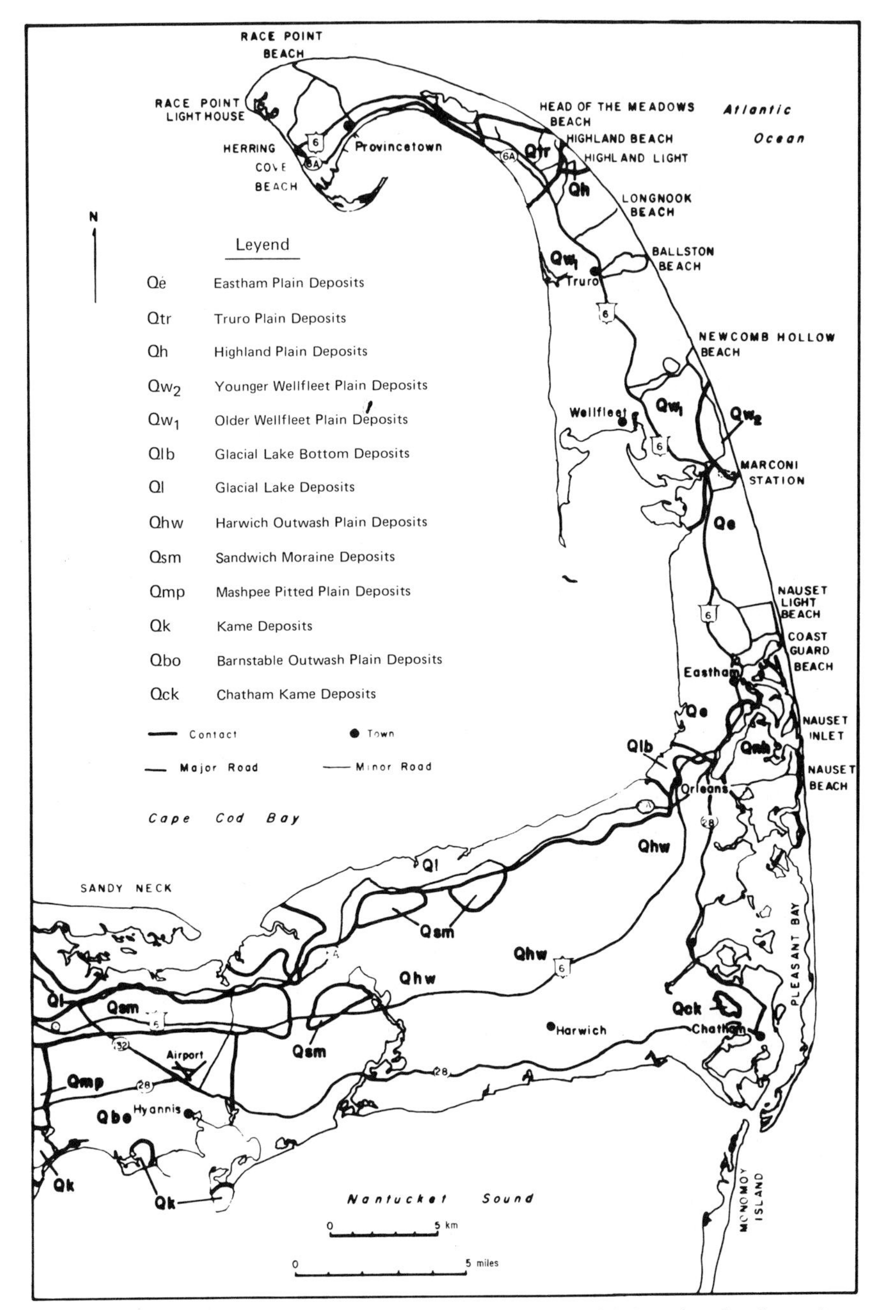

FIG. 2. Glacial geologic map of eastern Cape Cod with glacial deposits of end moraines, outwash plains, glacial lake deposits, and kame deposits. Outer Cape composed of outwash plain terraces, from north to south, of Truro plain, Highland plain, Wellfleet plain, and Eastham plain. (Sources: various U.S. Geological Survey maps.)

offshore from the present coast on the east. The outwash plain sediments themselves were deposited primarily from this south Channel lobe. The first of this system of south-to-north interlobate outwash plain deposits is the Eastham plain. The maximum altitude on the Eastham plain on uncollapsed glacial drift in the vicinity of Nauset Light is at an elevation of 25 m. Eastham plain outwash glacial deposits north of Nauset Light are mostly fine to very coarse sand with few cobbles, in contrast to the silt and boulders present in the deposits to the south. In general, these glacial deposits have a high content of felsic volcanic rocks (Oldale, 1968). Rising to the north above the Eashham plain is the Wellfleet plain of Grabau (1897) and Woodworth and Wigglesworth (1934). A glacial outwash plain with a gentle southwestward slope, large entrenched valleys, and numerous kettle holes, it is the oldest, highest, and most extensive of the Cape outwash plains. One-third the way down the cliff face at Marconi's Wireless Station site is the contact between the younger and older Wellfleet outwash plain glacial deposits. The older unit (Qw_1) is composed of fine to coarse sand. A younger second unit of the Wellfleet deposit (Qw_2) overlies the older Wellfleet deposit and occurs in a more limited area from just north of Cohoon Hollow to just south of the Marconi Station. It is similar to the older Wellfleet deposits, except that it is primarily gravelly sand. Highland Lighthouse at 40 m is situated on the Cape cliffs on the Highland outwash plain, a small triangular area north of the Wellfleet plain. It is primarily well-stratified sand and gravel, which earlier was thought to contain Wisconsinan glacial sands, Sangamon interglacial clays, and pre-Wisconsinan glacial basal sands in its prominent cliff outcrop (Woodworth and Wigglesworth, 1934). Now all deposits are found to be late Wisconsinan in age, with glacial lake clay deposits (Koteff *et al.*, 1967). Finally there is the Truro plain, most northern of these Cape Cod outwash plains. There are less ice collapse features on this plain, which is composed primarily of fine sand with foreset beds suggesting deltaic deposition in the meltwater lake adjacent to the Cape Cod Bay lobe.

The age of the glacial sediments that make up the Pleistocene of Cape Cod has been put at between 14,000 and 15,000 yr B.P. (Oldale *et al.*, 1968). This is estimated from radiocarbon dating of 15,300 $\pm$ 800 yr B.P. for tundra flora on Martha's Vineyard to the south and dates of 14,000 yr B.P. for marine clays north of Boston. An advance of ice over the area at about 20,000 yr B.P. is suggested from marine shells found in the Cape Cod glacial nonmarine sediments. These shells were probably picked up by the advancing ice from marine areas to the north and later deposited in the stratified drift by the meltwater streams as the ice sheet subsequently melted.

III. Coastal Geomorphology of Cape Cod

The wave-cut marine sea cliff extending for over 25 km along the eastern shoreline of Cape Cod is the most prominent coastal feature of the entire coast. It was eroded from a system of glacial outwash plains that increase in height toward the north, with the highest elevation in the vicinity of Highland Light at over 40 m. Along this marine sea cliff, these glacial outwash plains from north to south are the Truro, Highland, Older Wellfleet, Younger Wellfleet, and Eastham plains. Further to the south, behind Nauset and Pleasant bays, are the Nauset Heights and Harwich outwash plain deposits (Oldale, 1976). A series of "hollows" or gaps in this cliff face is the result of cliff erosion breaching or beheading former glacial outwash channels, called pamets, that once flowed from east to west across these outwash plains.

These wave-cut cliffs along the outer Cape are the remnants of a once extensive glacial land mass that extended offshore to the east. Reconstruction of the shape and extent of this former coastline can be developed from different lines of evidence. Shaler (1897) claimed that "slope extension" of the outwash plain surface indicates that this original shoreline could not be less than ½ mile or more than 4 miles. Davis (1896) earlier suggested that the "greatest retreat of the original shore to the present shore" was about 2½ miles, based on his analysis of fulcrum retreat and reconstructed the initial shoreline on this basis. He also suggested that this erosion would have occurred primarily within the past 3000 years, based on measured average rates of erosion. Johnson (1925), in a similar manner, calculated how long this outer section of Cape Cod might continue to exist. At Highland Light and south of Wellfleet, the width of the Cape is about 10,000 ft, while the wider section north of Wellfleet averages about 20,000 ft in width, with the maximum 25,000 ft. With an average rate of recession of 3 ft/yr, the narrow parts should erode through in about 3000 yr, and the wider parts in 6000–8000 yr (Fig. 3).

Recent sea-level curve data can also now be used to determine the original extent of the Cape Cod offshore land mass by multiplying the average rate of cliff erosion of 3 ft/yr (Zeigler, 1960) by the length of time the sea has been at its present level. Curray's (1965) sea-level rise curve indicates that about 3500 yr B.P. sea level was close to the present level and has been rising slowly but steadily since then. Application of these two values indicates that the original Cape Cod shoreline could have extended a minimum of some 2 miles offshore ($3500 \times 3 = 10{,}500$ ft) during the past 3500 yr. A prominent spit along Cape Cod Bay shoreline is Sandy

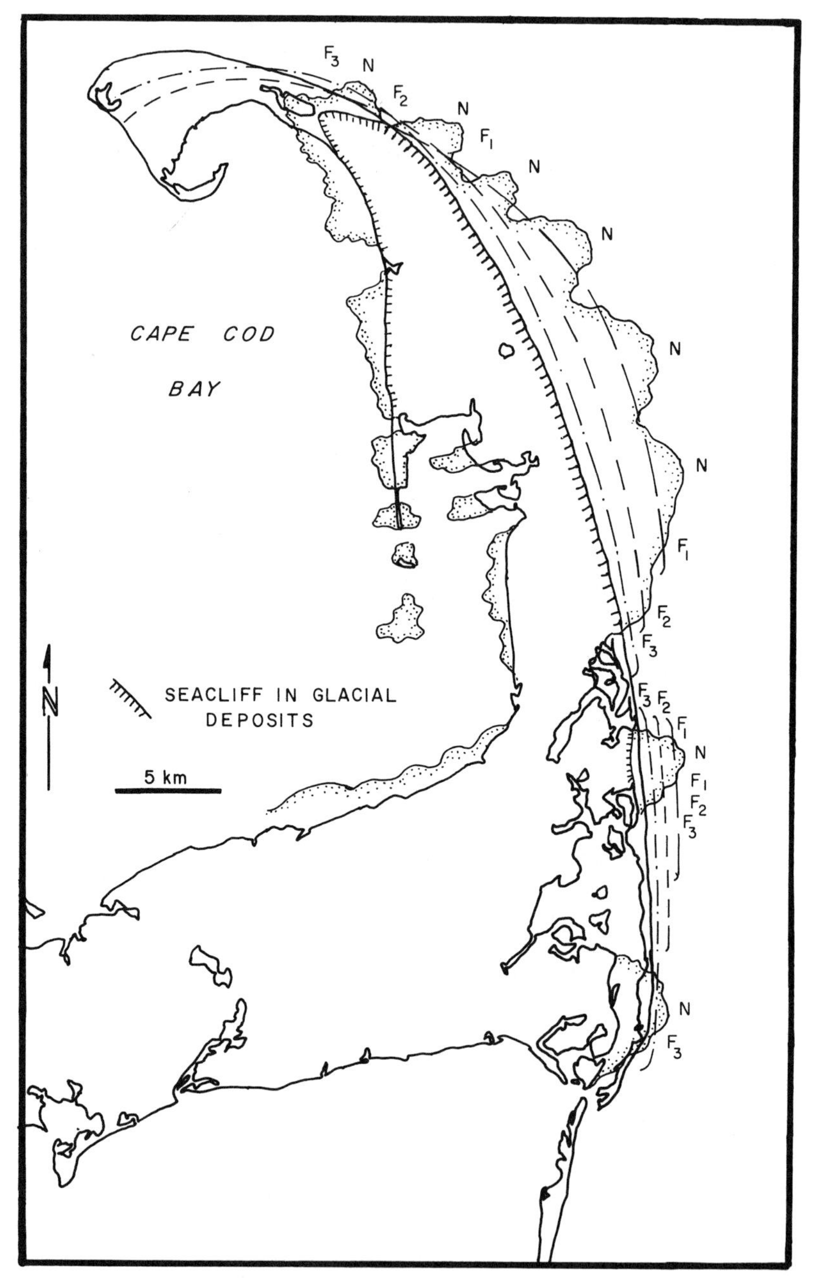

Fig. 3. Reconstructed development of present-day Cape Cod shoreline by wave erosion from initial irregular outwash plain submerged by rising sea level. Migrating northern fulcrums, F_1, F_2, and F_3, during development of Provinceland spits after Davis (1896). Nodal points, N, indicated for initial headlands (stippled pattern: initial outwash plain).

Neck, which extends eastward for over 5 miles and encloses Barnstable Harbor. Coring studies in the tidal salt marshes of this area, together with radiocarbon dating, indicate that the peat has grown upward at the rate of 0.003 ft/yr since 2100 B.P., but before that time it was at the rate of 0.01 ft/yr. These rates suggested to Redfield and Rubin (1963) that changes in the sea level and land level along the Cape Cod shoreline are perhaps due primarily to land subsidence rather than sea-level rise.

The most extensive beaches on the Cape Cod coast extend to the north and south of the above marine sea cliff. A complex of spits, barrier spits, and barrier island shoreline extend along this eastern shoreline. North from this sea cliff is the well-known extensive Provincelands Spit with Race Point and Herring Cove beaches. To the south of this sea cliff is the North Beach barrier spit, 4 km, which includes at its north end Coast Guard Beach. Nauset Bay, Nauset Marshes, and Nauset Harbor lie behind this barrier spit. South of Nauset Inlet is Nauset Beach, 20 km, another barrier-spit beach fronting on Pleasant Bay and extending to Chatham Inlet. South of this inlet is Monomoy Island, an island that is primarily depositional in nature.

Provincelands Spit is Cape Cod's most prominent depositional landform. It is a large foreland with numerous relict beach ridges and migrating dunes. It is easy to visualize each of these relict ridges as coastal "lines of growth." A wide trough, between the most recent relict beach ridges, is occupied by the Provincetown Airport. This trough probably represents a former very wide relict beach in front of the frontal or foredune ridge on the expanding spit, a situation similar to the wide beach in front of present-day Race Point Beach. Offshore from Race Point Beach there is an extensive permanent submarine longshore bar called Peaked Hill Bar. Peaked Hill Bar actually begins tangent to the beach at Highland Point to the east and extends westward, terminating at Race Point. This bar is offshore a distance of 600 m, and as the shore makes a sharp turn to the south at Race Point, the Peaked Hill Bar also turns, but gradually merges into the shoreline. Extensive migrating sand dunes on the Provincelands foreland spit are parabolic or U shaped, with the open end of the dune facing into the dominant prevailing wind, which along this section is from the northwest, and the dunes migrating to the southeast.

In a sense, the entire Cape Cod coast appears to resemble a giant mega-winged headland with extensive spits developed by longshore currents extending to both the north and south from the glacial marine seacliff. Similar and analogous, but much smaller, relict "winged headlands" have also developed by wave erosion of drumlins in Boston Harbor to the north. Longshore drift by wave refraction had developed spits ("wings") on either side of these headlands during the development of the Nantasket area (Johnson, 1925).

IV. Cape Cod's Fulcrum–Nodal Coastline

Fulcrum as a coastal term is often taken to indicate that point along a shoreline where a shoreline of erosion changes to a shoreline of deposition. In his classical study on the development of the northern Cape Cod shoreline, Davis (1896) pointed out that, during the growth of the Provincelands Spit and while the cliff shoreline is retreating, there must be a "neutral point of no change" referred to as a "fulcrum." With continuing cliff erosion, this fulcrum would shift or migrate in the direction of the distal or downdrift end of the spit. The point where the present spit shoreline intersects the cliffed shoreline, while presently a fulcrum, is not the original point of attachment of spit and cliff, which was seaward of the present shoreline. Based on present rates of cliff retreat, Davis estimated that the original Provincelands shoreline was at the first fulcrum point about 1200 yr B.P. At later stages of development, the fulcrum as it migrates may be located along the spit shoreline at the intersection of an earlier spit shoreline with that of later spit development. In some cases, the term "fulcrum" has been limited to indicate this intersection of a recurved spit, with the next stage of spit development producing a "compound recurved spit." Sandy Hook Spit at the north end of the New Jersey coastline is such a spit, with fulcrums along that shoreline and developing in a manner analogous to the Provincelands Spit (Davis, 1896). This development of a spit through a fulcrum point as an extension of a cliff or mainland shoreline under the action of longshore transport was considered by Davis as development of a "graded shoreline." Along some individual island shorelines, such as along the southern New Jersey coast, the location of fulcrum points is important to indicate the change from an eroding shoreline section to an adjacent accreting shoreline section. Recognition of this point is important for shoreline reconstruction studies, especially the placement of beach fill.

At the Provinceland section of Cape Cod, this northern fulcrum point appears to be located at the Head of the Meadow Beach. The meadow referred to is the salt meadow that separates the Truro plain from the Provincelands Spit that has grown from a junction to the south along the sea cliffs. This junction on the initial Cape Cod shoreline must have been southeast of the present junction at Head of the Meadow Beach. The cliff, above the present meadow, was originally a marine sea cliff above the open ocean until the first spit was built from the fulcrum to the east, forming the ocean shoreline of present-day Pilgrim Lake. Pilgrim Lake was originally a saltwater bay until its inlet to Cape Cod Bay closed in 1879. The present-day spit that forms the bay shore of Pilgrim Lake ("Pilgrim

Beach") developed from the low marine cliff (average height, 50 ft) to the south along the Cape Cod Bay shoreline (Fig. 4).

Shoreline nodal points, not to be confused with migrating fulcrum points, are an area in which the predominant direction of longshore transport changes (CERC, 1973). These diverging longshore currents may be due to wave refraction, especially at headlands, or current eddies. Nodal zones may also result from seasonal changes in the longshore transport direction. For example, along the Cape Cod coast the dominant (strongest) northeasterly winter winds produce a southerly sediment movement, while the prevailing (commonest) southerly summer winds move sediment to the north. The net annual sediment movement behaves as if there were a nodal point along this eastern shore. The nodal point, or that point at which this net flow of sand apparently changes from north to south along this coast, has not been well established. Hartshorn *et al.* (1967) suggested that this dividing line was somewhere near the center of the eastern Cape Cod shoreline. This might put its location just north of the Marconi Station. Schalk's (1938)) study of sediments along the beaches of Cape Cod suggests that the shoreline opposite the Pamet River might be the dividing point. Dominant-wave vector analysis might be useful in locating this nodal point. Dominant waves for this coast come from about the east–northeast, and this direction is perpendicular to a tangent to the shoreline just east of Gull Pond. Location of this nodal point is important in developing shoreline sediment budgets for shore protection studies as well as for understanding the affected beach morphology. Beaches within and outside a nodal zone may have different stable foreshore slopes and widths. In addition, movement of this nodal zone along a particular shoreline might also indicate migration of the fulcrum points elsewhere along the coast. Location of nodal points has been easiest along shorelines where there are numerous shore protection structures such as jetties and groins. Along the northern New Jersey coast, where there are numerous groins, the accumulation of longshore drift on the southern side of the groins, from Sandy Hook south to Point Pleasant at the head of Barnegat Bay is reversed south of this point, where groins accumulate sediment on their northern sides. Thus the nodal point is easily located at Point Pleasant. Unfortunately, we cannot use this common technique to locate the Cape Cod nodal point, because along the entire outer Cape shoreline from Race Point to the southern tip of Monomoy Island, there is not a single shore protection structure. Because it is a National Seashore Park, there is not a single jetty, groin, or even seawall along this coast to assist in determining direction or rates of longshore transport.

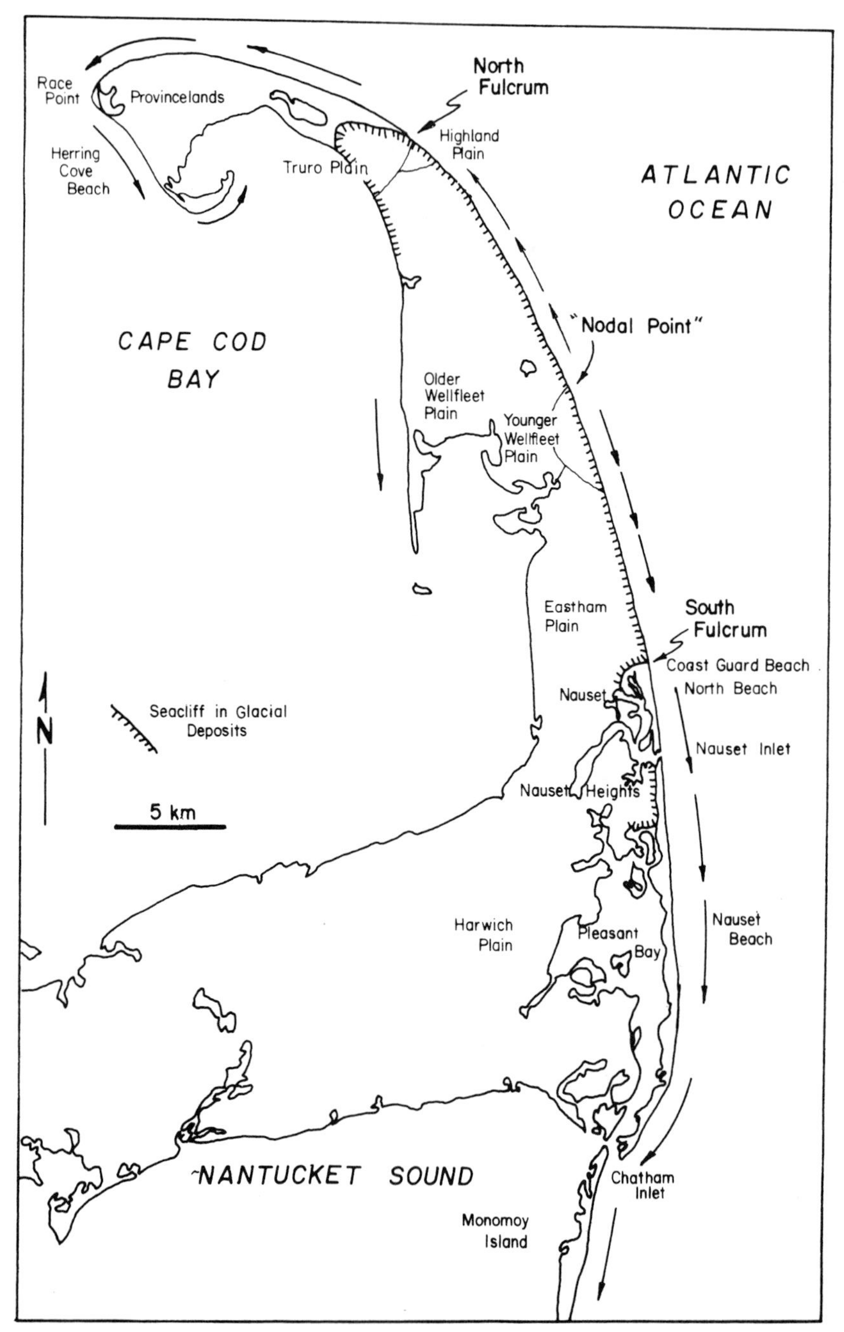

Fig. 4. Fulcrum–nodal point shoreline of present-day Cape Cod shoreline. Northern fulcrum at base of Provincelands Spit, "nodal point" zone south of Gull Pond, and southern fulcrum point in Coast Guard Beach area. Arrows indicate direction of resultant longshore drift.

V. Fulcrum Cliffline Erosion

Erosion of the Cape Cod marine cliffs is both an important coastal geomorphologic process and an environmental engineering problem. At Highland Lighthouse, the original site consisted of 10 acres but now only 4 acres are left. With a measured erosion rate of 5 ft/yr, the lighthouse is now less than 200 ft from the edge of the cliff. Further south, at Nauset Light, the cliff edge has retreated some 250 ft since the first lights were erected in 1885 (Leatherman, 1979). Continuing erosion is now cutting away at a parking lot, where an entire row of parking spaces has been lost to cliff erosion over a recent period of 5 yr. Dramatic evidence of this sea-cliff erosion is measurable at the Marconi Wireless Station's former antenna site near the cliff edge. It was first a simple circular system, which was destroyed in a gale in September 1901. Later four towers, in a square, were erected and began transmitting across the Atlantic in January 1903. In 1917, the station was closed and the towers were sold for scrap for the war. Since the placement of the towers in 1902, the cliff has eroded back 170 ft, and the two other concrete bases have crumbled to the beach below. Presently the foundation of the former electrical powerhouse on the edge of the cliff is undergoing erosion.

The most recent dramatic shoreline erosion in the Cape Cod National Seashore during the last decade has been at the Coast Guard Beach recreational bathing area. The beach takes its name from the former Coast Guard Station on the former glacial sea cliff bluff overlooking the beach, and it was among the earliest recreational beaches of the Seashore. Coast Guard Beach, at the north end of a barrier spit, has been built by the longshore movement of glacial sediments eroded from the Eastham plain cliffs in the north and carried to Nauset Harbor inlet on the south. This barrier spit is about 4 km in length and separates Nauset Bay (directly west) and Salt Pond Bay from the ocean. The entire spit has sometimes been referred to as North Spit or North Beach to distinguish it from Nauset Beach, a similar barrier spit south of Nauset Inlet that has also been built southward from the Nauset Heights outwash plain.

Cliff, dune, and beach erosion of a serious nature has been apparent at the northern end of Coast Guard Beach since 1969 (Fig. 5). In addition to the fact that this erosion has made the beach less suitable for recreational activities, there is the additional concern that such continuing erosion might breach the narrow sand spit at Coast Guard Beach. If such an inlet into Nauset Bay to the west actually were opened, the change in salinity to the estuarine waters would perhaps be fatal to the shellfish in these brack-

FIG. 5. Extensive erosion at Coast Guard Beach cliffs, southern fulcrum point, as well as adjacent beach, due to possible migration of fulcrum point zone. Note loss of parking spaces due to erosion of cliff edge and pavement on beach face below cliff.

ish waters. During the February 7 blizzard of 1978, a record blizzard since colonial times, major overwash wave action occurred at this beach. The entire frontal dune was eroded, and wave action destroyed a large National Seashore bathhouse as well as several other smaller beach houses along the beach immediately to the south. Four hundred meters of frontal dunes were eroded and washed westward over the bathhouse's paved parking lot. Prior to that time, however, there had been continuing erosion of both this frontal dune and the sea cliff to the north for the preceding 10 years. In an attempt to reduce this erosion, in the early 1970s large concrete slabs covered by sand fill were placed along the front of the cliffs and the frontal dunes. This material was emplaced in the spring (March), and the first storm of that year removed all the sand cover and displaced the slabs onto the beach. These slabs can still be seen about the Coast Guard Beach area. The beach bathhouse and parking lot were not rebuilt after the 1978 blizzard. People now reach the beach for limited use by a shuttle bus from the Orleans Museum. Plans have been made for a larger recreational beach between Coast Guard Beach and the Nauset Light Beach to the north. However, even this beach may show increased erosion, as discussed under fulcrum migration.

While the erosion of the sea cliff is understandable as a normal geomorphic coastal process, what was not understood is why the beach, which is part of a spit, a depositional landform, also should be undergoing erosion. In determining the rate of erosion or deposition along Coast Guard Beach, it is necessary to separate the short-term and long-term changes. Zeigler (1960), in comparing his surveys of 1957–1959 to the various surveys of Marindin (1891) for this area, indicated that "erosion is 4–6 ft/yr where Nauset Spit is being driven into the marshes behind it." It is possible that, if this is a long-term erosion, this eroded material either is moving into the bays through the inlets or is being transported even further south by longshore transport. One suggestion for the increased erosion of Coast Guard Beach during the 1970s is southward fulcrum migration (A. N. Strahler, 1970, personal communication).

Possible migration of this fulcrum zone in the Coast Guard Beach area might have been responsible for the extensive erosion starting in the early 1970s. A former fulcrum point at the intersection of the marine sea cliff and the Coast Guard Beach spit could start migrating southward. The erosive phase along the marine cliffs would then be transferred to the earlier deposited frontal dune shoreline along Coast Guard Beach. It would be a form of "erosional cannibalism," with erosion of previously deposited sands. Even if this is the case, the following questions remained:

(1) If the fulcrum migrated southward, why did the cliff also exhibit excessive erosion?

(2) Was possible sapping of the cliff toe and accelerated erosion due to leakage of water from the kettle-hole pond directly behind the cliff?

(3) In the mid 1970s (1975), Nauset Inlet to the south started to migrate northward. This should have caused the fulcrum to migrate back northward to the cliff face and stopped erosion of the frontal dunes. However, dune-face erosion continued through this time until the extensive 1978 erosion.

(4) If the fulcrum has now migrated northward, as suggested by the migration of Nauset Inlet, will this cause increased erosion along the proposed new beach as has occurred earlier along the Coast Guard Beach?

VI. Cape Cod Beach Sediment Patterns

In order to see whether the location of the shoreline fulcrum and nodal points could be detected by sediment size analysis, a short-term synoptic beach sampling program was planned for the entire Cape Cod shoreline. Previously, Schalk (1938) had conducted such a study, but logistics required that it be spread out over two seasons. Zimmerman (1963) conducted a detailed beach study along the barrier spit from Coast Guard Beach to Nauset Inlet with samples collected every 500 ft (150 m). The mean size of all foreshore samples was 1.061 ϕ (about 0.5 mm). Foreshore samples on Coast Guard Beach were the coarsest (0.4 ϕ to 0.5 ϕ) for the entire barrier spit, because these beach sands directly in front of the glacial sediment sea cliffs have not yet been subjected to the sorting action of the winds and currents. A definite trend can be noted, however, as the mean size decreases southward, due probably to the longshore currents that transport the sand southward from its source in the glacial material. A similar plot of standard deviation ("sorting") and kurtosis ("range of sizes") showed no discernable trend. A similar beach study was conducted along Nauset Beach to the south (Felsher, 1963). Six beach sampling stations were established at about 1½-mile (4-km) intervals. Felsher reported that, south along Nauset Beach, there is no significant decrease of median grain size; however, there is a significant change in sorting, with the better-sorted materials to the south.

Our regional sediment study was conducted in October 1970, to represent the end of the summer season of accumulation. Stations wer located 1 mile (1.6 km) apart from Chatham Point to Provincetown for a total of 46 stations (Fig. 6). During the 2-day sampling period, wind and wave conditions were constant and minimal to produce an effective "synoptic sampling." Three 25-cm-deep samples, 5 m apart, were collected just below the high tide line. They were split and combined into one sample in the laboratory and sieved to ¼ϕ intervals. This regional sediment size analysis indicates certain interesting patterns of distribution that relate directly to the coastal geomorphology as follows (Fig. 7).

A. Barrier Spit Unit

The southern barrier spit shorelines from Coast Guard Beach to Chatham Point do not show any decrease in size southward away from the source area of the marine cliffs. From Coast Guard Beach (station 16) to Nauset Inlet (station 13), the sediment does start to decrease in size away from the marine cliffs, as shown by Zimmerman (1963), but further south it increases in size to Nauset Inlet. Along Nauset Beach (stations 13–1), sediment size shows no strong decreasing size trend, as also shown by Felsher (1963), and exhibits the greatest variability of grain size of all the coastal morphologic units of the Cape Cod shoreline. Along this barrier

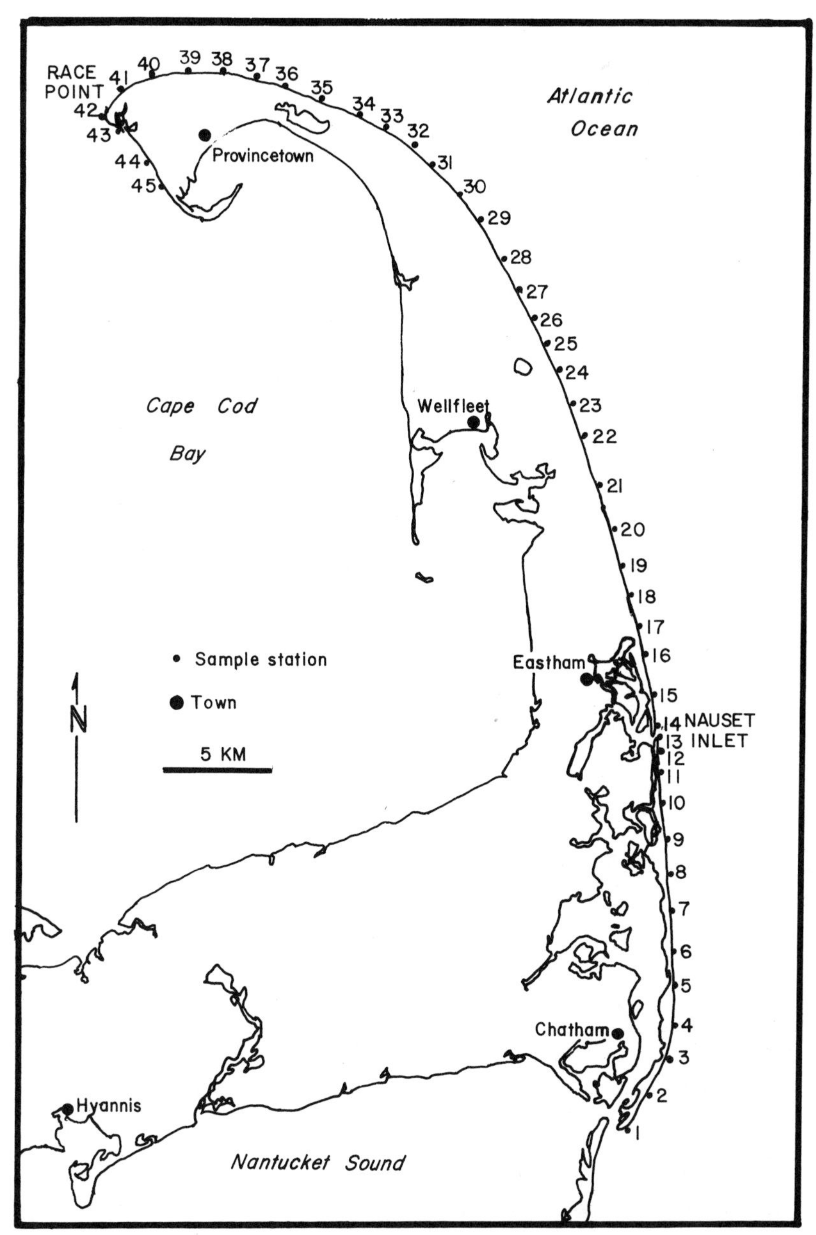

FIG. 6. Shoreline beach sediment sampling stations along outer Cape Cod. Core samples taken on foreshore at 300-m intervals under synoptic short-term sampling programs of constant summer wind–wave conditions. Sampling program to relate sediment parameters to fulcrum–nodal coastline model.

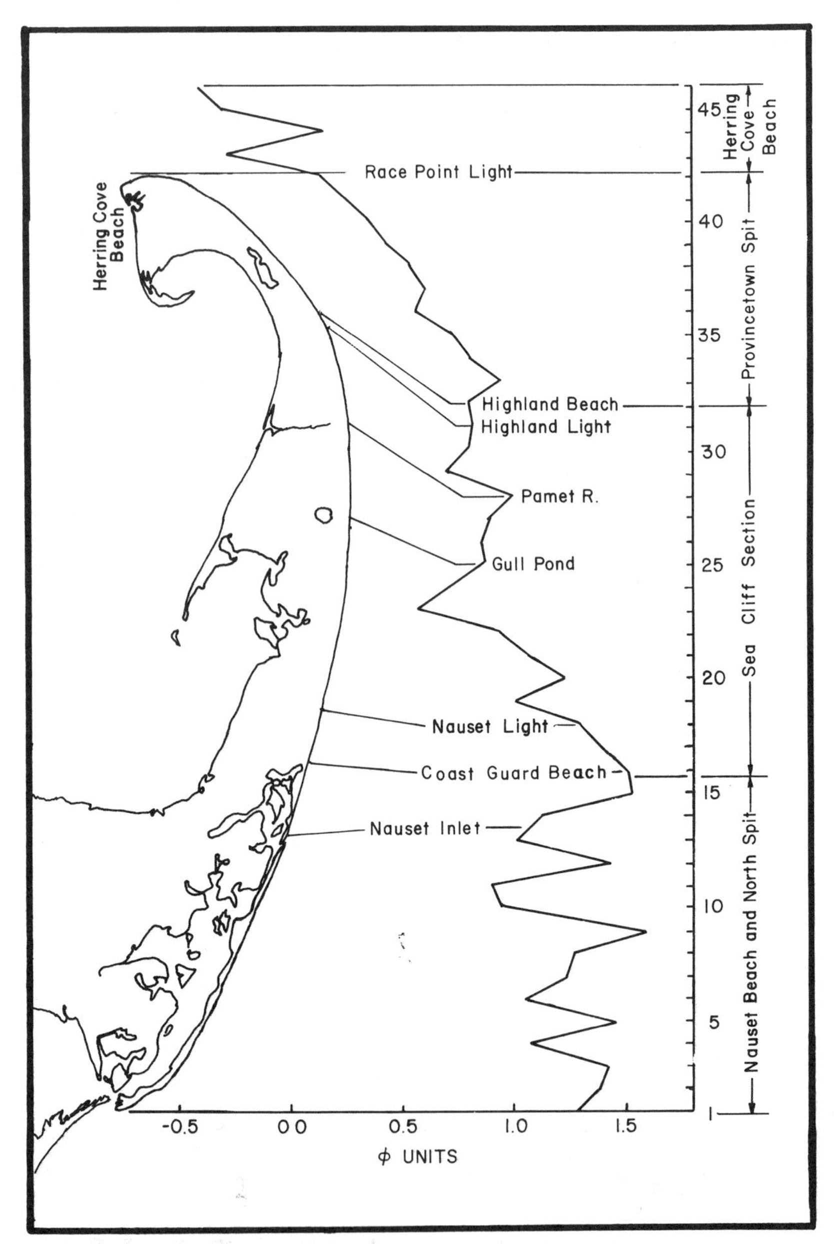

Fig. 7. Median grain-size distribution of beach sediments along outer Cape Cod. Location of fulcrum and nodal points indicated primarily by changes in overall trend pattern (see text). Note also general increase in sediment size from medium sand at southern barrier spits to very coarse sand at Race Point beaches.

spit morphologic unit, grain size ranges from 1.0 ϕ to 1.5 ϕ and averages about 1.2ϕ.

B. Sea Cliff Unit

Along the sea cliff morphologic unit, beach sediment increases in size northward, ranging from 1.5 ϕ to 0.8 ϕ. This is a progressive increase in grain size from the southern fulcrum point at Coast Guard Beach (1.5ϕ) to the northern fulcrum at Highland Beach (0.8 ϕ). The southern fulcrum point is shown as a transition from a highly variable grain size trend for the barrier spits to the less variable trend along the sea cliff, while the northern fulcrum is marked by a transition to the even less variable trend in grain size of the Provincetown Spit unit. The sediment nodal point appears to be south of Gull Pond, around station 24. There is greater change in average grain size with distance south from this point (0.8 ϕ/8 miles), while from Gull Pond north to Highland Beach (stations 25–32) there is much less change in average grain size for the same distance (0.1 ϕ/13 km).

C. Provincetown Spit Unit

This morphologic unit shows a very uniform increase in grain size from the north fulcrum at Highland Beach to Race Point Lighthouse. Grain size ranges from 1.0 ϕ to 0.1ϕ, with almost no variability around this increasing trend. This would not be unusual except that grain size *increases* toward the distal end of the Provincetown Spit. The coarsest grain size of the entire Cape Cod shoreline occurs along the Herring Cove Beach section of the Provincetown Spit. This is the recurved portion of the spit, and it has a grain size averaging between 0.0 and -0.5ϕ. Coarse gravel and pebble-size grains are also present along these beaches.

D. Regional Pattern

The most unusual aspect of the regional sedimentation pattern is that the average grain size does not decrease north and south from the sea cliff source area to the distal spits. Schalk (1938) in his study found the finest grain size along the cliffs, with median grain size *increasing* both to the north and south. Since his study extended over several sampling periods, this may be a factor of the sampling. In contrast, our study shows median grain size averaging medium sand size along the southern barrier spits, increasing from medium to coarse sand size along the sea cliff, and finally increasing through coarse to very coarse along the northern Provincetown Spit. Since the sampling was after the summer season of prevailing winds from the south, one would further expect seasonally the grain size to become *finer* to the north, downdrift, rather than coarser. So both the morphology and wave dynamics are in contradiction to the expected regional sediment pattern. If this sediment is transported by longshore drift from the highland sea cliffs, one would expect the sediment size to decrease away from the source area. Changes in the direction of longshore drift were thought to occur some 18,000 and 6,000 yr B.P. when sea level was lower. At that time, sand eroded from the sea cliffs was moved from the north to the south because Georges Bank shoals to the south blocked the southeasterly waves. As sea level rose, and Georges Bank to the east was submerged, more waves could reach Cape Cod from the east and southeast and the Provincetown Spit could then begin to form (Zeigler *et al.*, 1965). That reversal in littoral drift direction occurred many years ago and does not seem to explain Provincetown Race Point beaches unless a relict pattern is present.

Chesil Beach, a shingle–cobble beach along England's southern coast, is another beach that has beach sediments that increase in size in the direction of longshore drift (Carr, 1969). Increase in wave energy downdrift has been suggested as a means of increasing grain size (King, 1972) or by directional variations in wave energy along the beach (Lewis, 1938). Komar (1976) suggests that since Chesil Beach may lack new sediments it is a closed system, and that the present sediments are relict with a zero net littoral transport. The most recent work (Carr and Blackley, 1974) claims that the size sorting is due to absence of new pebbles, increasing wave energy downdrift, and "rejection" offshore of coarser material alongshore. For the similar Cape Cod regional trend, Schalk (1938) felt that the finer size material in the sea cliff source material (glacial outwash) was lost offshore as longshore drift transported the sediment away from the source area. However, the present study shows instead that the southern barrier spits do decrease somewhat in size downdrift, while the Provincetown Spit increases markedly downdrift. In fact, the entire Cape Cod coastline from Chatham Point to Race Point could be considered as showing an increasing grain size. For the very coarse sand with gravel and pebbles along Herring Cove Beach, the source could be truncating erosion of the relict spit ridges that make up the entire spit. For the increase in grain size of the Provincetown Spit toward its distal end, it is possible that waves under certain conditions could transport the finer sand back toward the fulcrum and leave the coarser material behind. This would leave coarser sands at the distal point. However, along this coast the more dominant winter waves are from the northern sector, and they should be able to transport both the fine and coarse sizes. In fact, it is difficult to visualize how the weaker prevailing summer waves from the south could transport both the coarser and fine materials northward, while the stronger winter waves from the north would transport back only the finer size. While answers to the regional sediment pattern distribution are complex, it is worthwhile to observe that the major geomorphic coastal units, as well as the shoreline fulcrums and node, do appear to be shown by sediment patterns.

VII. Barrier Beach Analogs of Cape Cod

The previous sedimentological study shows that the Cape Cod shoreline can be divided into discrete coastal geomorphologic units. In addition, these units indicate an analog similarity with other middle Atlantic barrier-island coasts and that the Cape Cod shoreline could be included with these shorelines.

The Middle Atlantic Coast barrier-island model (Fisher, 1967) consists of the following four shoreline segments for each coast (Fig. 8a).

I. A short cuspate "cape" or spit shoreline segment extending into relatively deep and open water. The width varies from narrow in the case of the spit to very wide for the cuspate cape. Relict beach ridges on both of these are common, with the most prominent ridges on the spit at their distal ends.

II. A longer mainland beach shoreline usually fronting on a coastal plain with associated narrow baymouth barrier beaches across coast perpendicular estuaries or shallow drowned river valleys. Inlets may exist through the baymouth barrier beaches, usually opposite these drowned river valleys.

III. A still longer barrier-island shoreline segment that may contain a barrier-spit shoreline attached to the adjoining mainland beach shoreline of segment II. The barrier island is usually wider than the barrier beaches of segment II. The lagoon behind the barrier island is coast-parallel, wide, and lacks shoal or marsh deposits except near the inlets. Inlets are few and again sometimes are located opposite small estuaries, usually shallow drowned river valleys along the lagoonal mainland shoreline. Lack of inlets is why this segment is referred to as the "continuous barrier-island" shoreline in this model.

IV. The final barrier-island shoreline segment is usually as long as segment III, but the individual barrier islands are shorter, wider, and more

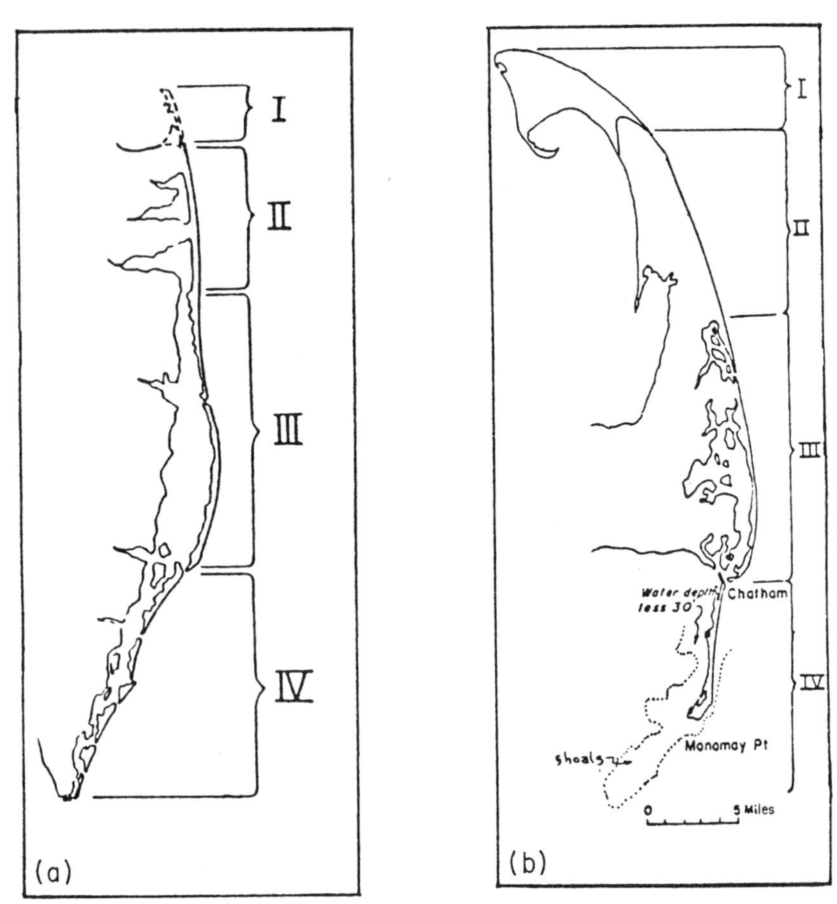

Fig. 8. (a) Coastal geomorphological model of middle Atlantic barrier-island shorelines, consisting of (I) spit or cape, (II) mainland beach, (III) barrier spit and islands, and (IV) barrier islands. (b) Coastal units of Cape Cod, consisting of (I) Provincelands Spit, (II) glacial sea cliffs, (III) North and Nauset beach barrier spits, and (IV) Monomoy Island and Handkerchief Shoals.

numerous. In contrast, the lagoon is narrower and is filled with more shoal and marsh deposits, due in part perhaps to the inlets, which are more numerous in this segment. Again, small drowned river-valley estuaries are present along the lagoonal mainland, often opposite the inlets separating the many barrier islands. The presence of these numerous islands is why this segment is referred to as the "discontinuous barrier-island shoreline."

The Cape Cod coastline can be related by this model to other Atlantic coast barrier shorelines. First, the Cape Cod coastline can be subdivided into four morphologic-unit shoreline segments (Fig. 8b). There is the northern Provincetown spit (I), the sea cliff, Truro–Wellfleet–Eastham headlands (II), barrier spit segments that are convex seaward in plan, North (Coast Guard) and Nauset beaches (III), and a "barrier-island" segment that is concave seaward in plan, Monomoy Island and Handkerchief Shoals to the south (IV). Segments I, II, and III of Cape Cod most closely resemble the model, with segment IV presenting some changes. Although isolated, Monomoy Island has developed by longshore drift similar to a barrier-spit development. And while it is not segmented by inlets, the February 1978 blizzard did open an inlet, and the island does seem susceptible to inlet breaching, more so than the Nauset Beach spit barrier. The inclusion of Handkerchief Shoals in the coastline model is for several reasons. Since it is less than 10 m below the surface, linear in plan, and does have the concave shape of the model, I feel that it may have developed in response to longshore drift from the Cape Cod shoreline and is maintained in part by this material. No field data is available on the sediment distribution pattern for this shoal, but it is possible that it might be a sediment sink for material eroding from Cape Cod.

The classical barrier-island shoreline of Long Island (Fig. 9a) and New Jersey (Fig. 9b) can be shown to possess these same model shoreline segments, as can the Delmarva barrier-island shoreline (Fig. 9c). Finally, the last barrier-island shoreline, south of Chesapeake Bay, the Virginia–North Carolina "Outer Banks" barrier-island chain (Fig. 9d) also has these segments on perhaps a slightly larger scale than the previous barrier-island coastlines. For the Long Island barrier-island coast, there are slight changes. Segment II is a mainland developed from glacial sediments (the Wisconsinan terminal moraine and outwash plain) rather than the coastal plain mainland of the other barrier-island coasts. In this sense, it resembles the Cape Cod coast and is a bridge between the Atlantic barrier coastlines and the Cape Cod coast barriers. In addition, for Long Island, segment I is represented not by a spit, but by relict beach-ridge deposition between glacial bluffs in the Montauk Point region. For the Delmarva barrier coast, Kraft *et al.* (1973), in a study of the subsurface Holocene stratigraphy of the Delmarva coast, were able to apply this model to paleosedimentary environments.

In addition, the development of the Cape Cod shoreline also bears directly on the origin of barrier-island coastlines in general. Fisher (1967, 1968, 1973; see also Saxena and Klein, 1974) suggested that the morphologic and stratigraphic features of certain Atlantic barrier islands indicate that they develop due primarily to longshore drift on shorelines of

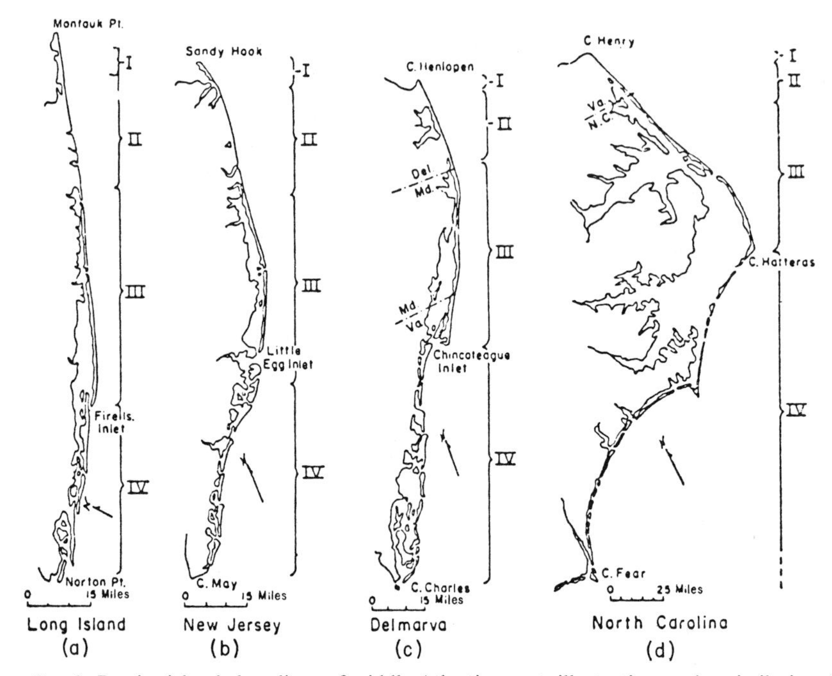

FIG. 9. Barrier-island shorelines of middle Atlantic coast, illustrating analog similarity of larger scale fourfold barrier-island coastal division to developed fourfold division of outer Cape Cod shoreline. (a) Long Island, (b) New Jersey, (c) Delmarva, and (d) North Carolina barrier-island chain coastlines.

submergence, not on shorelines of emergence. Thus, if the Cape Cod coastline, with headland zone and its north Provincetown Spit and south barrier spit, can also be shown to be related in its shoreline segments to those of ''Fisher's barrier-island model'' and to typical barrier islands along the Atlantic coast, then this is evidence by analog that these barrier islands developed primarily by longshore drifting, as has the Cape Cod coastline.

VIII. Conclusion

Cape Cod's outer coastline has developed from a complex system of interlobate glacial outwash plain terrace deposits. Rising sea level submerged the initial glacial outwash shoreline, and wave erosion developed an extensive prominent marine sea cliff. Dominant winter wind wave conditions from the northeast (''nor'easters'') together with prevailing summer wind wave conditions from the southeast have acted to develop the present coastline. Provincelands Spit has developed by deposition from a northern fulcrum point by erosion of the sea cliff. A southern fulcrum exists at the junction of Coast Guard Beach, a barrier spit, and the same sea cliff. Extensive recent erosion at this beach is probably due to recent migration of this fulcrum point southward. Thus the area of earlier erosion at the cliff has migrated into the former beach depositional area. Between the two fulcrum points, there exists a nodal point zone along the prominent sea cliff. There is no single nodal point, but it is the effective result of longshore transport to the south during the winter months and then longshore transport to the north during the summer months. A similar regional nodal zone separating opposing longshore current patterns has recently been shown to exist along the New Jersey coastline (Ashley *et al.*, 1986).

Regional beach sediment patterns along the Cape Cod shoreline indicate the location of this nodal zone, after the summer period, midway along the glacial sea cliffs shoreline segment. Just south of this nodal zone, seasonal beach cycles were found related to northeast storms that occur in late autumn, winter, and early spring (Miller and Aubrey, 1985).

Foreshore beach sediment size increases along the Provincelands Spit, moving north from the glacial sea cliff headland source area. A slight increase in beach grain size was found to the south from the Coast Guard sea cliff source area. Along the English coast, similar coarsening of sand size away from the cliff sources suggested that the finer sediment was being lost from the foreshore to offshore (Carr, 1981). Along this Cape Cod Provincelands Spit foreshore, dominant northeasterly winter waves probably selectively winnow out the finer sediments, leaving behind the coarser sediments.

Intensity of foreshore sedimentation processes is also indicated by the variability of the sediment size pattern (Fig. 6). Along the Provincelands Spit segment, grain size increases in a regular and rapid fashion, while along the sea cliff section to the south it increases at a rate half that of the previous spit segment and with a slight variation. Further south along the barrier spit and barrier beach segments, grain size shows no distinct trend and varies even more. This indicates that longshore transport processes are most active along the northern spit shore, less active along the sea cliff shoreline, and least active along the southern barrier sections. Reduction of longshore power may explain why there has been a northerly migration of Nauset Inlet along the southern barrier section in recent years (Leatherman and Zaremba, 1986).

A four-part regional pattern for the outer Cape Cod coastline consists of the northern Provinceland spits, the glacial outwash mainland marine cliff, the southern Nauset Beach barrier spits and barrier island, and the southernmost Monomoy Island–Handkerchief shoals complex. This morphological pattern can be related to a similar regional coastline pattern of the classical barrier-island coastlines of Long Island, New Jersey, Delmarva, and North Carolina. This analog similarity relationship indicates development under similar past and present marine dynamic conditions, as well as a similar history of development.

Recent inlet formation along the southern Cape Cod coastline is in keeping with the discontinuous character of the barrier islands of segment IV. The blizzard of 1978 in February opened an inlet along Monomoy Island that still shows no sign of closing almost 10 years later. More recently, again in February 1987, a northeasterly storm opened another inlet in the barrier island opposite the town of Chatham (south of sampling station 5). It also appears to be remaining open. This inlet-prone nature is predicated for the southernmost segment based on application of the mid-Atlantic coast barrier-island model.

References

Ashley, G. M., Halsey, S. D., and Buteux, C. B. (1986). New Jersey's longshore current pattern. *J. Coastal Res.* **2,** 453–463.

Carr, A. P. (1969). Size grading along a pebble beach: Chesil Beach, England. *J. Sediment. Petrol.* **39,** 297–311.

Carr, A. P. (1981). Evidence for the sediment circulation along the coast of East Anglia. *Mar. Geol.* **40,** 9–22.

Carr, A. P., and Blackley, M. W. L. (1974). Ideas on the origin and development of Chesil Beach, Dorset. *Proc. Dorset Natl. Hist. Archaeol. Soc.* **95,** 9–17.

CERC–Coastal Engineering Research Center (1973). "Shore Protection Manual," Vol. 3. Dep. Army, Corps Eng., Fort Belvoir, Virginia.

Curray, J. R. (1965). Late Quaternary history, continental shelves of the United States. *In* "The Quaternary of the United States" (H. E. Wright and D. G. Frey, eds.), pp. 723–735. Princeton Univ. Press, Princeton, New Jersey.

Davis, W. M. (1896). The outline of Cape Cod. *Am. Acad. Arts Sci. Proc.* **31,** 303–332.

Felsher, M. (1963). Beach studies on the outer beaches of Cape Cod, Mass. M.S. Thesis, Univ. of Massachusetts, Amherst.

Fisher, J. J. (1967). Origin of barrier island chain shorelines, Middle Atlantic states. *Geol. Soc. Am. Spec. Pap.* **115,** 66–67.

Fisher, J. J. (1968). Barrier island formation, discussion. *Geol. Soc. Am. Bull.* **79,** 1421–1426.

Fisher, J. J. (1972). "Field Guide to Geology of Cape Cod National Seashore," Fifth Annual Meeting American Association of Stratigraphic Palynologists. Dep. Geol., Univ. of Rhode Island, Kingston.

Fisher, J. J. (1973). Barrier island formation. *In* "Barrier Islands: Benchmark Papers in Geology" (M. L. Schwartz, ed.), pp. 227–232. Dowden, Hutchinson & Ross, Stroudsburg, Pennsylvania.

Grabau, A. W. (1897). The sand plains of Truro, Wellfleet, and Eastham. *Science* **5,** 334–35.

Hartshorn, J. H., Oldale, R. N., and Koteff, C. (1967). Preliminary report on the geology of the Cape Cod National Shore. *In* "Economic Geology in Massachusetts" (O. C. Forquhar, ed.), pp. 49–58. Grad. Sch., Univ. of Massachusetts, Amherst.

Johnson, D. W. (1919). "Shore Processes and Shoreline Development." Wiley, New York.

Johnson, D. W. (1925). "The New England–Acadian Shoreline." Wiley, New York.

King, C. A. M. (1972). "Beaches and Coasts," 2nd Ed. St. Martin's Press, New York.

Komar, P. D. (1976). "Beach Processes and Sedimentation." Prentice-Hall, Englewood Cliffs, New Jersey.

Koteff, C., Oldale, R. N., and Hartshorn, J. H. (1967). Geological quadrangle map of North Truro, *U.S. Geol. Surv. Quadrangle Map* GQ-599.

Kraft, J. C., Biggs, R. B., and Halsey, S. D. (1973). Morphology and vertical sedimentary sequence models in Holocene transgressive barrier systems. In "Coastal Geomorphology" Publications in Geomorphology, S.U.N.Y. Binghamton, New York.

Leatherman, S. P., ed. (1979). "Environmental Geologic Guide to Cape Cod National Seashore." Soc. Econ. Paleontol. Mineral. East. Sect. Univ. Mass., Amherst, Mass.

Leatherman, S. P., and Zaremba, R. E. (1986). Dynamics of a northern barrier beach: Nauset Spit, Cape Cod, Massachusetts. *Geol. Soc. Am. Bull.* **97,** 116–124.

Lewis, W. V. (1938). Evolution of shoreline curves. *Proc. Geol. Assoc.* **49,** 107–127.

Marindin, H. L. (1891). Cross-sections of the shore of Cape Cod, Mass., between the Cape Cod and Long Point light-houses. *U.S. Coast Geodet. Surv. Rep. 1891* Part II, 289–341.

Miller, M. C., and Aubrey, D. C. (1985). Beach changes on eastern Cape Cod, Massachusetts, from Newcomb Hollow to Nauset Inlet, 1970–1974. *CERC Misc. Pap.* **85-10.**

Oldale, R. N. (1968). Geologic map of the Wellfleet quadrangle. *U.S. Geol. Surv. Geol. Quadrangle Map* GQ-750.

Oldale, R. N. (1976). Notes on the generalized geologic map of Cape Cod. *U.S. Geol. Surv. Open-File Rep.* **76765.**

Oldale, R. N., Koteff, C., and Hartshorn, J. H. (1968). "Field Trip to Cape Cod, Massachusetts," Friends of the Pleistocene, 31st Annual Reunion.

Redfield, A. C., and Rubin, M. (1963). The age of salt marsh peat and its relation to recent changes in sea level at Barnstable, Mass. *Natl. Acad. Sci. Proc.* **48,** 1728–34.

Saxena, R. S., and Klein, G. (1974). "Sandstone Depositional Models for Exploration." New Orleans Geol. Soc. New Orleans, Louisiana.

Schalk, M. (1938). A textural study of the outer beach of Cape Cod, Massachusetts. *J. Sediment. Petrol.* **8,** 41–54.

Shaler, N. S. (1897). Geology of the Cape Cod District. *U.S. Geol. Surv. Ann. Rep.* **18,** 503–593.

Woodworth, J. B., and Wigglesworth, E. (1934). Geography and geology of the region including Cape Cod, Elizabeth Islands, Nantucket, Martha's Vineyard, No Mans Land and Block Island. *Mus. Comp. Zool. Mem.* No. 52.

Zeigler, J. M. (1960). "Beach Studies, Cape Cod, August, 1953–April, 1960," Rep. 60-20. Woods Hole Oceanogr. Inst., Woods Hole, Massachusetts.

Zeigler, J. M., Tuttle, D. S., Tash, H. J., and Giese, G. S. (1965). The age and development of the Provincelands, Outer Cape Cod, Massachusetts. *Limnol. Oceanogr.* **10,** R298–R311.

Zimmerman, B. (1963). The size analysis of the sediments of Nauset Harbor, Cape Cod, Massachusetts. M.S. Thesis, Univ. of Massachusetts, Amherst.

CHAPTER 12

Reworking of Glacial Outwash Sediments along Outer Cape Cod: Development of Provincetown Spit

STEPHEN P. LEATHERMAN

Laboratory for Coastal Research
and
Department of Geography
University of Maryland
College Park, Maryland

Cape Cod is a relatively young feature in terms of geologic time, having been formed by glaciers about 14,000 years ago. These original glacial deposits, consisting primarily of unconsolidated sands and gravels, have been reshaped during rising sea levels by winds, waves, and currents to form the distinctive outline of the Cape. What was initially a mass of glacial drift has evolved into many diverse geobiological environments, such as sandspits, kettle-hole ponds, and freshwater and salt-water marshes.

Provincetown Spit did not begin forming until 5500–6000 yr B.P., following the submergence of Georges Bank. Prior to that time, waves were blocked by this subaerial obstruction, and there was no appreciable northern component of littoral drift. Interpretation of the sedimentary deposits tended to support the hypothesis of Zeigler *et al.* (1965), who believed that each sequential spit built a short distance to the west and then hooked to the south. Core data did not support the Davis (1896) theory that the present-day dune ridges correspond to the former prograding sandpits. Shifting sand has effectively erased much of the surficial evidence relating to the origin of this complex recurved spit.

I. Introduction

Cape Cod is a product of the last Ice Age (Wisconsin glacial stage of the late Pleistocene epoch). During the last few hundred thousand years, only the last event (glacial stage) had a profound effect in the Cape Cod area, since it obliterated any earlier glacial features generated by the three previous periods of ice advance and retreat. A rapid warming of the world climate about 15,000 years ago initiated melting of the glacier covering New England. As the ice withdrew to the north, the original form of Cape Cod as a complex of end moraines and associated outwash plains was evident. Cape Cod was thus composed entirely of glacial drift, consisting of sand, gravel, silt, clay, and boulders. The underlying crystalline bedrock has had no influence on the present topography, since it lies from 80 ft (24.2 m) near Cape Cod Canal to over 900 ft (275 m) below sea level along the outer Cape (Oldale, 1969).

The outer Cape from Orleans to Truro (Fig. 1) consists of pitted outwash plains that slope gently westward. This sediment (largely sand and gravel) was deposited by the meltwater streams from the South Channel ice lobe to the east. The moraine that once existed eastward of these outwash plains has been completely eroded away by storm waves of the Atlantic Ocean during the last 6000 years, leaving a wave-cut cliff along most of the outer Cape Cod shoreline.

For the past 7000 years, Cape Cod has evolved through erosion of glacial sediments and accretion of coastal deposits (Davis, 1896). These processes have been governed by rising eustatic sea levels, attributable to the melting of glaciers. As sea level rose, the original irregular shoreline of the Cape was subjected to Atlantic Ocean storm waves. According to Davis (1896), the greatest retreat from the original shoreline to the present shoreline has been approximately 2.5 miles (4.0 km).

A. Sea-Level Rise

The Holocene coastal features of Cape Cod were formed by the deposition of eroded glacial material (Fig. 1). The Provincelands to the north, the Nauset–Monomoy spit system to the south, and tombolos of Wellfleet along Cape Cod Bay were all formed by the process of longshore sediment transport. These spits subsequently served as protective barriers for the development of salt marshes.

Holocene coastal deposits on Cape Cod—barrier beaches, salt marshes, and sand dunes—are closely related to the submergence and erosion of glacial drift caused by postglacial sea-level rise. Sea level is most directly

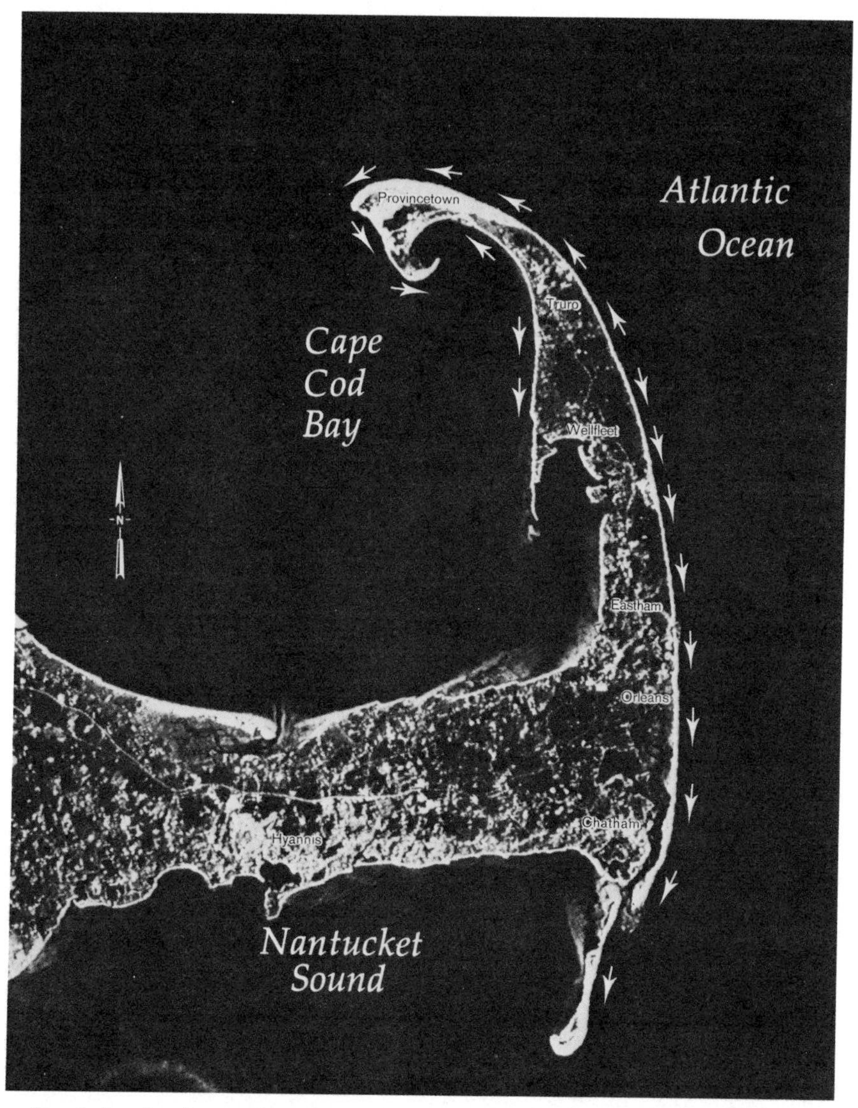

FIG. 1. Landsat image of Cape Cod, showing littoral drift directions and place names.

affected by the freezing and melting of glaciers, which is caused by global climatological changes (Donn and Shaw, 1963). The Holocene transition coincides with the sharp increase of surface temperatures about 15,000 years ago, which initiated the melting of the Wisconsinan glacier covering New England.

The change in sea level relative to land is due to the combined effects of the eustatic and isostatic components (Redfield, 1967). The eustatic component is representative of worldwide changes in sea level due to deglaciation, and is therefore the same in all parts of the ocean. The isostatic component refers to local land subsidence, and is regionally variable. In southern New England, sea level has been rising more rapidly than in relatively stable regions, due largely to a greater degree of local subsidence.

Several researchers have proposed sea-level rise curves for the northeastern United States based on radiocarbon dates on shells and/or peat. The curves most applicable to Cape Cod are those constructed by Redfield and Rubin (1962) and Oldale and O'Hara (1978). The curve of Redfield and Rubin is based on the radiocarbon age of samples of high-marsh saltwater peat recovered from various depths in Barnstable marsh. Because the vertical accretion of the high marsh is controlled by the rate of sea-level rise, their results are interpreted to indicate the change of the relative elevation of the sea and land that has occurred in the Cape Cod region during the past 4000 yr. They determined that sea level has been rising at a rate of 3.3×10^{-3} ft/yr (1.0×10^{-3} m/yr) since 2100 yr B.P. Prior to that time the average rate of sea-level rise was 10.0×10^{-3} ft/yr (3.0×10^{-3} m/yr).

Oldale and O'Hara's (1978) sea-level rise curve from 12,000 yr B.P. to the present is based on radiocarbon dates on shells and freshwater peat collected near the late Wisconsinan glacial maximum on the continental shelf. It is considered to be most accurate from about 10,000 to 7000 yr B.P., where it is extrapolated to join the Redfield and Rubin (1962) curve. Oldale and O'Hara showed that sea level rose at a rate of 5.6×10^{-2} ft/yr (1.7×10^{-2} m/yr) from 12,000 to 10,000 yr B.P. Between 10,000 and 6000 yr B.P. the rate of rise decreased gradually to about 9.8×10^{-3} ft/yr (3.0×10^{-3} m/yr), and remained at that rate until 2000 yr B.P. at which time there was an abrupt change in the rate of sea-level rise to 3.3×10^{-3} ft/yr (1.0×10^{-3} m/yr), as noted by Redfield and Rubin (1962).

Redfield (1967) found that this abrupt change in the rate of sea-level rise occurred at various times from 4000 to 2000 yr B.P. at different geographic locations. He asserts that at about 4000 yr B.P. there was a sudden reduction in the rate of eustatic rise in sea level. Local variations were therefore controlled by different local subsidences. There appears to have been about 13 ft (4 m) of subsidence between 4000 and 2000 yr B.P. for Cape Cod.

B. Shoreline Changes and Spit Development

In 1896, Davis attempted to reconstruct the original outline of Cape Cod by mentally reversing the processes at work on the present coastline

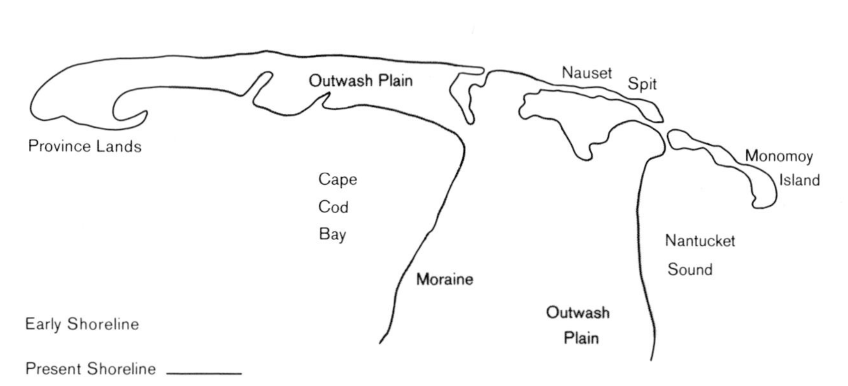

FIG. 2. Sediments eroded from glacial deposits were transported by longshore currents to form coastal features. [Adapted from Davis (1896).]

(Fig. 2). He projected that the original outer Cape had an irregular outline, which extended more than 2 miles beyond the present shoreline to the east and less than a mile to the west. Holocene features such as sandspits and tidal marshes did not yet exist. The changes that have occurred since ice retreat are a result of the erosive power of the sea and submergence of the land, the latter process being more important in embayments.

As sea level rose following deglaciation, Atlantic Ocean storm waves began reworking the glacial deposits of Cape Cod, gradually producing a graded (smoother) outline. The irregular outer shoreline of early postglacial Cape Cod approached an arcuate shape due to differential erosion of the glacial drift, forming an almost continuous marine scarp from Eastham to North Truro. The average rate of scarp retreat during historical times has been determined to be 2.2 ± 0.9 ft/yr (0.67 ± 0.3 m/yr), which compares favorably with the value of 2.4 ± 0.9 ft/yr (0.73 ± 0.3 m/yr) for the Zeigler *et al.* (1964) data. Profile 96 illustrates the long-term cliff retreat in the Wellfleet area based on field surveying (Fig. 3). While rates of change provide an indication of long-term cliff retreat, it actually tends to be an aperiodic, storm-related process (Leatherman *et al.*, 1981).

Prior to submergence of George's Bank, the dominant direction of littoral drift was to the south. Between 6000 and 7000 yr B.P. eroded glacial materials were transported by littoral drift to form Nauset Spit. About 6000 years ago, George's Bank was submerged, thus allowing a southeast wave approach and a pronounced northerly movement of littoral drift. This event signaled the formation of the Provincetown Peninsula, a complex recurved spit. This coastal landform was created by a series of prograding sandspits; however, the mechanism for its evolution is debatable. Two opposing theories have been proposed by Davis (1896) and Zeigler

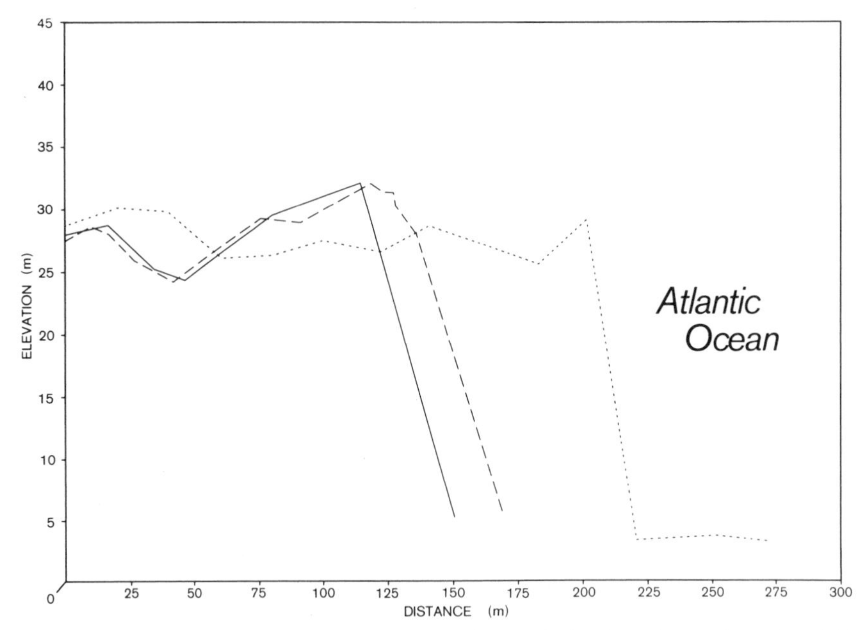

FIG. 3. Historical cliff retreat at Wellfleet-by-the-Sea during last 90 years based on field surveys: Marindin, 1887–1889 (······), Zeigler *et al.*, 1956–1962 (– – –), and Leatherman *et al.*, 1979–1981(——).

et al. (1965) concerning the developmental sequences of this unique geomorphic feature, as explained in the following section.

The tombolos of Wellfleet began forming as spits about 5000 years ago, when the glacial deposits along Cape Cod Bay were subject to erosional forces of northwesterly waves (Fig. 1). Elongation and recurvature of the Provincetown Spit provided protection to the bayside shoreline. By 2000 yr B.P., the peninsula was large enough to effectively block northwest wave approach and greatly reduce the transport of southeast littoral drift. Southwest waves eventually dominated the northeast shore of Cape Cod Bay, causing a reversal in littoral drift in the vicinity of Pamet Harbor (Giese, 1964). Conditions were thus favorable for the northward growth of Beach Point Spit (Pilgrim Beach), which enclosed East Harbor as it lengthened. In 1869, Beach Point was permanently closed by a dike, converting a tidal lagoon into the freshwater, eutrophic Pilgrim Lake of today.

In general, salt-marsh development began shortly after the formation of sandpits, which serve as protective barriers. Marsh development is contingent on rising water levels and a continuous supply of sediment to build up the elevation of mud flats to the level at which *Spartina alterniflora* (salt-marsh cordgrass) may colonize. Along the shores of Cape Cod, this sediment is derived from eroding sea cliffs and is transported into embayments by longshore currents.

The initiation of salt-marsh development in different locations on Cape Cod is quite variable. According to Redfield (1967), marsh has been developing behind Sandy Neck in Barnstable since about 4000 yr B.P. His study was based on the relation of age to depth of deposits, and he was able to describe and define the development of Sandy Neck. The outer Cape salt marshes, however, are much younger in age as the peat layers are generally less than a few meters thick.

II. Evolution of Provincetown Spit

Provincetown Spit is an excellent example of a large coastal landmass created by marine and aeolian processes. Although most of Cape Cod is composed of glacial sediments in the form of terminal moraines and outwash plains, Provincetown Peninsula was formed by deposition of sediments eroded from these glacial deposits (Fig. 2). This complex, recurved spit was created by a series of prograding sandspits, but the location, shape, and orientation of the original sandspits as well as their relationship to the extensive present-day system of parallel and curving dune ridges are still debatable. Two theories have been proposed to explain its evolution (Davis, 1896; Zeigler *et al.*, 1965).

Davis (1896) hypothesized that Provincetown Spit was formed by a number of prograding sandspits built from sediment eroded from the glacial deposits of Truro and Wellfleet and transported to the northwest by longshore currents. The formation of the peninsula began about 6000 years ago, when the first sandspit was built to the northwest of the glacial headland (Fig. 4). As these spits build northwest, they cut off areas of water in between as they curved and fused at the end of the peninsula. At the point of attachment to the mainland, the accreting spit would be quite narrow and subject to inlet breaching; this occurrence has been historically documented for Sandy Hook, New Jersey, a large recurved spit that Davis (1896) used for comparison to Provincetown. He also suggested that as the glacial sediments of the outwash plain eroded toward the southeast, the change in sea cliff position shifted the fulcrum of the accreting sandspit system (Fig. 5). This shift in the fulcrum was believed to result in the building of a new tangential spit to the north and seaward of the previous spit.

Messinger (1958) presented a modified version of the Davis (1896) orig-

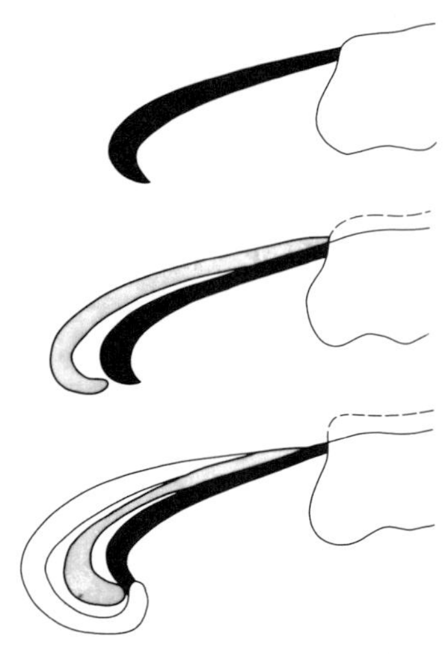

Fig. 4. Conceptual evolution of a spit according to Davis (1896).

inal theory. He mapped the location of the parallel dune ridges but did not provide any additional data to support this hypothesis. The Davis (1896) and Messinger (1958) interpretations were based entirely on the assumption that the present-day parallel and curving dune ridges correspond to the location of the shoreline at different stages in the development of the peninsula. They therefore believed that each of the sandspits that formed the peninsula built out from east to west over the entire length of the peninsula before being subsequently cut off by newly forming sandspits to the north (Fig. 4). In this fashion the peninsula's length was established

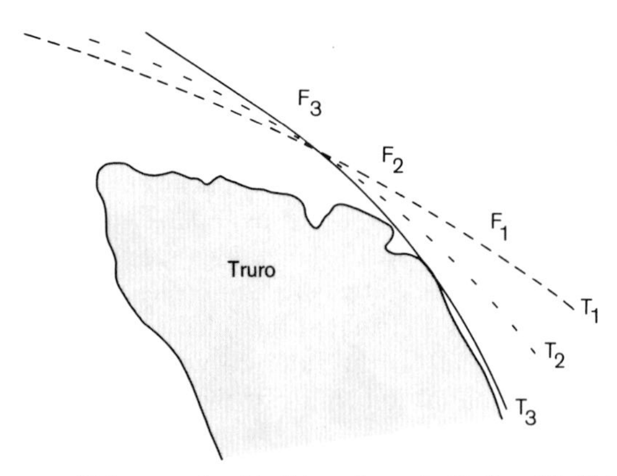

Fig. 5. Movement of fulcrum point (F_1–F_3) northward with time (T_1–T_3) as Truro erodes and sediment accretes to form Provincetown Spit. [From Davis (1896), with permission.]

as early as the completion of the first sandspit. Subsequent spits forming to the north would increase the width of Provincetown Spit. Messinger (1958) further expressed the belief that the peninsula formed around a preexisting island or landmass off the tip of Cape Cod. Well log analysis by Zeigler *et al.* (1965) indicated that this was not the case. Also, Messinger's (1958) explanation of the beach–dune ridges required an oscillating sea level during development, but this trend is not reflected in the sea-level curve of Redfield and Rubin (1962).

Zeigler *et al.* (1965) presented a different interpretation of the origin of the peninsula, disputing the assumption of Davis (1896) and Messinger (1958) that the present-day parallel dune ridge systems correspond to the location of the original prograding sandspits. According to Zeigler *et al.* (1965), each sequential spit built a short distance to the west and then hooked to the south (Fig. 6). In this way the present-day width of the Provincetown Spit at that point was essentially established, while the peninsula as a whole slowly built further to the west with the addition of each new spit. Thus, the westermost end of the peninsula was only recently formed.

Zeigler *et al.* (1965) theorized that there was a change in deposition patterns approximately 2000 years ago. Instead of the series of tight hooks that originally formed the peninsula, deposition tended to extend along the whole northerly coast seaward to the north (Fig. 7). This shift, which was believed to correspond to an abrupt change (slowing) in the rate of sea-level rise, resulted in the widening of the end of the Provincetown hook. Therefore, it follows that there is no relation between the present-

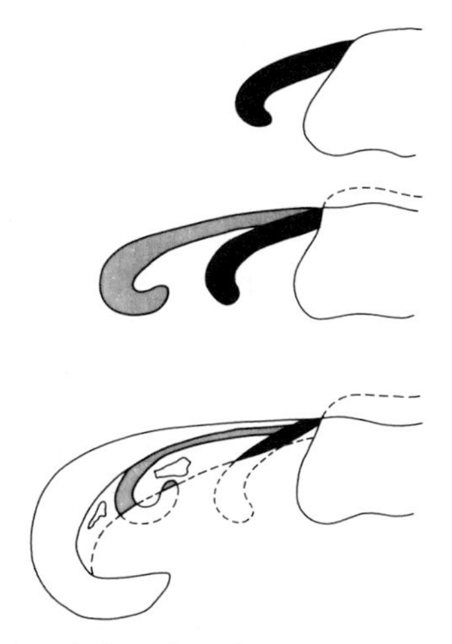

Fig. 6. Conceptual evolution of a spit according to Zeigler *et al.* (1965).

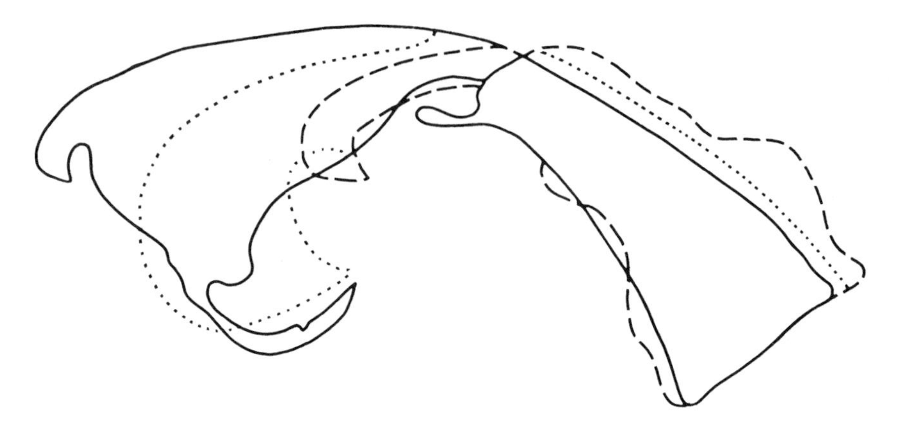

FIG. 7. Hypothesized evolution of Provincetown Spit: present (——), middle (⋯⋯), and early (– – –) (Zeigler *et al.*, 1965).

day parallel dune ridges and the location of the previous shorelines (sandspits), except for the most recent ridge enclosing Hatches Harbor (Fig. 8).

III. Methodology

From an aerial vantage point, the swale cranberry meadows and ericaceous shrub swamps correlate well with the form and orientation of the dune ridge system of the Provincetown Peninsula. If these swales were derived from areas of salt water, cut off by the prograding sandspits as postulated by Davis (1896), then cores in these areas should encounter plant remains and sediment indicative of a marine origin. The swale cores should contain a basal vegetative layer of salt-marsh peat, followed by a successive sequence of plant community remains. The vegetative change would reflect changes in the marine influence as the areas became more terrestrial in nature. Even if the cores were not deep enough to intercept the originally formed salt-marsh vegetation, the vertical sequence of wetland plant remains should indicate the development of this swale from a body of salt water. Therefore, coring along the swales should show if the parallel dune–ridge systems correspond to the location of the original sandspit shoreline.

Radiocarbon dating of organic materials recovered at depth by coring aided in defining the rate of sandspit formation and determined the age

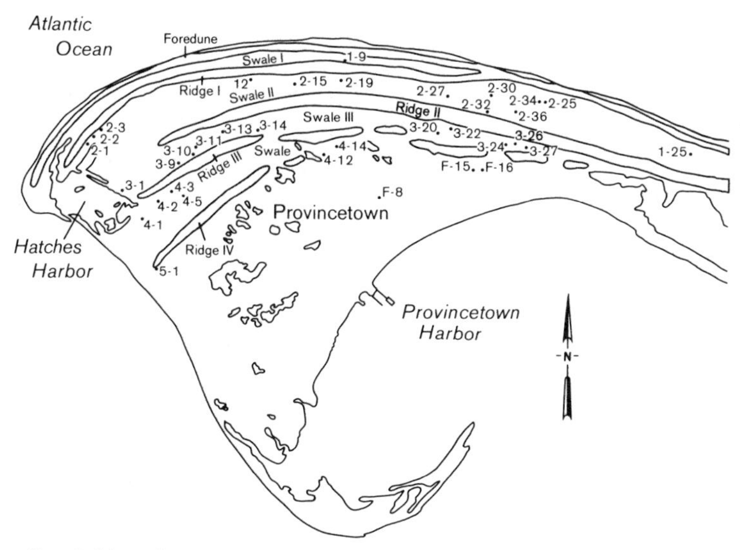

FIG. 8. Map of apparent ridge and swale topography of the Provincelands.

of the different prograding spits. By dating subsurface deposits found in cores from widely scattered areas of Provincetown Spit (different east and west locations and north and south swales), it was possible to determine the relative age of these deposits. For example, if swales on opposite sides of a dune ridge contained sediments of the same relative age, then the Davis (1896) hypothesis would seem incorrect, since this dune ridge would not represent a true sandspit.

Forty-one cores were taken in the four swales (Fig. 8), as previously defined by Messinger (1958). The pile-driving technique limited coring penetration to a maximum of 3 m. Radiocarbon dates were obtained for sediments in four cores. Coring sites were selected at widely scattered locations throughout the Provincelands and numbered separately for each swale, in increasing order from west to east. The four swales were numbered from I to IV, bounded by the dune ridges (Fig. 8). The two northermost swales can be traced eastward toward the terminus of the North Truro outwash plain at Pilgrim Heights.

IV. Results

According to the Davis (1896) hypothesis, swale IV would be the oldest and swale I the youngest. Coring sites to the west end (low numbers) would be the youngest and those to the east (high numbers) the oldest. Finally, the sites to the extreme western end of a more southern swale (swale IV) would be older than sites to the extreme eastern end of the subsequent swale formed to the north (Fig. 8). According to the hypothesis of Zeigler *et al.* (1965), the relative ages of the coring sites are different. In general, sites to the western end of the peninsula would be the youngest, regardless of the swale in which they were located. All the sites (roughly north-to-south line across the peninsula) would be about the same age. Finally, eastern coring sites would always be older than western sites, regardless of their north–south location.

Analysis of the core data tends to dispute the Davis (1896) interpretation of the origin of Provincetown Peninsula. Most of the 41 cores revealed thin surface organic layers underlain by wind-deposited (aeolian) sediments (Fig. 9). Inspection of 17 cores revealed vegetation or sedimentary information anomalous to the Davis (1896) interpretation, but that could be adequately explained considering the spit orientation and age sequence hypothesized by Zeigler *et al.* (1965).

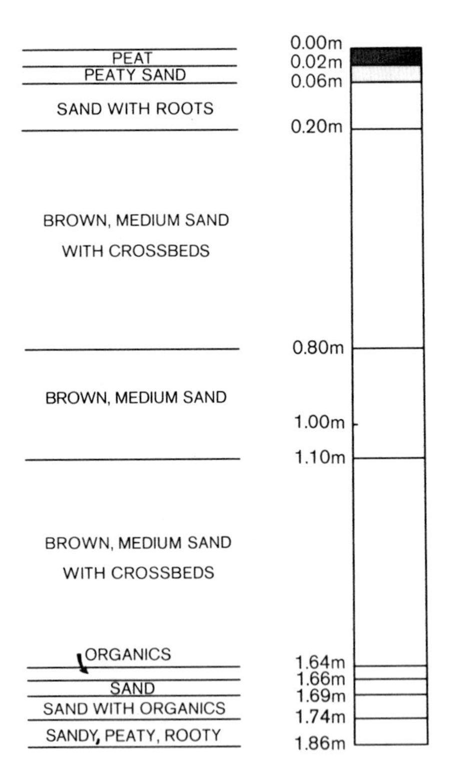

Fig. 9. Core log of PL 3-22.

No salt-marsh deposits were penetrated by the Provincetown Spit cores. Coring through cranberry meadows and other plant communities, located at various sites in all four swales, yielded only a thin surface organic layer underlain by wind-deposited sediments (for example, see core diagram PL 1-9, Fig. 10). Cores taken in Race Run, where recent marine influence has been documented (Woodworth and Wigglesworth, 1934), exhibited marine deposits underlying surface cranberry layers (core diagram PL 2-15, Fig. 10).

At three different coring sites, extensive compacted woody peat deposits, overlain by up to 1.8 m of wind-deposited sand, were found (Fig. 11, PL 4-3B). This subsurface peat probably formed under a dense mixed-shrub swamp similar to swamps characteristic of many areas on the peninsula today. Other scattered cores revealed nonwoody subsurface peat covered by wind-deposited sand (Fig. 11, PL 1-25). These subsurface organic layers appeared to be peat, similar to that found in present-day freshwater marshes.

Although the evidence is not conclusive, the general orientation and characteristics of the former freshwater marshes and swamps are better explained by the Zeigler *et al.* (1965) interpretation. Moreover, the positions of the swales did not help in determining the location of these former wetlands. In fact, very similar peat deposits were located in different swales, instead of along the same swale as would be predicted from

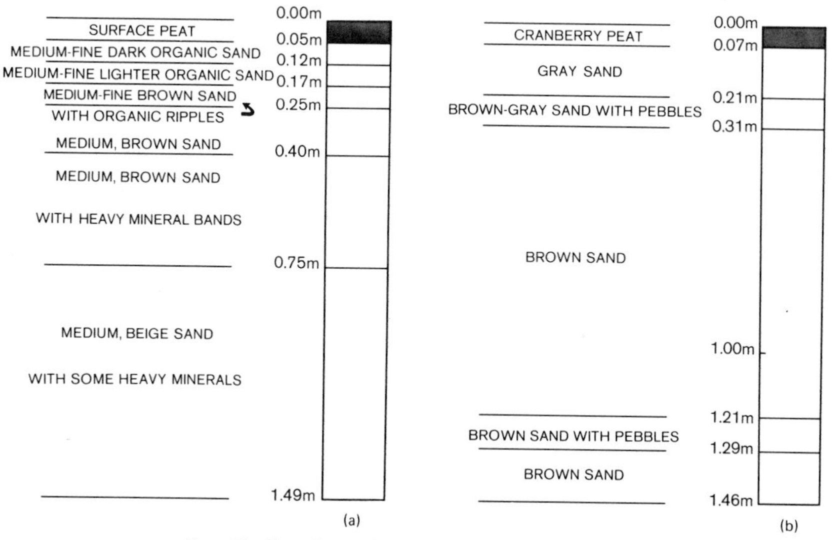

Fig. 10. Core logs of (a) PL 1-9 and (b) PL 2-15.

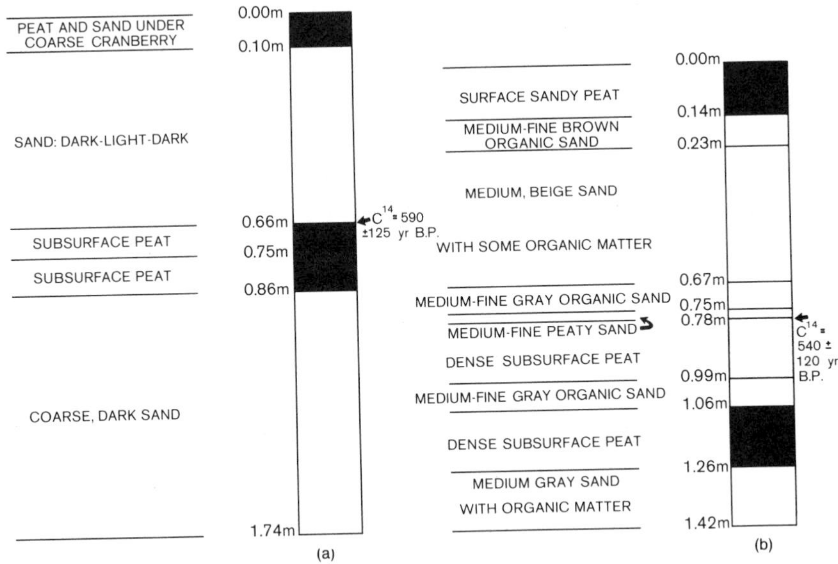

FIG. 11. Core logs of (a) PL 4-3B and (b) PL 1-25.

the Davis (1896) interpretation. For example, very similar dense, compacted, subsurface peat deposits were found in cores PL 2-32 and PL 3-26 (Fig. 12). Comparison of these sites on a topographic map showed that PL 3-26 is located just south of PL 2-32, separated only by a dune ridge (Fig. 8). The relative elevations of these two coring sites were obtained by field surveying with transit and rod. Assuming a sea-level rise of 0.5 ft/100 yr (15 cm/100 yr), the age difference in these freshwater peats should be 1400 yr, since the elevational difference is approximately 7 ft (2.1 m). In fact, the peat in core PL 3-26 is only 200 years older than that found in core PL 2-32 (Fig. 12). If there are deeper peat layers at coring site PL 2-32, it would follow that these sediments would be older than the 3.4 ft (1.05 m) peat layer in this core (PL 2-32) and perhaps older than the 10.5 ft (3.2 m) peat in core PL 3-26. This finding indicates that these areas are nearly equivalent in age, instead of exhibiting age differences of approximately 1000 yr, and that dune ridge II is merely a surficial feature rather than an original sandspit. A deeper core at site PL 2-32, obtained with a vibracorer, showed another interval of aeolian sand overlying yet another freshwater organic layer (Fig. 13).

Sedimentary evidence from the cores was also best explained by the hypothesis of Zeigler *et al.* (1965). For example, very coarse sand and pebbles (good evidence of marine deposition) were found less than a meter

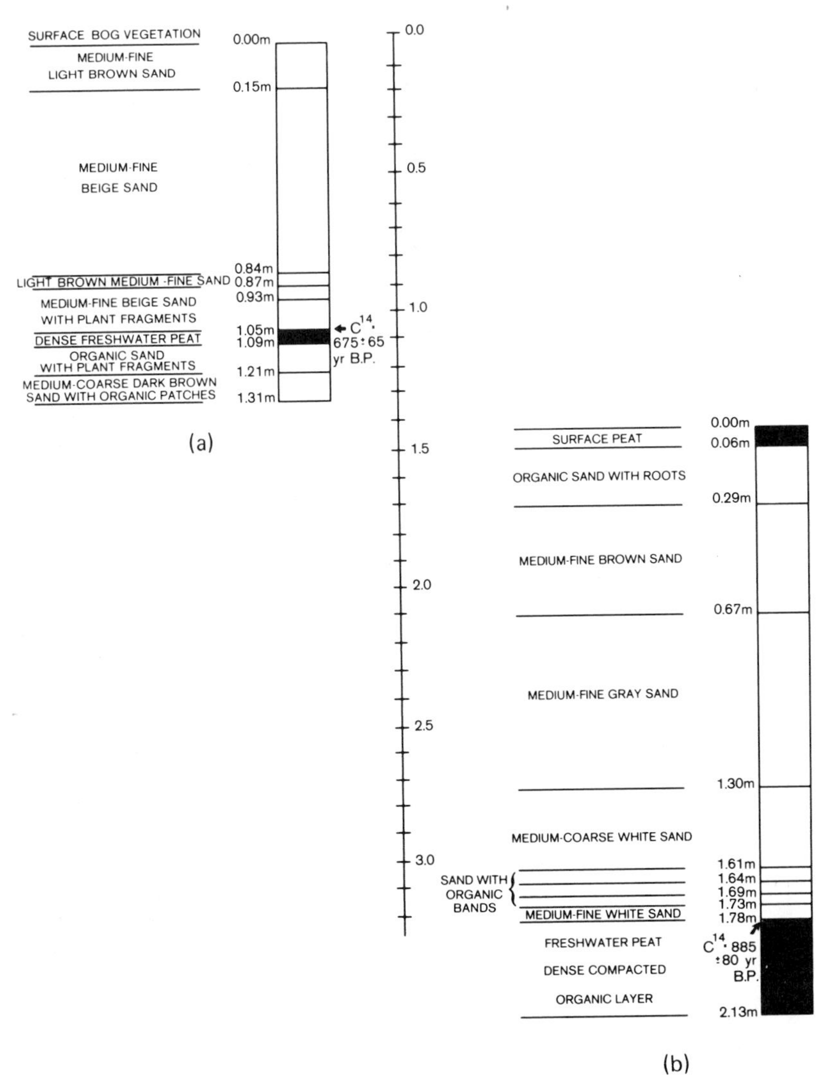

FIG. 12. Core logs of (a) PL 2-32 and (b) PL 3-26.

below the surface at coring sites PL 3-10 (Fig. 14) and PL 3-14. According to the Davis (1896) hypothesis, this site is in one of the older swales and therefore was formed quite a long time ago when sea level was far below its present level. It follows that marine deposits would not be found as close to the surface if this area is in fact quite old. Only by applying Zeigler's interpretation is it possible to explain these marine deposits.

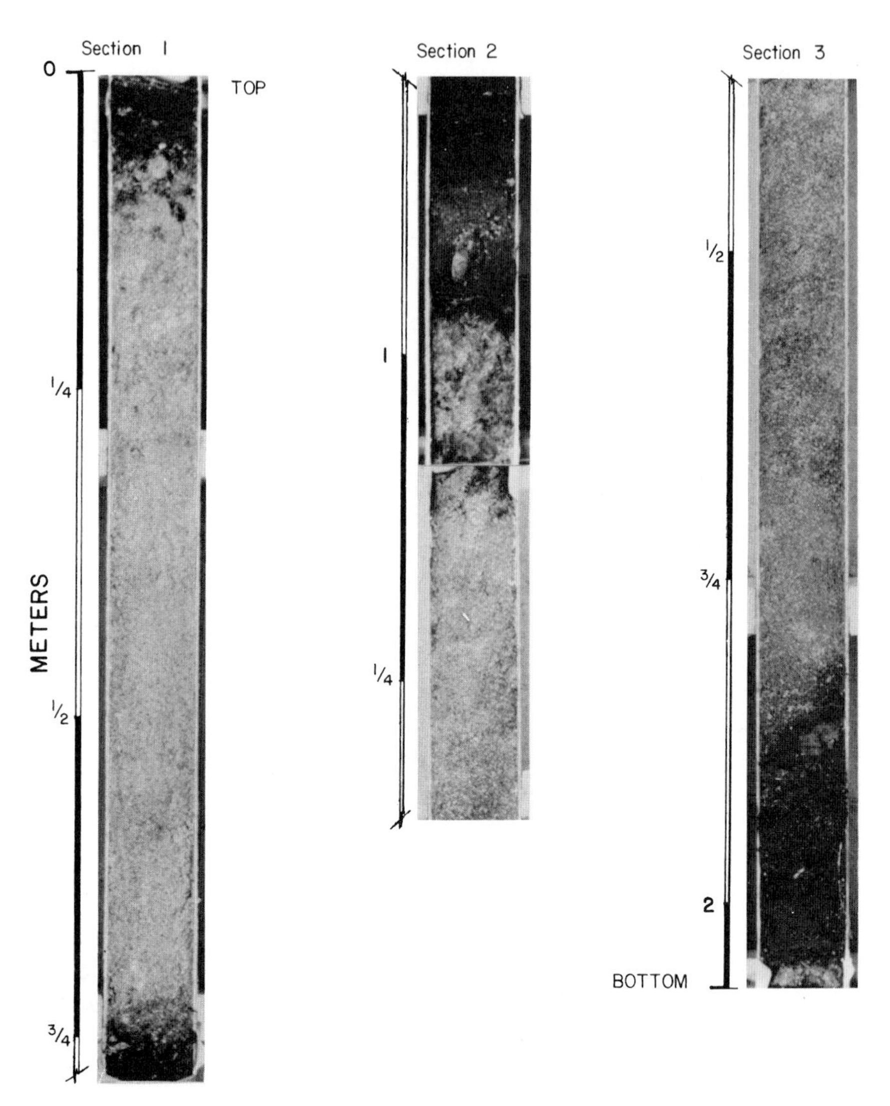

Fig. 13. Deeper (over 2 m) core at site PL 2-32 revealed three freshwater organic layers separated by aeolian sand deposits.

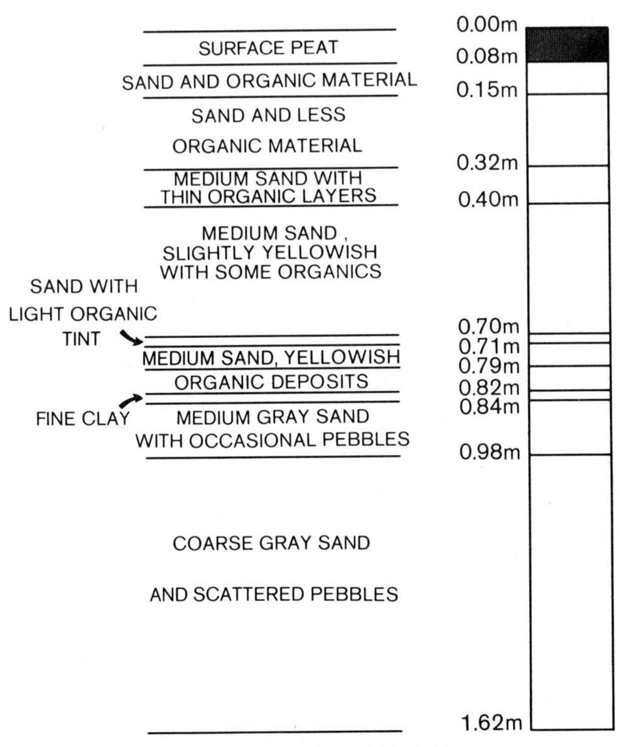

Fig. 14. Core log of PL 3-10.

V. Summary and Conclusions

Limitations of the coring technique did not allow us to obtain the data necessary to fully address the questions posed by this investigation. Cores deeper than 3 m will be necessary, considering the overall age of the peninsula and the rise of sea level over the past 4000 to 6000 yr. Analysis of sediments obtained from swale cores indicated sand deposition by wind rather than water, except for the Race Run–Hatches Harbor area (a historically known waterway). These findings showed that the Provincetown Peninsula is a very complex geological area, characterized by unstable and constantly shifting dunes. Cores taken in forested swales between relatively stabilized dune ridges indicated that in the recent past even these areas were subjected to extensive dune migration. Although not conclusive, these findings are better explained by the mechanism of Zeigler *et al.* (1965) for the formation of Provincetown Peninsula than by the Davis (1896) theory.

This study also confirmed the conclusions drawn by McCaffrey and Leatherman (1979) regarding the massive disruption of sand-dune ecosystems by the Pilgrims. The core data showed quite clearly that dune migration was rampant throughout the Provincetown area at an earlier time. Many freshwater bogs and swamps were destroyed by sand encroachment and eventual burial. It appears that there was at least one period of extreme aeolian sand movement through the area; this time probably corresponded to the arrival and settlement of the area by the Pilgrims.

Acknowledgments

This research was supported by the National Park Service. Mark Benedict helped carry out the field work, and much information for this chapter was derived from our technical report. Background information was also obtained from a technical report co-authored with Patty O'Donnell. The graphics were prepared by Roy Ashley and Joel Miller of the Department's Cartographic Services Laboratory. The typing was completed by Pat Leedham and Dorie Taylor.

References

Benedict, M. A., and Leatherman, S. P. (1978). "Preliminary Investigation of the Geobotanical Evolution of Provincetown Peninsula," NPS Res. Unit Report No. 38. Univ. of Massachusetts, Amherst.

Davis, W. M. (1896). The outline of Cape Cod. *Proc. Am. Acad. Arts Sci.* **31,** 303–332.

Dillon, W. P., and Oldale, R. N. (1978). Late Quaternary sea-level curve, reinterpretation based on glaciotectonic influence. *Geology* **6,** 56–60.

Donn, W. L., and Shaw, D. M. (1963). Sea level climate of the past century. *Science* **142,** 1166–1167.

Giese, G. S. (1964). Coastal orientations of Cape Cod Bay. M.S. Thesis in Oceanogr., Univ. of Rhode Island, Kingston.

Leatherman, S. P., Giese, G., and O'Donnell, P. (1981). "Historical Cliff Erosion of Outer Cape Cod," NPS Res. Unit Rep. No. 53. Univ. of Massachusetts, Amherst.

Marindin, H. L. (1891). Cross-sections of the shore of Cape Cod between Chatham and Highland Lighthouse. *U.S. Coast Geodetic Surv. Rep.*, 409–457.

McCaffrey, C., and Leatherman, S. P. (1979). Historical land-use practices and dune instability in the Province Lands. *In* "Environmental Geologic Guide to Cape Cod National Seashore" (S. P. Leatherman, ed.), UM/NPSCRU, pp. 207–222. Univ. of Massachusetts, Amherst.

Messinger, C. (1958). A geomorphic study of the dunes of the Provincetown peninsula, Cape Cod, Massachusetts. M.S. Thesis in Geol., Univ. of Massachusetts, Amherst.

O'Donnell, P., and Leatherman, S. P. (1980). "Generalized Maps and Geomorphic Reconstruction of Outer Cape Cod between 12,000 BP and 500 BP," NPS Res. Unit Rep. No. 48. Univ. of Massachusetts, Amherst.

Oldale, R. N. (1969). Seismic investigations on Cape Cod, Martha's Vineyard, and Nantucket, Mass., and a topographic map of the basement surface from Cape Cod Bay to the Islands. *U.S. Geol. Surv. Prof. Pap.* **65-B,** B122–B127.

Oldale, R. N., and O'Hara, C. J. (1978). New radiocarbon dates from the inner continental shelf off southeastern Massachusetts and a local sea level rise curve for the last 12,000 years. *U.S. Geol. Surv. Open File Rep.*

Redfield, A. C. (1967). Post glacial change in sea level in the western North Atlantic Ocean. *Science* **157,** 687–692.

Redfield, A. C., and Rubin, M. (1962). The age of salt marsh peat and its relation to recent changes in sea level at Barnstable, Massachusetts. *Proc. Natl. Acad. Sci. USA* **48,** 1728–1735.

Woodworth, J. B., and Wigglesworth, E. (1934). Geography and geology of the region including Cape Cod, Elizabeth Islands, Nantucket, Martha's Vineyard, No Mans Land and Block Island. *Harv. Univ. Mus. Comp. Zool. Mem.* **52,** 328 p.

Zeigler, J. M., Tuttle, S. D., Giese, G. S., and Tasha, H. J. (1964). Residence time of sand composing the beaches and bars of outer Cape Cod. *Proc. Conf. Coastal Eng., 9th Am. Soc. Civ. Eng.* pp. 403–416.

Zeigler, J. M., Tuttle, S. D., Tasha, H. J., and Giese, G. S. (1965). The age and development of the Province Lands hook, outer Cape Cod, Massachusetts. *Limnol. Oceanogr.* **10,** R98–R311.

CHAPTER 13

Development of the Northwestern Buzzards Bay Shoreline, Massachusetts

DUNCAN M. FITZGERALD,
CHRISTOPHER T. BALDWIN,
NOOR AZIM IBRAHIM,
AND
DAVID R. SANDS

Department of Geology
Boston University
Boston, Massachusetts

The western Buzzards Bay coast consists of a series of headlands and intervening drowned valleys which are fronted by various types of barriers. The general topography of the region is a product of late Tertiary fluvial erosion and repeated Pleistocene glaciations. The depth and width of the valleys coupled with the availability of sediment have dictated the morphological genesis of individual embayments. Where valleys are narrow and/or shallow, coastal ponds fronted by barrier beaches have formed. Larger and deeper valleys are protected by spit systems and maintain tidal inlets due to greater bay tidal prisms. Differences in the dynamic histories of the two largest embayments in this area (Slocum River and Westport River estuaries) are explained in terms of availability and proximity of glaciogenic and other sediment sources as well as contrasting exposures to wave processes.

I. Introduction

The northern shore of Buzzards Bay, Massachusetts, is dominated by a series of north–south oriented peninsulas and embayments that become more northwest–southeast trending toward the Cape Cod Canal (Fig. 1). The coastline can be divided into two distinct morphological units. The western half of the coast, from the Sakonnet River in Rhode Island eastward to Apponagansett Bay in Massachusetts, consists of elongated bays fronted by barrier spits and tidal inlets, or in some cases barrier beaches (Fig. 2). Several of the bays in this area have small rivers discharging into them. The eastern half of the shoreline, east of Apponagansett Bay to the Cape Cod Canal, is characterized by long peninsulas (2–5 km in length) and, unlike the coast to the west, contains few wave-built depositional

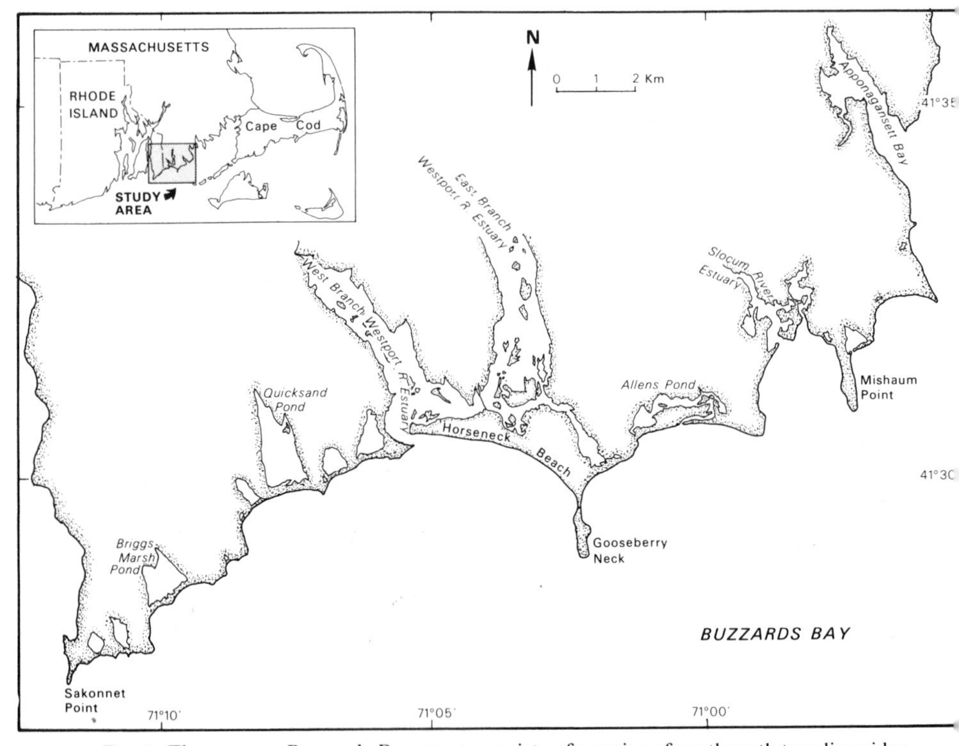

Fig. 1. The western Buzzards Bay coast consists of a series of northsouth trending ridge and valley systems. The flooding of the valleys and subsequent barrier formation during the Holocene transgression has produced coastal ponds, estuaries, and tidal inlets.

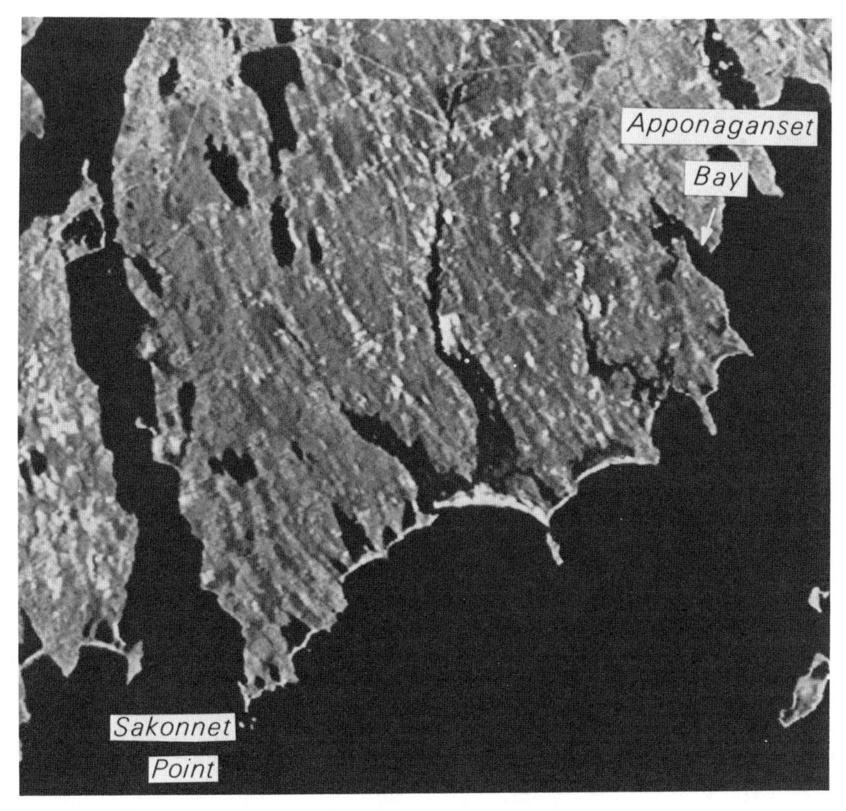

FIG. 2. High-altitude ERTS aerial photograph of the western Buzzards Bay coast. The study area is located between Sakonnet Point and Apponaganset Bay.

features. Eroding headlands, narrow sand and gravel pocket beaches, and peripheral marsh areas dominate this coastline. Landward-migrating, small sand and gravel spit systems also occur at some of the island and headland areas.

This chapter is concerned with the western portion of the northern Buzzards Bay shoreline. The morphologic development of this coast is a product of a number of interactive physical processes operating on pre-existing ridge and valley topography. This landscape owes its origin to the structural framework of the region that has been accentuated by glacial excavation. Glaciers were also responsible for delivering large amounts of till and outwash sediment to the coast. In a regime of rising sea level, these deposits have been subsequently reworked by storms, waves, and tidal processes. This has led to the formation of a number of barrier beaches, spits, and beach ridge barriers in front of many of the Holocene drowned valleys. It is the intent of this chapter to describe the morphology of this shoreline and to assess in detail the processes responsible for its genesis.

II. Physical Setting

The northern coast of Buzzards Bay is a microtidal shoreline with tidal ranges increasing toward the Cape Cod canal from a mean of 1.1 m at Sakonnet Point to 1.3 m at Great Hill [National Ocean Survey (NOS), 1986]. During spring tide conditions, tidal ranges at these locations increase to 1.3 and 1.6 m, respectively. This region experiences semidiurnal tides, with inequalities reaching a maximum of 0.3 m.

Wave energy along the study area is low and highly variable. Three years of data (incomplete) taken at a wave gage situated 9.0 km due south of Gooseberry Neck (Fig. 1) indicate that the deepwater mean significant wave height in this region is 0.9 m and the mean wave period is 7.5 s (Thompson, 1977). Wave heights along the open shoreline are generally less than 0.9 m and are considerably smaller in the embayments ($H < 0.3$ m) (Sands and FitzGerald, 1984). The study area is most susceptible to storms that track west of Buzzards Bay. These storms, like the 1938 hurricane (storm of record for this coast), generate southerly winds and waves. Cape Cod and the Elizabeth Islands limit the size of waves approaching the coast from east and southeasterly directions. Long Island serves a similar function for waves approaching from the westerly quadrant. A wind rose constructed for Block Island (U.S. Army Corps of Engineers, 1986) indicates that the prevailing winds come from the southwest while the dominant northeast winds blow offshore. The study area spans the tide-dominated, mixed-energy setting in the shoreline classification of Hayes (1979) and Nummedal and Fischer (1978) (Fig. 3). However, it should be noted that due to the overall low energy of this coast, infrequent large-magnitude storms (50-yr storm) have exerted a strong influence on the development of this shoreline.

III. Bedrock and Glacial Geology

Tectonically, the northern coast of Buzzards Bay is part of the Proterzoic southeast New England platform (Barosh and Hermes, 1981). The

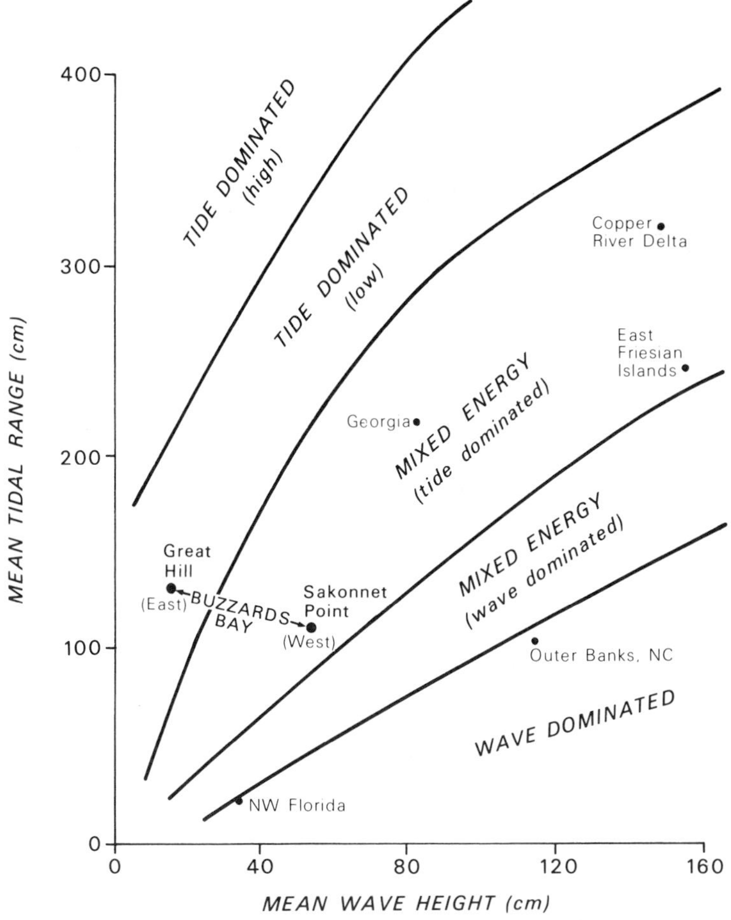

FIG. 3. The study area plots in the mixed-energy (tide-dominated) environment in the Hayes (1979, Fig. 15) and the Nummedal and Fischer (1978, Fig. 18) coastline classification. The eastern section of the Buzzards Bay shoreline is tide dominated due to the limited fetch.

rocks of this area belong to the gneissic terrane, consisting primarily of biotite granites and hornblende gneisses (Zen, 1983). During the late Cretaceous and early Tertiary the bedrock surface was eroded and unconformably overlain by coastal plain and continental shelf sediments that extended along most of New England (Kaye, 1983). These strata were eroded during the late Tertiary, forming southerly flowing drainage systems. Seismic data reveal that the fluvially formed ridge and valleys extend several kilometers offshore and are now covered by reworked Pleistocene

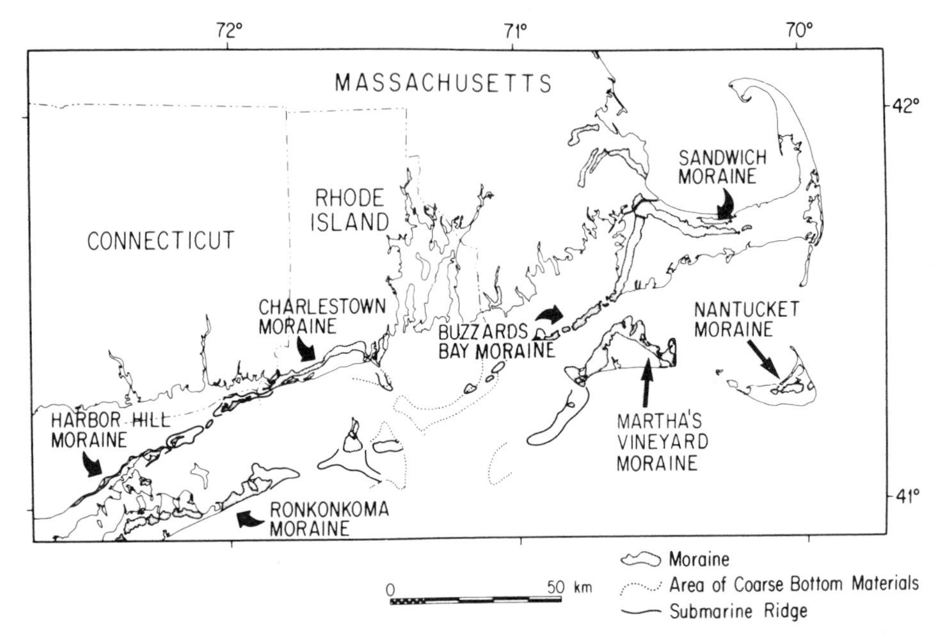

FIG. 4. Map of the coastal end moraines deposited during the late Wisconsinan stage of glaciation. [From Oldale (1982, Fig. 2); after Schafer and Hartshorn (1965, Fig. 2).]

and older sediments (McMaster and Asraf, 1973). In the study area the ridges have an average spacing of 1–3 km.

More recently, the coastal landscape has been sculpted by glacial processes during the Pleistocene epoch. Stratigraphic evidence including outcrops and core data from Martha's Vineyard demonstrates that this region has experienced at least four major episodes of glaciation, ranging from Nebraskanan to Wisconsinan in age (Kaye, 1964). The maximum extent of the Wisconsinan ice sheet is marked by a discontinuous terminal moraine extending from southern Long Island through Block Island to Martha's Vineyard and Nantucket Islands. The ice retreated from this region approximately 15,300 to 14,250 yr B.P. (Larson, 1982), leaving behind the Sandwich, Buzzards Bay, and Charlestown recessional moraines (Fig. 4).

IV. Sediment Supply

Like most shorelines in New England, except those near large river systems (e.g., the barriers near the Merrimack River, Massachusetts, and

Kennebec River, Maine), the coast of western Buzzards Bay receives most of its supply of sediment from glacial sources. Sand and minor amounts of gravel eroded from these nearshore deposits have been delivered to beaches via longshore transport. Along the North Carolina coast it has been suggested by Hine *et al.* (1979) that as the rate of sea-level

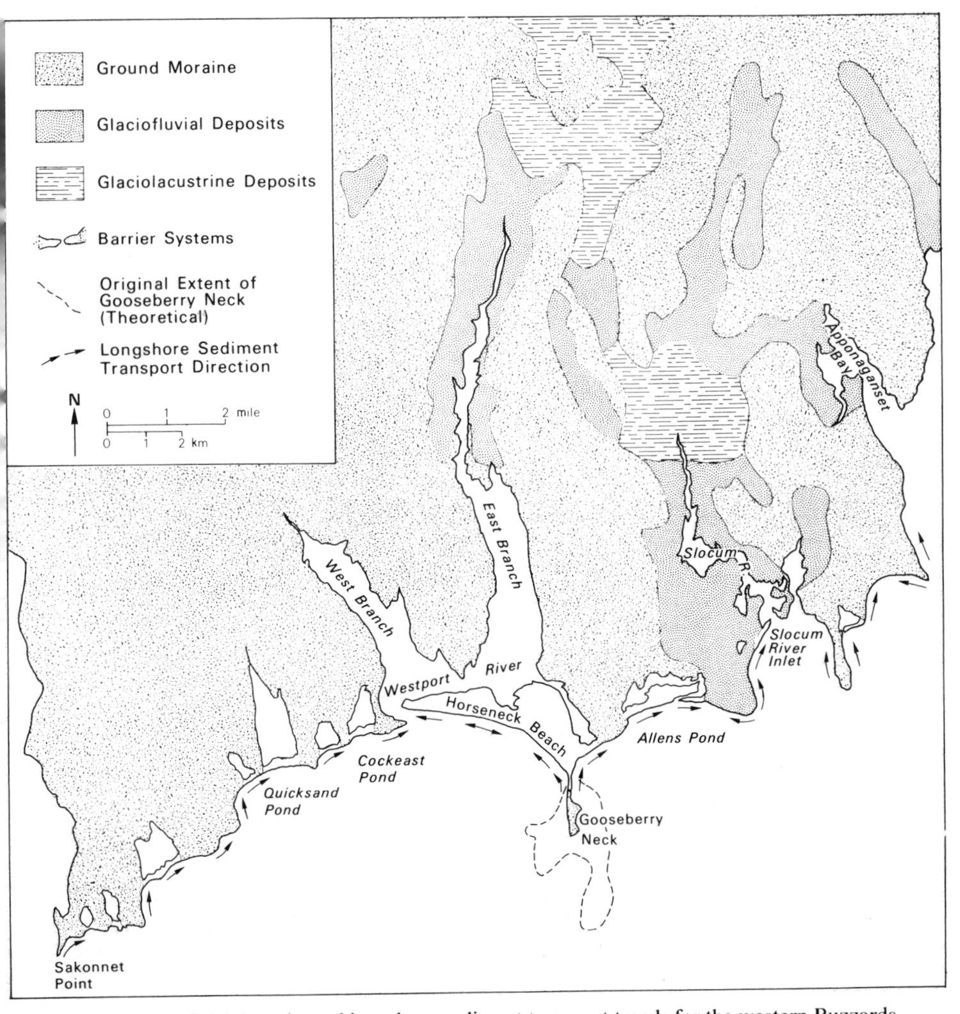

FIG. 5. Surficial deposits and longshore sediment transport trends for the western Buzzards Bay coast. Surficial data, in part, are from Stone and Piper (1982). Longshore transport directions are based on wave refraction analysis, morphologic evidence, process data, and grain-size trends (Sands and FitzGerald, 1984; Magee and FitzGerald, 1980).

rise slowed during the Holocene still stand, the onshore movement of sediment contributed significantly to the progradation of the barrier systems. Similarly, the marine reworking and onshore transport of drowned glaciofluvial deposits may have also been an important supply of sediment to the western Buzzards Bay coast. The primary sources of sand along this coast include the large till island of Gooseberry Neck and the glaciofluvial and glaciolacustrine deposits associated with the east branch of the Westport River and the Slocum River embayment (Fig. 5) (Stone and Piper, 1982). Minor amounts of sand, commonly less than 15–20% of the matrix, have also been derived from the erosion of till-covered headlands (Fig. 6).

Although Gooseberry Neck appears to be a rather small source of sediment when compared to the volume of sand contained in Horseneck Beach and the spits fronting Allens Pond, it is believed that in the past it was many times its present size and, when combined with a possibly more vigorous storm regime (Lamb, 1977), may have been a major local source (see Fig. 5, dashed line). The extent of this source deposit is marked approximately by the 5-m isobath and contains several areas exposed at mean low water (e.g., Hens and Chickens), including an extensive boulder retreat lag that mantles the sea floor and surrounds the island. Gooseberry Neck is 10–15 m high with low, cliffed seaward margins and is composed of boulders, cobbles, and gravel in a sandy silt matrix. The linear nature

FIG. 6. Till scarp near Quicksand Pond. Note the boulder pavement at the base of the slope. (Photograph by Jamie Greacen.)

of the deposit and numerous bedrock outcrops offshore suggest that the island may be rock-cored.

Although no direct observational evidence exists, it seems likely that other, now submerged, or indeed, completely removed glaciogenic sources may have been present in this area. Other rock core remnants and associated boulder remaneés such as those adjacent to Twomile ledge and Mishaum ledge may be the only indicators. In essence, it would seem reasonable to postulate that Gooseberry Neck is representative of a number of similar high sediment input events that characterized different stages of the trangression of this complex glaciated topographic surface.

V. Shoreline Components and Processes

The western Buzzards Bay coast can be separated into a number of morphological components including small bedrock outcrops and eroding till headlands, barrier beaches, spit systems, tidal inlets, and beach-ridge barriers. Behind the barriers are ponds, estuaries, and marsh and tidal creek systems. The morphology of these features and the sedimentation processes affecting them are discussed below. Their distributions are shown in Fig. 7.

A. Bedrock and Till Headlands

Bedrock, although comprising a relatively small percentage of the coast (Fig. 7), has had an important influence in the overall development of the shoreline. Bedrock outcrops at the Acoaxet Point (Fig. 14a) and at Potomska Point (Fig. 12g) have stabilized the positions of Westport River and Slocum River inlets and bedrock islands along the western barrier beaches have led to the formation of intertidal tombolos (Fig. 8). Ibrahim (1986) in a stratigraphic study of Horseneck Beach has also shown that bedrock has strongly influenced beach barrier formation by providing anchoring sites for sedimentation and in organizing the back-barrier drainage.

The till-covered ridges of the study area, although forming slight promontories (Fig. 7), have been retreating at the same approximate rate as the intervening barriers. Mishaum Point and Gooseberry Neck are major exceptions to this trend. As till headlands erode and contribute sand and gravel to the adjacent barriers and offshore region, low till scarps less than 3 m in height are commonly formed. Where the till has a relatively high boulder content, the till scarps become somewhat protected by a boulder retreat lag (Fig. 6).

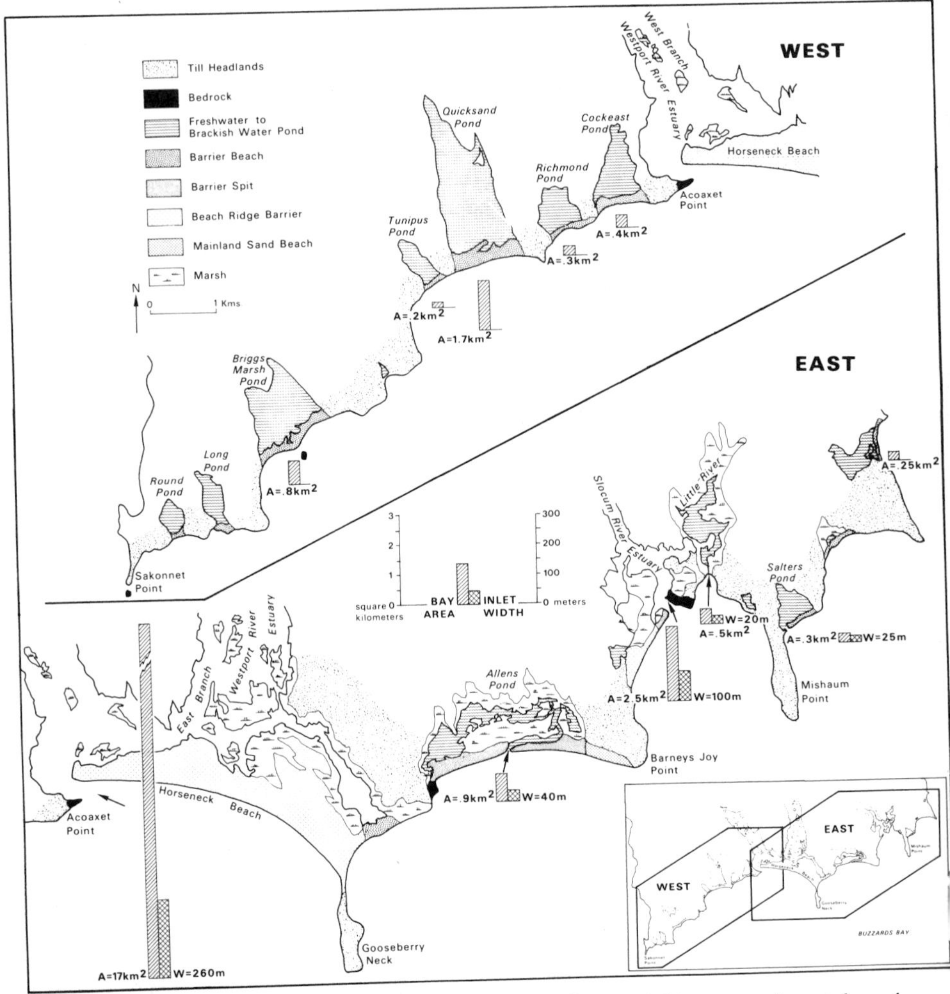

FIG. 7. Coastal geomorphic map of western Buzzards Bay. A histogram of coastal pond and estuary bay areas is plotted against their respective tidal inlet widths (indication of inlet size).

B. Coastal Ponds and Barrier Beaches

Barrier beaches and coastal ponds are located primarily west of Westport River inlet and east of Slocum embayment where headlands are spaced closely together (mean spacing = 1.0 km) and the intervening valleys have been flooded. Several of the ponds along this section of coast have developed during the past 100 years. For example, an 1885 U.S. Coast and Geodetic Survey coastal chart indicates that both Salters Pond and Briggs Marsh Pond were once entirely marsh areas. Sea-level rise, barrier construction, and occasional salt-water inundation transformed these marshes to brackish water ponds. The formation of other ponds, such as Quicksand and Richmond ponds, predates historical records.

There is a clear correspondence between the small area of the ponds and the absence of permanent, well-developed tidal inlets along the fronting barriers (Fig. 7). Because of the low tidal range (TR) of this region (TR = 1.1 m), the diminutive bay areas produce small tidal prisms that are incapable of maintaining a tidal inlet (O'Brien, 1931, 1969). However, at many of the ponds the significant freshwater inflow that comes from melting snow and ice during late winter and from precipitation during early spring produces pond outlets. These narrow discharge channels (width = 5–20 m) are formed when pond waters overtop the lowest point along the barrier, usually the site of a previous outlet, and flow out to the ocean.

The depth to which the outlet channel is scoured is dependent on the hydraulic head that is developed during the overtopping process, and the width and stratigraphy of the barrier. Channel depths seldom exceed mean low water and are normally intertidal (Fig. 8). In some cases, the highly porous nature of the gravel beaches permits the discharge of pond waters through the barrier system. As the outlets become fully developed, the ensuing tidal exchange between ocean and pond establishes tidal inlet processes. These processes include recurved spit formation, inlet migration, and flood tidal delta sedimentation and are illustrated at Briggs Marsh and Quicksand ponds in Fig. 9. Boothroyd *et al.* (1985) described similar inlet processes along the Rhode Island barrier coast. It should be noted that tidal inlets also develop along this section of shoreline during intense storms as a product of wave erosion, overwash activity, and storm surge flooding. Regardless of how the inlets are formed, they are ephemeral features that close by wave-generated longshore sediment transport and through the construction of flood-tidal deltas, which impede flood tidal currents (Fig. 9). The inlets that begin as seasonal late-winter channels are usually closed by mid-summer.

The barriers that front the coastal ponds have relatively low elevations (height < 4 m) and are comprised of both sand and gravel beaches. Where sand is abundant, low vegetated dune ridges have formed, while in gravel-dominated regions a series of storm ridges is present. Along the landward side of most of the barriers a rim of high salt marsh has developed on overwash deposits (Fig. 9). Although overwash is neither an active process along all of the barrier beaches on a yearly basis, nor in some sites even during a 10- to 20-yr period, major storms like the 1938 hurricane overwash the entire shoreline. Storms of this magnitude (50- to 100-yr storm event) erode the barrier beaches depositing much of the sediment in extensive

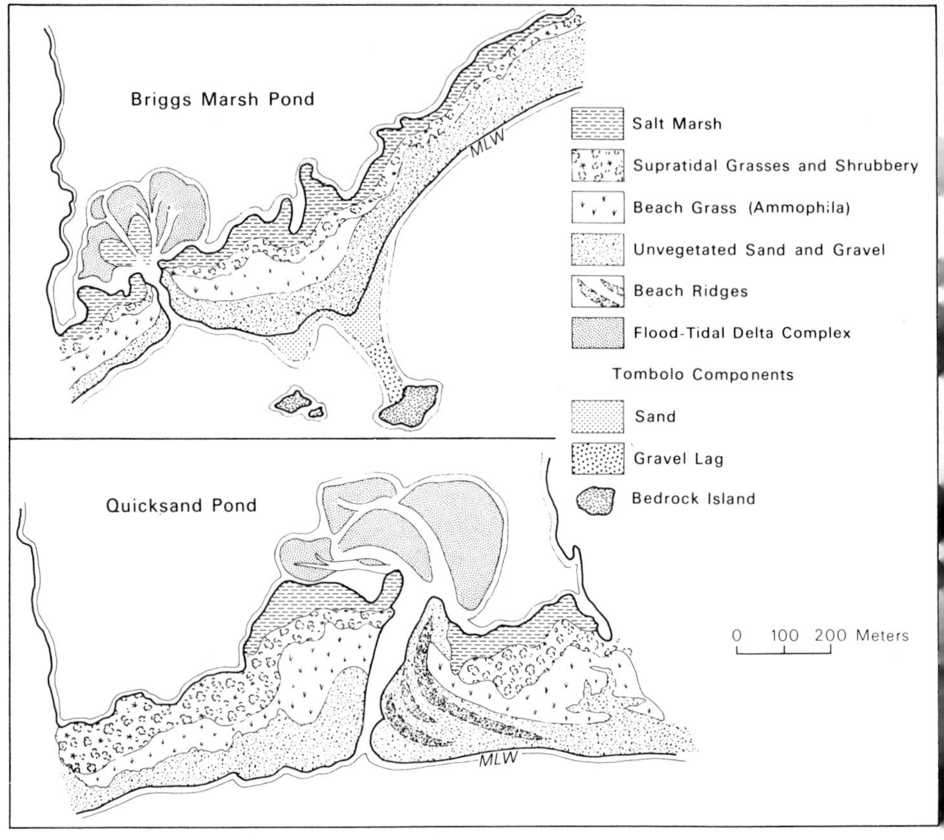

FIG. 8. Sedimentologic components of the barriers fronting Briggs Marsh and Quicksand ponds. [Modified after Greacen *et al.* (1983).] The inlets are ephemeral in nature.

lobate washover fans. The ongoing transgression of this shoreline is also accomplished through the construction of flood-tidal deltas.

C. Spit Systems

The major spit systems in the study area include those fronting Allens Pond and bordering Slocum River inlet (Fig. 7). Morphological changes that have occurred to these spits during the past 50 years have been interpreted from a detailed analysis of vertical aerial photographs, U.S. Coast and Geodetic Survey coastal charts, and shoreline surveys (Figs. 10–13). Beach morphology, general grain size trends on the beach face, and longshore currents are summarized for the spits along Allens Pond in Fig. 11. Longshore currents were measured during varying wind conditions over

FIG. 9. Ephemeral inlet at Quicksand Pond in 1983. (Photograph by Jamie Greacen.)

seven half-tidal cycles, and the data shown in Fig. 11 are representative of waves produced by the southwesterly prevailing winds (7–9 m/s). Similar data for Slocum Spit are given in Fig. 13. In a comparison of the data sets for the two spits, it is seen that the two regions have experienced very different depositional histories.

1. Allens Pond Spit

The historical data for the Allens Pond spits demonstrate that the tidal inlet migrates to the west (Fig. 10). The average rate of inlet migration is more than 100 m/yr but decreases markedly as the inlet reaches a far westerly position and almost closes. In fact, the inlet has closed or was about to close on several occasions when the Town of South Dartmouth prevented this from happening by dredging a new inlet channel at the eastern end of the eastern spit. This is done to protect the shellfish in the pond, which require saline conditions and ocean-water circulation.

Despite the historical evidence of westerly inlet migration and spit building, other data demonstrate that the dominant longshore transport direction along this shoreline is to the east (Fig. 11). Morphological data that support an easterly direction include (1) the logarithmic spiral-shaped shoreline east of the bedrock outcrop at the western end of the beach (Yasso, 1964) and (2) the wider, higher, and increasingly better development of dunes (excepting the eastern spit) in an easterly direction along the Allens Pond shoreline. Likewise, average grain sizes along the beach face exhibit a fining trend to the east, decreasing from gravel-sized material at the western end of the beach to medium sands along the eastern end

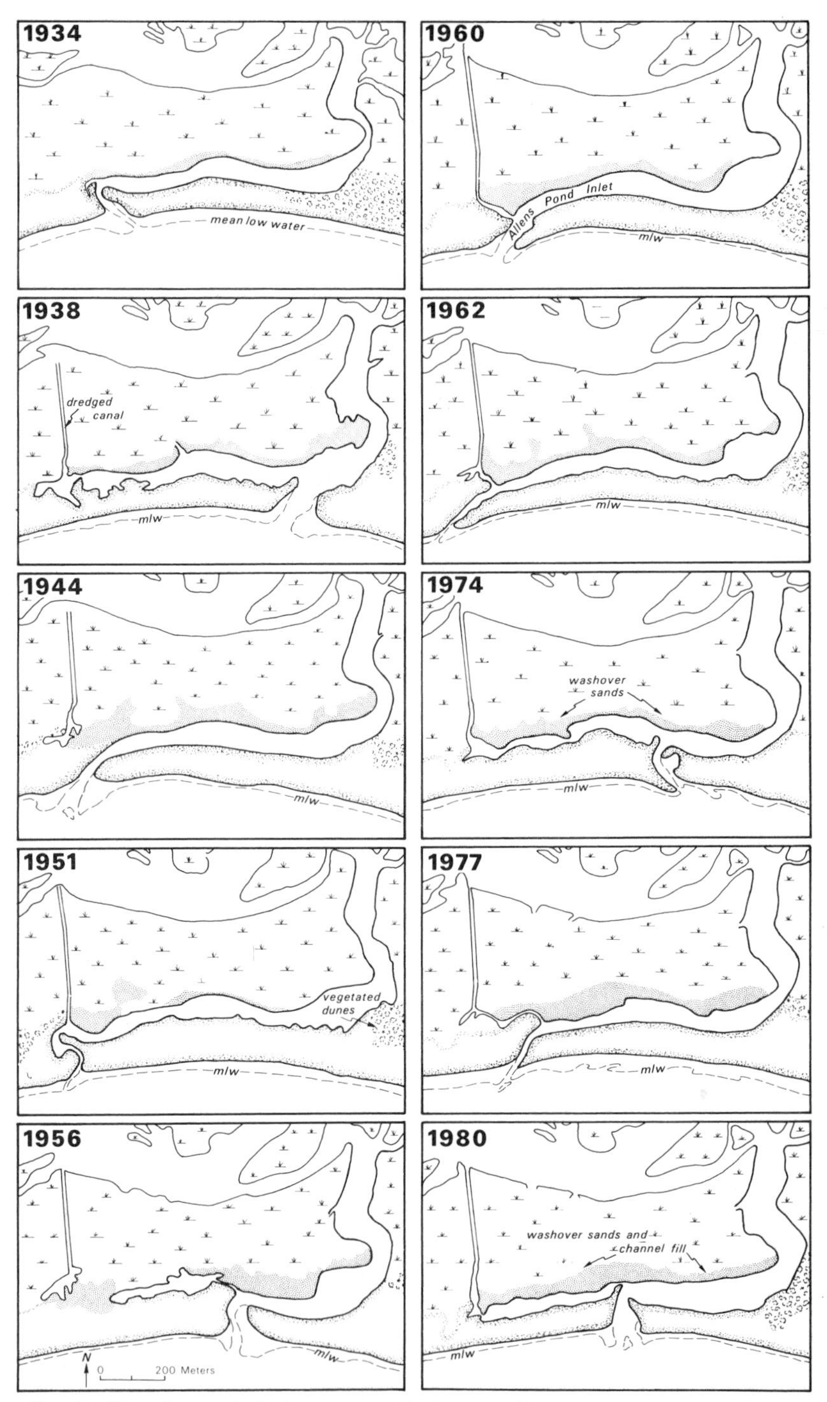

FIG. 10. Historic morphologic changes of Allens Pond inlet determined from U.S. Coast and Geodetic Survey coastal charts and vertical aerial photographs.

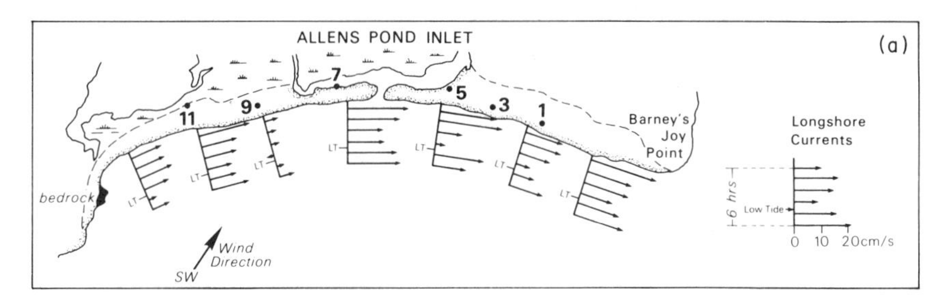

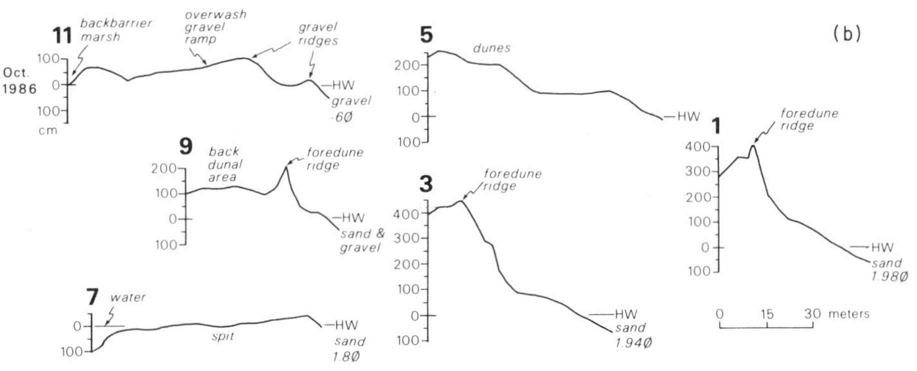

FIG. 11. (a) Longshore current and (b) morphologic and sedimentologic information for the Allens Pond region. Longshore currents were recorded on November 19, 1983, during southwesterly wind conditions. These data indicate a dominant easterly longshore transport trend.

of the beach (Fig. 11b). Finally, longshore current measurements indicate a dominant easterly transport direction. This is the same direction that would be predicted due to the prevailing southwesterly wind regime.

Thus, it can be concluded that the inlet is migrating against the dominant longshore sediment transport direction. This behavior of some tidal inlets has been recognized and explained by FitzGerald and FitzGerald (1977) and Aubrey and Spener (1984) as a product of ebb discharge in the back-barrier tidal creeks plus inlet thalweg migrational constraints. At Allens Pond updrift inlet migration is caused by the configuration of the back-barrier main channel, which runs parallel to the eastern spit before turning southward at the inlet mouth (Fig. 10). Ebb flow in this channel is directed at the western inlet shoreline, causing erosion of the western channel bank in a manner similar to that of the cut bank side of river meanders. The westerly accreting spit is the point bar side of the meander bend, receiving its sediment from the sand that has been eroded at the inlet and transported seaward to the westerly moving longshore transport system.

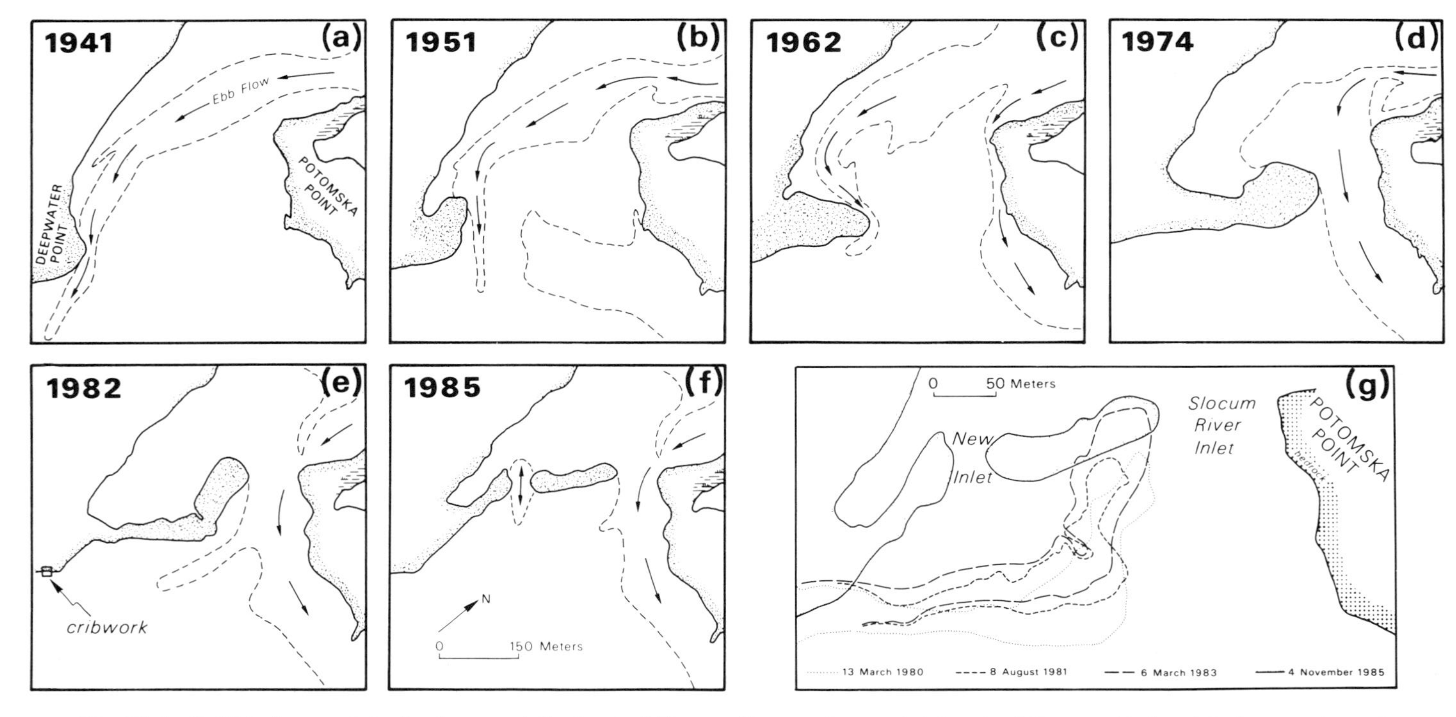

FIG. 12. (a–f) Historic morphologic changes at the head of Slocum River embayment determined from vertical aerial photographs and shoreline surveys. (g) Composite overlays of recent shoreline configurations. [From FitzGerald *et al.* (1986), with permission.]

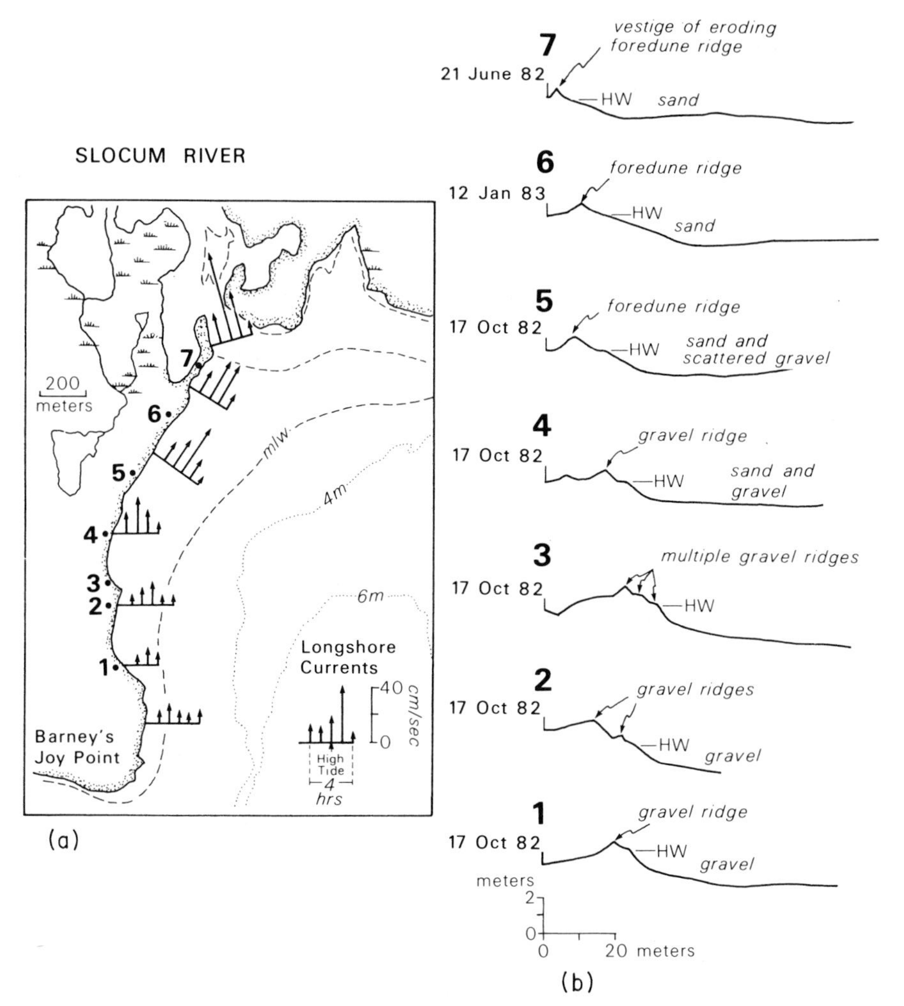

FIG. 13. (a) Longshore current and (b) morphologic and sedimentologic information for Slocum River embayment. Longshore current measurements were recorded on August 25, 1982. These data demonstrate that sediment is transported toward the inlet. [Data from Sands and FitzGerald (1984).]

2. Slocum Spit

Like the Allens Ponds spit system, the updrift shoreline of Slocum Spit exhibits similar morphologic and sedimentologic patterns (Fig. 13b). A strong trend of northeasterly transport is indicated by the replacement of multiple gravel ridges along the outer embayment shoreline by sand dunes next to Slocum River inlet. Due to the configuration of this embayed shoreline, any ocean wave from the southerly quadrant produces northeasterly longshore currents along the western beach. Note that the strength of the currents increases toward the inlet as the wave-generated currents are augmented by flood tidal flow (Fig. 13a).

The constructional and destructional history of Slocum spit during the past 44 years is a consequence of a period of abundant sand supply to the area (Fig. 12a–d) followed by sediment starvation (Fig. 12d–g). Prior to 1941, the entrance to the inner embayment was wide and shallow, and the main channel to the estuary was located along Deepwater Point. Sometime between 1941 and 1951, Slocum Spit began accreting eastward across the bay, causing a deflection of the embayment channel. This process continued until the mid-1970s, when the channel had become constricted next to Potomska Point and was only 100 m wide. Although tidal current flushing prevented further inlet narrowing at this point, the end of the spit continued building landward into the bay.

After the mid-1970s the supply of sand to the spit was drastically reduced, resulting in erosion of the beach and frontal dune ridge. As seen in historical map overlays (Fig. 12g), by 1980 the spit had thinned to half its original width, permitting frequent storm overwash. This caused a rapid landward migration of the spit, which ultimately lead to its breaching during a spring tide in November 1984. It now appears that the spit will become a subtidal bar or, if migration continues, part of the rear embayment beach. A more detailed account of the history of the spit can be found in Sands and FitzGerald (1984) and FitzGerald *et al.* (1986).

The formation of Slocum Spit followed by its transgression during a time span of only 45–50 yr indicates that a discrete sediment supply was probably responsible for its development. This sand is believed to have entered the embayment as a result of the 1938 hurricane, when storm waves from the southeast transported large quantities of sediment around Barney's Joy Point. Once the sediment entered the embayment, low wave energy gradually transferred the sand onshore and along the beach toward Deepwater Point. Additional factors may also be involved in this periodic supply of sediment to the embayment. Erosion of the shoreline in the vicinity of Deepwater Point has revealed the framework of a wooden crib that was once full of boulders and presumably built to prevent sand from filling the embayment channel. The destruction of this feature and possibly others during catastrophic storms would have released additional sand to the littoral system and aided spit development.

D. Tidal Inlets

The prominent tidal inlets along this section of coast are Westport River and Slocum River inlets. Minor inlets include Allens Pond inlet, discussed in the previous section, Little River inlet located in Slocum embayment, and the inlet to Salters Pond. As was shown earlier, the area of the bay can be used to approximate the size of the corresponding tidal inlet (Fig. 7). This relationship, which is based on an inlet's tidal prism versus its cross-sectional area (O'Brien, 1931, 1969) explains well the progressively smaller widths (sizes) of Westport River, Slocum River, and Little River inlets. However, this relationship does not explain why Allens Pond and Salter's Pond maintain tidal inlets whereas the adjacent Quicksand Pond and Briggs Marsh Pond do not, despite their having substantially larger bay areas and larger potential tidal prisms than the other two. This apparent anomaly is due to differences in wave energy. The ponds west of Westport River inlet are exposed to waves coming from an unlimited southerly fetch direction, including those generated by the prevailing southwesterly winds and extratropical southwesterly storms. In contrast, Little River inlet is highly sheltered, located at the head of Slocum embayment, while Salter's Pond inlet is protected by Mishaum Point and the offshore Elizabeth Islands. Consequently, these inlets experience lower wave energy than the barrier beaches fronting the coastal ponds to the west and therefore less sediment is delivered to their inlet channels.

The morphology and representative hydraulic data for Westport River and Slocum River inlets are given in Fig. 14. Detailed discussions of these inlets can be found in Magee and FitzGerald (1980) and Sands and FitzGerald (1984), respectively. The Westport River inlet channel cross section is highly asymmetric, with a deep thalweg bordered by a relatively shallow spit platform next to Horseneck Point (Fig. 14). This geometry has produced a segregation of tidal flow at the inlet whereby the deeper portion of the channel is dominated by ebb tidal flow and the spit platform by flood currents. This pattern of flow produces a counterclockwise recirculation of sand among the inlet channel, ebb-tidal delta, and adjacent beach. Similar sediment gyres have been described by Bruun and Gerritsen (1959), Dean and Walton (1975), and FitzGerald *et al.* (1976).

At Slocum River inlet, Sands and FitzGerald (1984), using regression analysis, demonstrated that the channel becomes increasingly flood dominant with increasing tidal range. During spring tide conditions, maximum flood currents exceed the ebb currents by 30–40 cm/s. Sand tracer experiments and the growth of a large flood-tidal delta inside the bay corroborate the landward transport of sediment through the inlet that the tidal current pattern implies (Sands and FitzGerald, 1984). The ultimate filling of this estuary is dependent on the rate of sand deposition in the bay versus the rate of sea-level rise. Because the embayment is presently in a sediment-starved condition, it is probable that the inner embayment will remain open and that if another spit should develop in this region it will form at a more landward position.

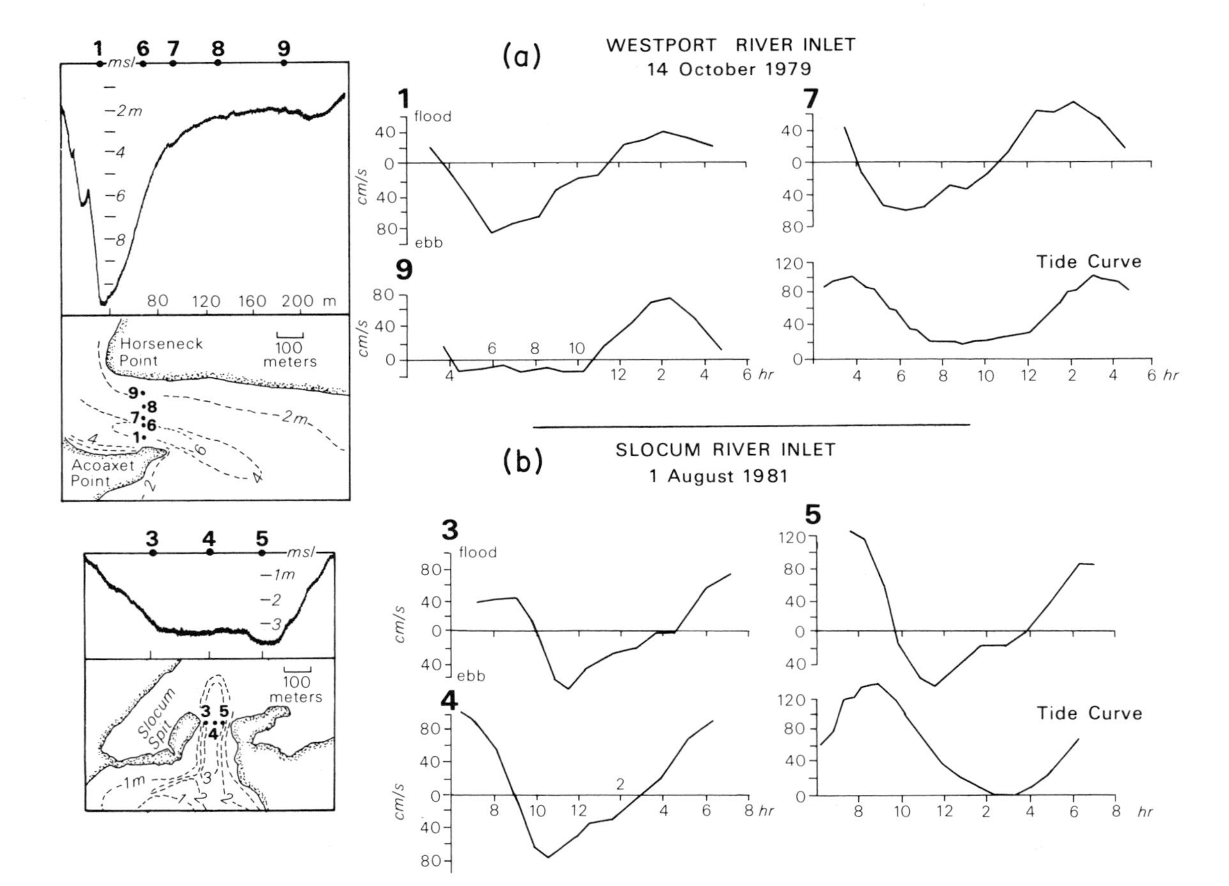

FIG. 14. Morphology and tidal current time series for (a) Westport River and (b) Slocum River inlets. [Data from Magee and FitzGerald (1980) and Sands and FitzGerald (1984).]

E. Beach-Ridge Barriers

1. Slocum Embayment

The shoreline southwest of Slocum Spit is one of two major sites of beach-ridge development in the study area (Fig. 15). The beach ridges in Slocum embayment front a small brackish water pond. Their formation caused a 300-m seaward extension and reorientation of the shoreline. Successive ridges are slightly skewed in a clockwise manner such that the present shoreline is swash aligned, parallel to the crests of refracted swell waves. Bakinowski (1987), using carbon-14 dates, has documented that beach ridge construction began after an estuarine and transgressive phase in the embayment (4170 ± 200 yr B.P.) and lasted until sometime shortly after 1010 ± 155 yr B.P.

2. Horseneck Beach

The Horseneck Beach ridge complex represents the largest accumulation of sand along the Buzzards Bay coast. The supply of sand to this region, which was derived from Gooseberry Neck and reworked glaciofluvial deposits, was sufficient to partially close off the Westport River estuary. Geomorphic evidence (Magee and FitzGerald, 1980) and stratigraphic data (Ibrahim, 1986) reveal that the barrier-building process at Horseneck Beach occurred in stages involving the closure of at least two inlets (Fig. 16). Magee and FitzGerald (1980) identified a paleotidal inlet in the middle of the barrier that is flanked on both sides by recurved spits. Ibrahim (1986) dated (carbon-14) the inlet fill sequence and showed that it closed approximately 485 yr B.P. during the main beach-ridge progradational phase. Historical accounts also document that "The Let" (Fig. 16) was once a passageway to the east branch of the Westport River and that it closed about 150–200 years ago. The formation of the barrier beach in front of "The Let" resulted in the connection of Horseneck barrier island to the mainland. The driving force behind the closure of these inlets was sedimentation in the estuary, causing a decrease in the bay tidal prism, which led to a smaller equilibrium inlet cross-sectional area.

The sedimentological development of Horseneck Beach has been interpreted from the analysis of 40 vibracores, 20 bore holes, 5 km of ground-penetrating radar, and eight topographic surveys across the island, and is reported in a detailed investigation by Ibrahim (1986). Three strike secions of the barrier complex from Ibrahim's study are presented in Fig. 17. From these profiles, it can be seen that the thickness of the barrier varies from 4 to 10 m and consists primarily of fine sands with some coarse sand and cobble-sized gravel layers. The gravel deposits increase in abun-

Fig. 15. Aerial photograph of beach ridges along (a) Slocum River embayment and (b) Horseneck Beach.

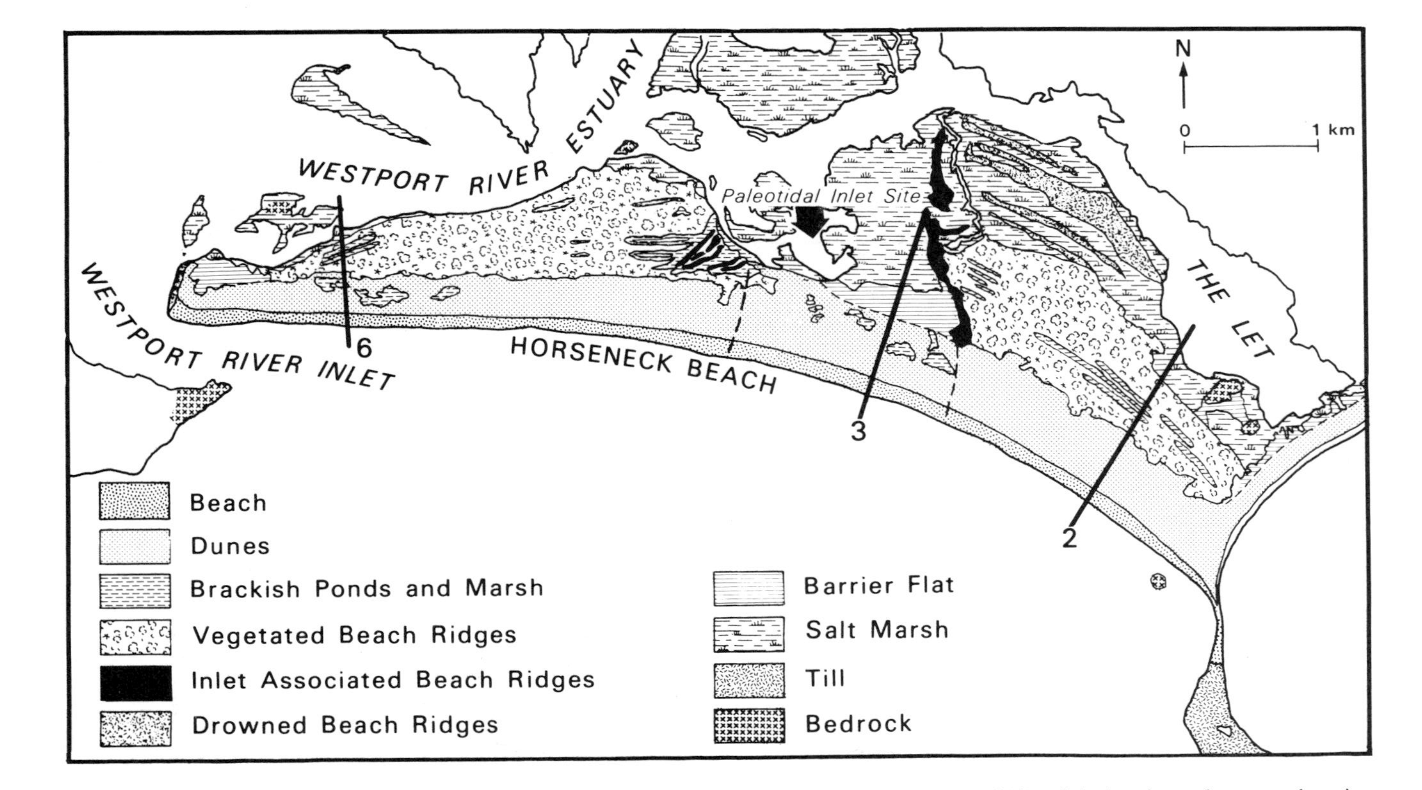

FIG. 16. Geomorphic map of Horseneck Beach, illustrating the site of the paleotidal inlet in the middle of the barrier and transect locations. [Modified from Ibrahim (1986, Fig. 8).]

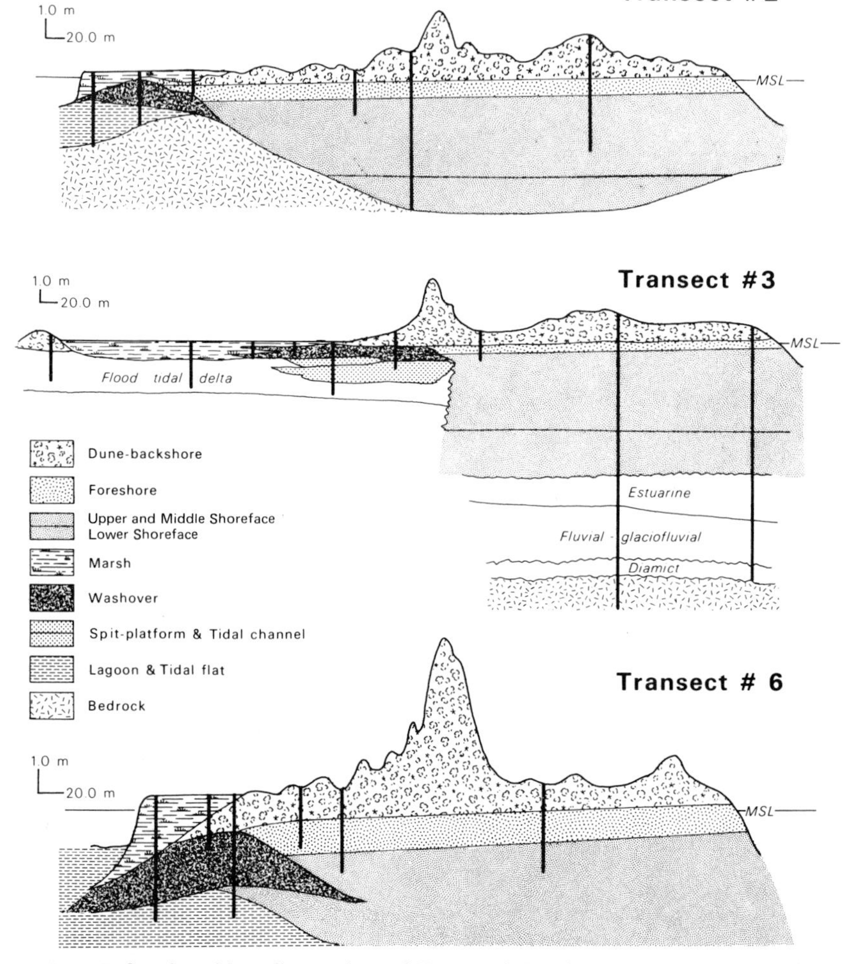

FIG. 17. Stratigraphic strike sections of Horseneck Beach. Locations shown in Fig. 16. [After Ibrahim (1986).]

dance toward the eastern end of the island and were likely derived from Gooseberry Neck. The barrier–beach face sands are underlain by estuarine sediments comprised of a coarsening upward sequence (3–8 m thick) composed of silts and clays grading to fine sands. Underlying the estuarine deposits are glacial deposits 3–6 m in thickness consisting of glaciofluvial sands and gravels and diamict sediments.

The filling of Westport River estuary, including the formation of Horseneck Beach, can be separated into five sedimentation stages based on the environment of erosion and deposition (Fig. 18a–d) (Ibrahim, 1986).

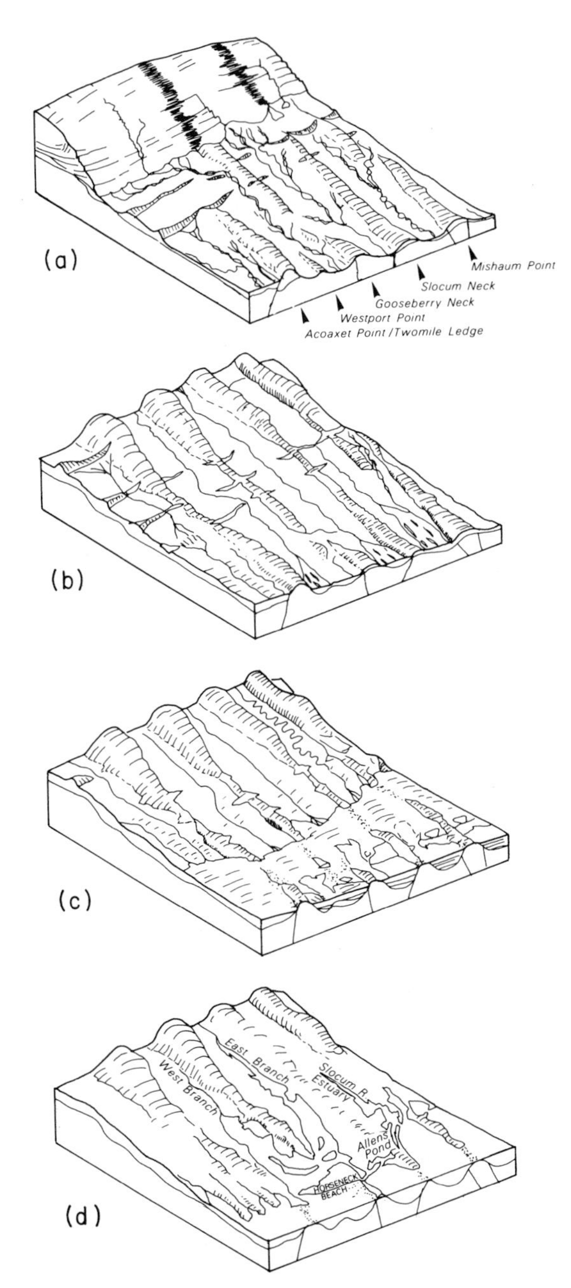

Fig. 18. Model of the sedimentological development of the Horseneck Beach region. Approximate time periods represent (a) stage 1, 15,000 yr B.P., deglaciation of the region; (b) stage 2, 14,000–7000 yr B.P., fluvial reworking; (c) stage 3, 7000–2800 yr B.P., estuarine sedimentation and transgressive phase of marine reworking; and (d) stage 4, 2800 yr B.P.–present, regressive period of beach-ridge construction.

The dates assigned to these stages have been determined from the overall stratigraphy of the region, the depth to the Holocene trangressive unconformity, carbon-14 dates, and Oldale and O'Hara's (1980) Holocene sea-level curve. This curve, which was constructed using numerous radiocarbon dates, covers the period from approximately 12,000 to 5000 yr B.P. and has been extrapolated to meet Redfield and Rubin's (1962) sea-level curve for the period from 3400 yr B.P. to the present.

The first stage occurred sometime after 15,300 ± 800 yr B.P., when the Buzzards Bay lobe began retreating from Martha's Vineyard (Larson, 1982). During its recession, the ice left several end moraines (i.e., Elizabeth Islands; Fig. 4) and an extensive cover of lodgement till (Fig. 18a). The retreating ice also formed meltwater streams that flowed through valleys, such as Slocum River and Westport River embayment, depositing glaciofluvial sediments. These glacial deposits form the basal unit of the sedimentary sequence in Westport River estuary and overlie a scoured bedrock surface.

During the second stage (14,000 to 7000 yr B.P.), sea level was rising quickly and the outer continental shelf was undergoing rapid transgression. At this time the inner continental shelf including Westport River embayment was exposed to fluvial processes and aeolian reworking of the glacial deposits (Fig. 18b). A fluvial erosional unconformity has been identified in seismic records from Buzzards Bay by Odale and O'Hara (1980) but has not been detected in subsurface data at Horseneck Beach (Ibrahim, 1986).

The third stage, which marked the beginning of estuarine sedimentation, began approximately 7000 yr B.P. in the deeper portions of Westport River embayment and lasted until about 5000 yr B.P. McMaster (1984) has postulated that estuarine conditions existed in the deeper passages of Narragansett Bay as early as 7500 yr B.P. During this period the Westport River embayment was protected from open-ocean processes by headlands at Gooseberry Neck and Twomile ledge, which at that time was connected to Acoaxet Point (Fig. 18c). These headlands were likely the anchoring sites of the first transgressive barrier beaches in this region. This type of coastal morphology is similar to that described by Boyd *et al.* (this volume, Chapter 4), for the drumlin coast of Nova Scotia.

The fourth stage encompassed the period from 5000 to 2800 yr B.P. when the study area was exposed to open-marine conditions. During this time a low, narrow, highly transgressive barrier migrated onshore in response to rising sea level (Fig. 18c). As this barrier rolled over itself, wave processes, storms, and coastal currents reworked part of the underlying estuarine and glaciofluvial sediments, forming an erosional unconformity. The roll-over process was achieved through overwash activity, deposition of flood-tidal deltas, and aeolian sedimentation. The earlier age of this stage was based on the correlation between the depth to the transgressive unconformity and Odale and O'Hara's (1980) sea-level curve. This same unconformity has been identified from seismic records and core logs at a number of inner continental shelf localities, including Buzzards Bay (Odale and O'Hara, 1980), Rhode Island Sound (Needel *et al.*, 1983) and Narragansett Bay (McMaster, 1984). The later date was approximated by correlating the depth to the contact between the washover facies and back-barrier unit to the sea-level curve.

The final stage, which has continued from 2800 yr B.P. to the present, accounted for the formation of the Horseneck Beach ridges (Fig. 18d). This period marked a change in barrier dynamics from a transgressive mode to a regressive phase that included the closure of inlets and a widening of the barrier to as much 1.2 km. As seen in Fig. 17, the transgressive washover deposits are found in the landward section of each of the barrier transects juxtaposed with the regressive beach-ridge deposits. This facies relationship has been identified at several other regressive barrier island studies, including Kiawah Island, South Carolina (Moslow and Colquhoun, 1981), along Bogue Banks, North Carolina (Steele, 1980), and the Netherland mainland barriers (Van Straaten, 1965). This stage of beach ridge construction coincided with a decrease in the rate of sea-level rise and a probable increase in supply of sediment to the shoreline. Hine *et al.* (1979) labeled this period the "late Holocene sedimentation interval" and postulated that the progradation of Bogue Banks along the North Carolina coast occurred during this time.

VI. Discussion and Summary

The structural grain of southeastern New England has had a pronounced influence on the development of the Buzzards Bay shoreline. Not only is it responsible for the north–southward trending ridge and valley topography of the region, but it also controlled the distribution of glacial deposits by confining glaciofluvial sedimentation to the existing valley systems. The overall morphology of the study area is primarily a product of the size of the drowned valleys and amount of sediment available for barrier construction. Where valleys are narrow and/or shallow, coastal ponds fronted by low, transgressive barriers have developed. The presence of inlets at these sites is dependent on the potential tidal prism of the bay and the exposure of the shoreline to wave energy. Coastal ponds that are sheltered from direct wave attack maintain open inlets when their tidal prisms, theoretically, would appear to dictate otherwise.

Where valleys are deep and wide, major estuaries have developed. However, differences in sediment supply to these areas have produced contrasting shoreline morphologies. The East and West Branches of the Westport River estuary are protected by an expansive beach ridge complex, while Slocum River estuary to the east is fronted by an open, shallow-water embayment. A large till deposit, several times the present size of Gooseberry Neck, and offshore glaciofluvial sediments nourished the building of the regressive beach ridges of Horseneck Beach. In the Slocum embayment region there were no large till deposits for barrier construction. In fact, the sediment responsible for the accretionary features that do exist in Slocum embayment, including the beach ridges and the sand platform, was probably derived, in part, from Gooseberry Neck through longshore transport. This process no longer occurs on a continuous basis today, due to the much smaller amount of sediment presently being eroded from Gooseberry Neck and the extension of the Barney's Joy Point promontory as the adjacent Allens Pond shoreline has migrated landward.

Stratigraphic and morphologic data indicate that since deglaciation the Buzzards Bay shoreline has undergone a transgressive stage followed by a regressive phase and that presently, most of the coast is once again being transgressed. The first transgression began approximately 7000 yr B.P. and lasted until 2800 yr B.P., during which time fine-grained sediments were deposited in the deep estuaries while elsewhere glaciofluvial sediments were being reworked by marine processes. During the end of this period, proto-Horseneck Beach was formed. Like many other beach-ridge barriers throughout the world, during its initial stages Horseneck Beach was a highly transgressive barrier comprised chiefly of washover deposits. As the rate of sea-level rise slowed, the supply of sediment became more abundant and beach-ridge construction proceeded at Horseneck Beach and in Slocum River embayment. During this stage it is probable that the barriers fronting the coastal ponds were wider and had more substantial dune systems.

As the onshore movement of sediment waned and sea level continued to rise, most of the Buzzards Bay shoreline began to erode. This final stage was reached at different times along the study area, depending on the available sediment supply, but it presently encompasses most of the coast.

Acknowledgments

The work presented in this chapter was supported by grants and contracts from Massachusetts Coastal Zone Management; the Towns of Westport and South Dartmouth, Massachusetts; PETRONAS of Malaysia; and by the Lloyd Center for Environmental Studies.

The arduous field work and data analysis of several graduate students, including Andrew Magee, Andrew Bakinowski, Jamie Greacen, Alan Saiz, and Cynthia Thomlinson, greatly contributed to the ideas presented in this chapter. Polychronis Tzedakis, Stephanos Stafilakis, and Alan Cathcart provided the process measurements presented for Allens Pond, and James Pollock and Christine Prekezes provided the beach profile and sedimentologic data for the same area. Peter Goldschmidt edited the manuscript. The figures in this paper were drafted by Eliza McClennen of the Boston University Cartographic Services Lab.

REFERENCES

Aubrey, D., and Spencer, W. D. (1984). Updrift migration of tidal inlets. J. Geol. **92,** 531–545.

Bakinowski, A. (1987). Subsurface investigation and historical reconstruction of the beach ridge complex of Slocum River Embayment, South Dartmouth, MA. Master's Thesis, Dep. Geol., Boston Univ., Boston, Massachusetts.

Barosh, P. J., and Hermes, O. D. (1981). General setting of Rhode Island and tectonic history of southern New England. *In* "Guidebook to Field Studies in Rhode Island and Adjacent Areas," New England Intercollegiate Geological Conference (J. C. Boothroyd and O. D. Hermes, eds.), pp. 1–16.

Boothroyd, J. C., Friederich, N. E., and McGinn, S. R. (1985). Geology of microtidal lagoons, RI. *Mar. Geol.* **63,** 35–76.

Bruun, P., and Gerritsen, F. (1959). Natural bypassing at coastal inlets. *J. Waterways Harbors Div., ASCE, Pap.* 2301, 75–107.

Dean, R. G., and Walton, T. L. (1975). Sediment transport processes in the vicinity of tidal inlets with special reference to sand trapping. *In* "Estuarine Research" (L. E. Chronin, ed.), pp. 129–150. Academic Press, New York.

FitzGerald, D. M., and FitzGerald, S. A. (1977). Factors influencing tidal throat geometry. *Proc. Coastal Sediments, ASCE* pp. 563–581.

FitzGerald, D. M., Nummedal, D., and Kana, T. (1976). Sand circulation patterns at Price Inlet, SC. *Proc. Coastal Eng. Conf., 15th, Am. Soc. Civ. Eng.* pp. 1868–1880.

FitzGerald, D. M., Saiz, A., Baldwin, C. T., and Sands, D. M. (1986). Spit breaching at Slocum River Inlet: Buzzards Bay, MA. *Shore Beach* **54,** 11–17.

Greacen, J., Rahilly, J., and Tomlinson, C. (1983). An examination of the coastal variation: Sakonnet Point, RI to New Bedford, MA. Unpubl. Rep., Coastal Environ. Res. Group, Dep. Geol., Boston Univ., Boston, Massachusetts.

Hayes, M. O. (1979). Barrier island morphology as a function of tidal and wave regime. *In* "Barrier Islands: From the Gulf of St. Lawrence to the Gulf of Mexico" (S. P. Leatherman, ed.), pp. 1–28. Academic Press, New York.

Hine, A. C., Snyder, S. S., and Nuemann, A. C. (1979). "Coastal Plain and Inner Shelf Structure, Stratigraphy and Geologic History: Bogue Banks Area, NC," Tech. Rep. North Carolina Sci. Technol. Comm., Chapel Hill.

Ibrahim, N. A. (1986). Sedimentological and morphological evolution of a coarse-grained regressive barrier beach, Horseneck Beach, MA. Master's Thesis, Dep. Geol., Boston Univ., Boston, Massachusetts.

Kaye, C. A. (1964). Outline of Pleistocene geology of Martha's Vineyard, MA. *U.S. Geol. Surv. Prof. Pap.* **501-C,** 134–139.

Kaye, C. A. (1983). "The Autochthonous and Allochthonous Coastal Plain Deposits of Martha's Vineyard and Marshfield–Scituate Area, Southeastern Massachusetts," Atlantic Coastal Plain Geological Association Fieldtrip Guidebook.

Lamb, H. H. (1977). "Climatic History in the Future," Vol. 2. Methuen, New York.

Larson, G. J. (1982). Nonsynchronous retreat of ice lobes from southeastern Massachusetts. *In* "Late Wisconsinan of New England" (G. J. Larson and B. D. Stone, eds.), pp. 101–115. Kendall/Hunt, Dubuque, Iowa.

McMaster, R. L. (1984). Holocene stratigraphy and depositional history of the Narragansett Bay system, RI, USA. *Sedimentology* **31,** 777–792.

McMaster, R. L. and Asraf, A. (1973). Subbottom basement drainage system of inner continental shelf of southern New England. *Geol. Soc. Am. Bull.* **84,** 187–190.

Magee, A. D., and FitzGerald, D. M. (1980). "Investigation of the Shoaling Problems at Westport River Inlet and Sedimentation Processes at Horseneck and East Horseneck Beaches," Tech. Rep. No. 3. Coastal Environ. Res. Group, Dep. Geol., Boston Univ., Boston, Massachusetts.

Moslow, T. F., and Colquhoun, D. J. (1981). Influence of sea-level change on barrier island evolution. *Oceanis* **7,** 439–454.

National Ocean Survey (1986). "Tide Tables East Coast of North and South America." U.S. Dep. Commer., Washington, D.C.

Needel, S. W., O'Hara, C. J., and Knebel, H. J. (1983). Quaternary geology of the Rhode Island inner shelf. *Mar. Geol.* **53,** 41–53.

Nummedal, D., and Fischer, I. A. (1978). Process–response models for depositional shorelines: German and Georgia Bights. *Proc. Coastal Eng. Conf., 16th, Am. Soc. Civ. Eng.* pp. 543–562.

O'Brien, M. P. (1931). Estuary tidal prisms related to entrance areas. *Civ. Eng.* **1,** 738–739.

O'Brien, M. P. (1969). Equilibrium flow areas of inlets on sandy coasts. *J. Waterways Harbors Div., ASCE,* **95,** 43–52.

Oldale, R. N. (1982). Pleistocene stratigraphy of Nantucket, Martha's Vineyard, the Elizabeth Islands and Cape Cod, Massachusetts. *In* "Late Wisconsinan of New England" (G. J. Larson and B. D. Stone, eds.), pp. 1–34. Kendall/Hunt, Dubuque, Iowa.

Oldale, R. N., and O'Hara, C. J. (1980). New radiocarbon dates from the inner continental shelf of southeastern Massachusetts and a local sea-level curve for the past 12,000 years. *Geology* **8,** 102–106.

Redfield, A. C., and Rubin, M. (1962). The ages of salt marsh peat and its relation to recent changes in sea-level at Barnstable, Massachusetts. *Proc. Natl. Acad. Sci. USA* **48,** 1728–1735.

Sands, D. R., and FitzGerald, D. M. (1984). "Morphology, Hydraulics and Sediment Transport Patterns at Slocum River Inlet: South Dartmouth, MA," Tech. Rep. 8. Coastal Environ. Res. Group, Dep. Geol., Boston Univ., Boston, Massachusetts.

Schafer, J. P., and Hartshorn, J. H. (1965). The Quaternary of New England. *In* "The Quaternary of the United States" (H. E. Wright, Jr. and D. G. Frey, eds.), pp. 113–128. Princeton Univ. Press, Princeton, New Jersey.

Steele, G. A. (1980). Stratigraphy and depositional history of Bogue Banks, North Carolina. Master's Thesis, Dep. Geol., Duke Univ., Durham, North Carolina.

Stone, B. D., and Piper, J. D. (1982). Topographic control of the deglaciation southern portion of the Connecticut Valley of Massachusetts. *In* "Late Wisconsinan of New England" (G. J. Larson and B. D. Stone, eds.), pp. 145–166. Kendall/Hunt, Dubuque, Iowa.

Thompson, E. F. (1977). "Wave Climate at Selected Locations along the U.S. Coasts," Tech. Rep. 77-1. U.S. Army Coastal Eng. Res. Cent., Washington, D.C.

U.S. Army Corps of Engineers (1986). "Detailed Project Report and Environmental Assessment: Small Beach Erosion Control Project at Okinawa, CT." N. Engl. Div., Waltham, Massachusetts.

Van Straaten, L. M. J. U. (1965). Coastal barrier deposits in north and south Holland in particular in the areas of Scheveningen and Ijlmuiden. *Meded. Rijks Geol. Dienst, Nieuwe Set. (Neth.)* **17,** 41–75.

Yasso, W. E. (1965). Plan geometry on the headland-bay beaches. *J. Geol.* **73,** 702–714.

Zen, E. (1983). "Bedrock Geologic Map of Massachusetts." U.S. Geol. Surv., Reston, Virginia.

Index

T

U

V

W